T0201323

ROUGH-FUZZY PATTERN RECOGNITION

Wiley Series on
Bioinformatics: Computational Techniques and Engineering

Bioinformatics and computational biology involve the comprehensive application of mathematics, statistics, science, and computer science to the understanding of living systems. Research and development in these areas require cooperation among specialists from the fields of biology, computer science, mathematics, statistics, physics, and related sciences. The objective of this book series is to provide timely treatments of the different aspects of bioinformatics spanning theory, new and established techniques, technologies and tools, and application domains. This series emphasizes algorithmic, mathematical, statistical, and computational methods that are central in bioinformatics and computational biology.

Series Editors: **Professor Yi Pan** and **Professor Albert Y. Zomaya**
pan@cs.gsu.edu   zomaya@it.usyd.edu.au

ROUGH-FUZZY PATTERN RECOGNITION

Applications in Bioinformatics and Medical Imaging

PRADIPTA MAJI
Machine Intelligence Unit, Indian Statistical Institute, Kolkata, India

SANKAR K. PAL
Center for Soft Computing Research, Indian Statistical Institute, Kolkata, India

IEEE PRESS

A JOHN WILEY & SONS, INC., PUBLICATION

Published by John Wiley & Sons, Inc., Hoboken, New Jersey
Published simultaneously in Canada

For general information on our other products and services or for technical support, please contact our Customer Care Department within the United States at (800) 762-2974, outside the United States at (317) 572-3993 or fax (317) 572-4002.

Wiley also publishes its books in a variety of electronic formats. Some content that appears in print may not be available in electronic formats. For more information about Wiley products, visit our web site at www.wiley.com.

Library of Congress Cataloging-in-Publication Data:

Maji, Pradipta, 1976–
  Rough-fuzzy pattern recognition : applications in bioinformatics and medical imaging / Pradipta Maji, Sankar K. Pal.
    p. cm. – (Wiley series in bioinformatics ; 3)
    ISBN 978-1-118-00440-1 (hardback)
  1. Fuzzy systems in medicine. 2. Pattern recognition systems. 3. Bioinformatics. 4. Diagnostic imaging–Data processing. I. Pal, Sankar K. II. Title.
  R859.7.F89M35 2011
  610.285–dc23

                                                                    2011013787

Printed in the United States of America

10 9 8 7 6 5 4 3 2 1

To our parents

CONTENTS

FOREWORD

It is my great pleasure to welcome the new book *Rough-Fuzzy Pattern Recognition: Applications in Bioinformatics and Medical Imaging* by the prominent scientists Professor Sankar K. Pal and Professor Pradipta Maji.

Soft computing methods allow us to achieve high-quality solutions for many real-life applications. The characteristic features of these methods are tractability, robustness, low-cost solution, and close resemblance with human-like decision making. They make it possible to use imprecision, uncertainty, approximate reasoning, and partial truth in searching for solutions. The main research directions in soft computing are related to fuzzy sets, neurocomputing, genetic algorithms, probabilistic reasoning, and rough sets. By integration or combination of the different soft computing methods, one may improve the performance of these methods. Among the various integrations realized so far, neuro-fuzzy computing (combing fuzzy sets and neural networks) is the most visible one because of its several real-life applications.

Both fuzzy and rough set theory represent two different approaches to analyzing vagueness. Fuzzy set theory addresses gradualness of knowledge, expressed by the fuzzy membership, whereas rough set theory addresses the granularity of knowledge, expressed by the indiscernibility relation. In 1999, together with Professor Sankar K. Pal, we edited the book *Rough-Fuzzy Hybridization* published by Springer. Since then, great progress has been made in the development of methods based on a combination of these approaches, both on foundations and applications. It is proved that by combining the rough-set and fuzzy-set approaches it is possible to achieve a significant improvement in the performance of methods. They are complementary to each other rather than competitive.

The book is based on a unified framework describing how rough-fuzzy computing techniques can be formulated for building efficient information granules, especially pattern recognition models. These granules are induced from some elementary granules by (hierarchical) fusion. The elementary granules can be induced using rough-set-based methods and/or fuzzy-set-based methods, and also the aggregation of granules can be based on a combination of such methods. In this way, one can consider different approaches such as rough-fuzzy, fuzzy-rough or fuzzy rough-fuzzy. For example, using rough-set-based methods one can efficiently induce some crisp patterns, which can be next fuzzified for making them soft. This can help, for example, in searching for the relevant fusion of granules. Analogously, the discovered fuzzy patterns may contain too many details and then, by using rough-set-based methods, one can obtain simpler patterns with satisfactory quality. Such patterns can be next used efficiently in approximate reasoning, for example, in the discovery of more complex patterns from the existing ones. In all these methods, the rough-set approach and the fuzzy-set approach work synergistically in efficient searching under uncertainty and imprecision for the target granules (e.g., classifiers for complex concepts) with a high quality. Note that in the fusion of granules, inclusion measures also play an important role because they allow us to estimate the quality of the granules constructed.

In a perfect way, the book introduces the reader to the fascinating and successful cooperative game between the rough-set approach and fuzzy-set approach. In this book, the reader will find a nice introduction to the rough-fuzzy approach and fuzzy-set approach. The discussed methods and algorithms cover all major phases of a pattern recognition system (e.g., classification, clustering, and feature selection). The book covers existing results and also presents new results. It was proved experimentally that the performance of the developed algorithms based on a combination of approaches is much better than the performance of algorithms based on approaches taken separately. This is shown in the book for several tasks such as feature selection of real valued data, case selection, image processing and analysis, data mining and knowledge discovery, selection of vocabulary for information retrieval, and decision rule extraction. The high performance of the developed methods is especially emphasized for real-life applications in bioinformatics and medical image processing. The balance of theory, algorithms, and applications will make this book attractive to many readers. The reader will also find in the book a discussion on various challenging issues relevant to application domains and possible ways of handling them with rough-fuzzy approaches.

This book, unique in its character, will be very useful to graduate students, researchers, and practitioners. The authors deserve the highest appreciation for their outstanding work. This is a work whose importance is hard to exaggerate.

ANDRZEJ SKOWRON

Institute of Mathematics
Warsaw University, Poland
December, 2011

PREFACE

Soft computing is a collection of methodologies that work synergistically, not competitively, that, in one form or another, reflects its guiding principle: exploits the tolerance for imprecision, uncertainty, approximate reasoning, and partial truth to achieve tractability, robustness, low cost solution, and close resemblance with human-like decision making. It provides a flexible information processing capability for representation and evaluation of various real-life, ambiguous and uncertain situations and therefore results in the foundation for the conception and design of high machine intelligence quotient systems. At this juncture, the principal constituents of soft computing are fuzzy sets, neurocomputing, genetic algorithms, probabilistic reasoning, and rough sets.

One of the challenges of basic soft computing research is how to integrate its different constituting tools synergistically to achieve both generic and application-specific merits. Application-specific merits point to the advantages of integrated systems not achievable using the constituting tools singly.

Rough set theory, which is considered to be a newer soft computing tool compared with others, deals with uncertainty, vagueness, and incompleteness arising from the indiscernibility of objects in the universe. The main goal of rough set theoretic analysis is to synthesize or construct approximations, in terms of upper and lower bounds of concepts, properties, or relations from the acquired data. The key notions here are those of information granules and reducts. Information granules formalize the concept of finite precision representation of objects in real-life situations, and the reducts represent the core of an information system, both in terms of objects and features, in a granular universe. Its integration with fuzzy set theory, called rough-fuzzy computing, has motivated researchers to design a much stronger paradigm of reasoning and handling uncertainties associated with the data compared with those of the individual ones. The generalized theories of rough-fuzzy sets (when a fuzzy set is defined over crisp granules), fuzzy-rough sets (when a crisp set is defined over fuzzy granules), and fuzzy rough-fuzzy sets (when a fuzzy set is defined over fuzzy granules) have been applied successfully to problems such as classification, clustering, feature selection of real valued data, case selection, image processing and analysis, data mining and knowledge discovery, selecting vocabulary for information retrieval, and decision rule extraction.

In rough-fuzzy pattern recognition, fuzzy sets are used for handling uncertainties arising from ill-defined or overlapping nature of concepts, classes or regions, in terms of membership values, while rough sets are used for granular computing and handling uncertainties, due to granulation or indiscernibility in the feature space, in terms of lower and upper approximations of concepts or regions. On the one hand, rough information granules are used, for example, in defining class exactness, and encoding domain knowledge. On the other hand, granular computing, which deals with clumps of indiscernible data together rather than individual data points, leads to computation gain, thereby making it suitable for mining large data sets. Furthermore, depending on the problems, granules can be class dependent or independent. Several clustering algorithms have been formulated where the incorporation of rough sets resulted in a balanced mixture between restrictive or hard clustering and descriptive or fuzzy clustering. Judicious integration of the concept of rough sets with the existing fuzzy clustering algorithms has made the latter work faster with improved performance. Various real-life applications of these models, including those in bioinformatics, have also been reported during the last five to seven years. These are available in different journals, conference proceedings, and edited volumes. This scattered information causes inconvenience to readers, students, and researchers.

The current volume is aimed at providing a treatise in a unified framework describing how rough-fuzzy computing techniques can be judiciously formulated and used in building efficient pattern recognition models. On the basis of the existing as well as new results, the book is structured according to the major phases of a pattern recognition system (for example, clustering, classification, and feature selection) with a balanced mixture of theory, algorithm, and applications. Special emphasis is given to applications in bioinformatics and medical image processing.

The book consists of nine chapters. Chapter 1 provides an introduction to pattern recognition and data mining, along with different research issues and challenges related to high dimensional real-life data sets. The significance of soft computing in pattern recognition and data mining is also presented in Chapter 1. Chapter 2 presents the basic notions and characteristics of two soft computing tools, namely, fuzzy sets and rough sets. These are followed by the concept of information granules, the emergence of a rough-fuzzy computing paradigm and their relevance to pattern recognition. It also provides a mathematical framework for generalized rough sets incorporating the concept of fuzziness in defining the granules as well as the set. Various roughness and uncertainty measures with properties are reported. Different research issues related to rough granules are stated.

Chapter 3 mainly centers on a generalized unsupervised learning (clustering) algorithm, termed as rough-fuzzy-possibilistic c-means. While the concept of lower and upper approximations of rough sets deals with uncertainty, vagueness, and incompleteness in class definition, the membership function of fuzzy sets enables efficient handling of overlapping partitions. It incorporates both probabilistic and possibilistic memberships simultaneously to avoid the problems of noise sensitivity of fuzzy c-means and the coincident clusters of possibilistic c-means. The concept of crisp lower bound and fuzzy boundary of a class enables efficient selection of cluster prototypes. The algorithm is generalized in the sense that all the existing variants of c-means algorithms can be derived from this as a special case. Superiority in terms of computation time and performance is demonstrated. Several quantitative indices are reported for evaluating the performance on various real-life data sets.

Chapter 4 provides various supervised classification methods based on class-dependent f-granulation and rough set theoretic feature selection. The significance of granulation for better class discriminatory information and neighborhood rough sets for better feature selection is demonstrated. Extensive experimental results with quantitative indices are provided on both fully and partially labeled data sets. Future directions on the use of this concept in other computing paradigms are also provided.

Selection of nonredundant and relevant features of real valued data sets is a highly challenging problem. Chapter 5 addresses this issue. Methods described here are based on fuzzy-rough sets by maximizing the relevance and minimizing the redundancy of the selected features. Various new concepts such as fuzzy equivalence partition matrix, representation of Shannon's entropy for fuzzy approximation spaces, f-information measures to compute both relevance and redundancy of features, and feature evaluation indices are stated along with experimental results.

While several experimental results on both artificial and real-life data sets, including speech and remotely sensed multi-spectral image data, are provided in Chapters 3, 4, and 5 to demonstrate the effectiveness of the respective rough-fuzzy methodologies, the next four chapters are concerned only with certain specific applications, namely, in bioinformatics and medical imaging. Problems considered in bioinformatics include selection of a minimum set of bio-basis strings with maximum information for amino acid sequence analysis (Chapter 6), grouping functionally similar genes from microarray gene expression data through clustering (Chapter 7), and selection of relevant genes from high dimensional microarray data (Chapter 8). Selection of bio-basis strings is done by devising a relational clustering algorithm, called rough-fuzzy c-medoids. It judiciously integrates rough sets, fuzzy sets, and amino acid mutation matrices with hard c-medoids algorithms. The selected bio-basis strings are evaluated with newly defined indices in terms of a homology alignment score. Gene clusters thus identified may contribute to revealing underlying class structures, providing a useful tool for the exploratory analysis of biological data. The concept of fuzzy equivalence partition matrix, based on the theory of fuzzy-rough sets, is shown to be effective for selecting relevant and nonredundant continuous valued genes from high dimensional microarray gene expression data.

Problems of segmentation of brain MR images for visualization of human tissues during clinical analysis is addressed in Chapter 9 using rough-fuzzy clustering with new indices for feature extraction. The concept of discriminant analysis, based on the maximization of class separability, is used to circumvent the problems of initialization and local minima of rough-fuzzy clustering. Different challenging issues in the respective application domains and the ways of handling them with rough-fuzzy approaches are also discussed.

The relevant existing conventional/traditional approaches or techniques are also included wherever necessary. Directions for future research in the concerned topic are provided in each chapter. Most of the materials presented in the book are from our published works. For the convenience of readers, a comprehensive bibliography on the subject is also appended in each chapter. Some works in the related areas might have been omitted because of oversight or ignorance.

This book, which is unique in its character, will be useful to graduate students and researchers in computer science, electrical engineering, system science, medical science, bioinformatics, and information technology, both as a textbook and a reference book for some parts of the curriculum. Researchers and practitioners in industry and R&D laboratories working in the fields of system design, pattern recognition, machine learning, image analysis, vision, data mining, bioinformatics, and soft computing or computational intelligence will also be benefited.

Finally, the authors take this opportunity to thank Mr. Michael Christian and Dr. Simone Taylor of John Wiley & Sons, Inc., Hoboken, New Jersey, for their initiative and encouragement. The authors also gratefully acknowledge the support provided by Prof. Malay K. Kundu, Dr. Saroj K. Meher, Dr. Debashis Sen, Ms. Sushmita Paul, and Mr. Indranil Dutta of Indian Statistical Institute in the preparation and proofreading of a few chapters of the manuscript. The book was written when one of the authors, Prof. S. K. Pal, held a J. C. Bose National Fellowship of the Government of India.

PRADIPTA MAJI
SANKAR K. PAL

Kolkata, India

ABOUT THE AUTHORS

Pradipta Maji received his BSc degree in Physics, MSc degree in Electronics Science, and PhD degree in the area of Computer Science from Jadavpur University, India, in 1998, 2000, and 2005, respectively.

Currently, he is an assistant professor in the Machine Intelligence Unit, Indian Statistical Institute, Kolkata, India. He was associated with the Center for Soft Computing Research: A National Facility, Indian Statistical Institute, Kolkata, India, from 2005 to 2009. Before joining the Indian Statistical Institute, he was a lecturer in the Department of Computer Science and Engineering, Netaji Subhash Engineering College, Kolkata, India, from 2004 to 2005. In 2004, he visited the Laboratory of Information Security & Internet Applications (LISIA), Division of Electronics, Computer and Telecommunication Engineering, Pkyoung National University, Pusan, Korea. During the period of September 2000 to April 2004, he was a research scholar at the Department of Computer Science and Technology, Bengal Engineering College (DU) (currently known as Bengal Engineering and Science University), Shibpur, Howrah, India. From 2002 to 2004, he also served as a Research and Development Consultant of Cellular Automata Research Laboratory (CARL), Kolkata, India. His research interests include pattern recognition, computational biology and bioinformatics, medical image processing, cellular automata, and soft computing. He has published more than 60 papers in international journals and conferences and is a reviewer for many international journals.

Dr. Maji received the 2006 Best Paper Award of the International Conference on Visual Information Engineering from The Institution of Engineering and Technology, UK, the 2008 Microsoft Young Faculty Award from Microsoft Research Laboratory India Pvt., the 2009 Young Scientist Award from the National Academy of Sciences, India, and the 2011 Young Scientist Award from the Indian National Science Academy, and was selected as the 2009 Associate of the Indian Academy of Sciences.

Sankar K. Pal is a distinguished scientist of the Indian Statistical Institute and a former Director. He is also a J.C. Bose Fellow of the Government of India. He founded the Machine Intelligence Unit and the Center for Soft Computing Research, a national facility in the institute in Calcutta. He received his PhD in Radio Physics and Electronics from the University of Calcutta in 1979, and another PhD in Electrical Engineering along with DIC from Imperial College, University of London, in 1982. He joined his institute in 1975 as a CSIR Senior Research Fellow and later became a Full Professor in 1987, Distinguished Scientist in 1998, and the Director for the term 2005–2010.

He worked at the University of California, Berkeley, and the University of Maryland, College Park, in 1986–1987; the NASA Johnson Space Center, Houston, Texas, in 1990–1992 and 1994; and the US Naval Research Laboratory, Washington, DC, in 2004. Since 1997 he has served as a distinguished visitor of the IEEE Computer Society (USA) for the Asia-Pacific region, and held several visiting positions at universities in Italy, Poland, Hong Kong, and Australia.

Prof. Pal is a Fellow of the IEEE, USA, the Academy of Sciences for the Developing World (TWAS), Italy, International Association for Pattern Recognition, USA, International Association of Fuzzy Systems, USA, and all the four National Academies for Science/Engineering in India. He is a coauthor of 17 books and more than 400 research publications in the areas of pattern recognition and machine learning, image processing, data mining and web intelligence, soft computing, neural nets, genetic algorithms, fuzzy sets, rough sets, and bioinformatics.

He received the 1990 S.S. Bhatnagar Prize (which is the most coveted award for a scientist in India), and many prestigious awards in India and abroad including the 1999 G.D. Birla Award, 1998 Om Bhasin Award, 1993 Jawaharlal Nehru Fellowship, 2000 Khwarizmi International Award from the Islamic Republic of Iran, 2000–2001 FICCI Award, 1993 Vikram Sarabhai Research Award, 1993 NASA Tech Brief Award (USA), 1994 IEEE Transactions Neural Networks Outstanding Paper Award (USA), 1995 NASA Patent Application Award (USA), 1997 IETE-R.L. Wadhwa Gold Medal, the 2001 INSA-S.H. Zaheer Medal, 2005–2006 Indian Science Congress-P.C. Mahalanobis Birth Centenary Award (Gold Medal) for Lifetime Achievement, 2007 J.C. Bose Fellowship of the Government of India, and 2008 Vigyan Ratna Award from Science & Culture Organization, West Bengal.

Prof. Pal is currently or has in the past been an Associate Editor of *IEEE Trans. Pattern Analysis and Machine Intelligence* (2002–2006), *IEEE Trans. Neural Networks* [1994–1998 & 2003–2006], *Neurocomputing* (1995-2005), *Pattern Recognition Letters*, *Int. J. Pattern Recognition & Artificial Intelligence*, *Applied Intelligence*, *Information Sciences*, *Fuzzy Sets and Systems*, *Fundamenta Informaticae*, *LNCS Trans. on Rough Sets*, *Int. J. Computational Intelligence and Applications*, *IET Image Processing*, *J. Intelligent Information Systems*, and *Proc.*

INSA-A; Editor-in-Chief, *Int. J. Signal Processing, Image Processing and Pattern Recognition*; Series Editor, *Frontiers in Artificial Intelligence and Applications*, IOS Press, and *Statistical Science and Interdisciplinary Research*, World Scientific; a Member, Executive Advisory Editorial Board, *IEEE Trans. Fuzzy Systems*, *Int. Journal on Image and Graphics*, and *Int. Journal of Approximate Reasoning*; and a Guest Editor of *IEEE Computer*.

1

INTRODUCTION TO PATTERN RECOGNITION AND DATA MINING

1.1 INTRODUCTION

Pattern recognition is an activity that human beings normally excel in. The task of pattern recognition is encountered in a wide range of human activity. In a broader perspective, the term could cover any context in which some decision or forecast is made on the basis of currently available information. Mathematically, the problem of pattern recognition deals with the construction of a procedure to be applied to a set of inputs; the procedure assigns each new input to one of a set of classes on the basis of observed features. The construction of such a procedure on an input data set is defined as pattern recognition.

A pattern typically comprises some features or essential information specific to a pattern or a class of patterns. Pattern recognition, as per the convention, is the study of how machines can observe the environment, learn to distinguish patterns of interest from their background, and make sound and reasonable decisions about the categories of the patterns. In other words, the discipline of pattern recognition essentially deals with the problem of developing algorithms and methodologies that can enable the computer implementation of many recognition tasks that humans normally perform. The objective is to perform these tasks more accurately, faster, and perhaps more economically than humans and, in many cases, to release them from drudgery resulting from performing routine recognition tasks repetitively and mechanically. The scope of pattern recognition also encompasses tasks at which humans are not good, such as reading bar codes. Hence, the goal

Rough-Fuzzy Pattern Recognition: Applications in Bioinformatics and Medical Imaging,
First Edition. Pradipta Maji and Sankar K. Pal.
© 2012 John Wiley & Sons, Inc. Published 2012 by John Wiley & Sons, Inc.

of pattern recognition research is to devise ways and means of automating certain decision-making processes that lead to classification and recognition.

Pattern recognition can be viewed as a twofold task, consisting of learning the invariant and common properties of a set of samples characterizing a class, and of deciding that a new sample is a possible member of the class by noting that it has properties common to those of the set of samples. The task of pattern recognition can be described as a transformation from the measurement space $\mathcal{M}$ to the feature space $\mathcal{F}$ and finally to the decision space $\mathcal{D}$; that is,

$$\mathcal{M} \rightarrow \mathcal{F} \rightarrow \mathcal{D}, \tag{1.1}$$

where the mapping $\delta : \mathcal{F} \rightarrow \mathcal{D}$ is the decision function, and the elements $d \in \mathcal{D}$ are termed as *decisions*.

Pattern recognition has been a thriving field of research for the past few decades [1–8]. The seminal article by Kanal [9] gives a comprehensive review of the advances made in the field until the early 1970s. More recently, a review article by Jain et al. [10] provides an engrossing survey of the advances made in statistical pattern recognition till the end of the twentieth century. Although the subject has attained a very mature level during the past four decades or so, it remains green to the researchers because of continuous cross-fertilization of ideas from disciplines such as computer science, physics, neurobiology, psychology, engineering, statistics, mathematics, and cognitive science. Depending on the practical need and demand, various modern methodologies have come into being, which often supplement the classical techniques [11].

In recent years, the rapid advances made in computer technology have ensured that large sections of the world population have been able to gain easy access to computers on account of the falling costs worldwide, and their use is now commonplace in all walks of life. Government agencies and scientific, business, and commercial organizations routinely use computers, not only for computational purposes but also for storage, in massive databases, of the immense volumes of data that they routinely generate or require from other sources. Large-scale computer networking has ensured that such data has become accessible to more and more people. In other words, we are in the midst of an information explosion, and there is an urgent need for methodologies that will help us to bring some semblance of order into the phenomenal volumes of data that can readily be accessed by us with a few clicks of the keys of our computer keyboard. Traditional statistical data summarization and database management techniques are just not adequate for handling data on this scale and for intelligently extracting information, or rather, knowledge that may be useful for exploring the domain in question or the phenomena responsible for the data, and providing support to decision-making processes. This quest has thrown up a new phrase, called *data mining* [12–14].

The massive databases are generally characterized by the numeric as well as textual, symbolic, pictorial, and aural data. They may contain redundancy, errors, imprecision, and so on. Data mining is aimed at discovering natural structures within such massive and often heterogeneous data. It is visualized as being capable of knowledge discovery using generalizations and magnifications of existing and new pattern recognition algorithms. Therefore, pattern recognition plays a significant role in the data mining process. Data mining deals with the process of identifying valid, novel, potentially useful, and ultimately understandable patterns in data. Hence, it can be viewed as applying pattern recognition and machine learning principles in the context of voluminous, possibly heterogeneous data sets [11].

One of the important problems in real-life data analysis is uncertainty management. Some of the sources of this uncertainty include incompleteness and vagueness in class definitions. In this background, the possibility concept introduced by the fuzzy sets theory [15] and rough sets theory [16] have gained popularity in modeling and propagating uncertainty. Both the fuzzy sets and rough sets provide a mathematical framework to capture uncertainties associated with the data [17]. They are complementary in some aspects. The generalized theories of rough-fuzzy sets and fuzzy-rough sets have been applied successfully to feature selection of real-valued data [18, 19], classification [20], image processing [21], data mining [22], information retrieval [23], fuzzy decision rule extraction, and rough-fuzzy clustering [24, 25].

The objective of this book is to provide some results of investigations, both theoretical and experimental, addressing the relevance of rough-fuzzy approaches to pattern recognition with real-life applications. Various methodologies are presented, integrating fuzzy logic and rough sets for clustering, classification, and feature selection. The emphasis of these methodologies is given on (i) handling data sets which are large, both in size and dimension, and involve classes that are overlapping, intractable and/or having nonlinear boundaries; (ii) demonstrating the significance of rough-fuzzy granular computing in soft computing paradigm for dealing with the knowledge discovery aspect; and (iii) demonstrating their success in certain tasks of bioinformatics and medical imaging as an example. Before describing the scope of the book, a brief review of pattern recognition, data mining, and application of pattern recognition algorithms in data mining problems is provided.

The structure of the rest of this chapter is as follows: Section 1.2 briefly presents a description of the basic concept, features, and techniques of pattern recognition. In Section 1.3, the data mining aspect is elaborated, discussing its components, tasks involved, approaches, and application areas. The pattern recognition perspective of data mining is introduced next and related research challenges are mentioned. The role of soft computing in pattern recognition and data mining is described in Section 1.4. Finally, Section 1.5 discusses the scope and organization of the book.

1.2 PATTERN RECOGNITION

A typical pattern recognition system consists of three phases, namely, *data acquisition*, *feature selection or extraction*, and *classification or clustering*. In the data acquisition phase, depending on the environment within which the objects are to be classified or clustered, data are gathered using a set of sensors. These are then passed on to the feature selection or extraction phase, where the dimensionality of the data is reduced by retaining or measuring only some characteristic features or properties. In a broader perspective, this stage significantly influences the entire recognition process. Finally, in the classification or clustering phase, the selected or extracted features are passed on to the classification or clustering system that evaluates the incoming information and makes a final decision. This phase basically establishes a transformation between the features and the classes or clusters [1, 2, 8].

1.2.1 Data Acquisition

In data acquisition phase, data are gathered via a set of sensors depending on the environment within which the objects are to be classified. Pattern recognition techniques are applicable in a wide domain, where the data may be qualitative, quantitative, or both; they may be numerical, linguistic, pictorial, or any combination thereof. Generally, the data structures that are used in pattern recognition systems are of two types: object data vectors and relational data. Object data, sets of numerical vectors of m features, are represented as $X = \{x_1, \ldots, x_i, \ldots, x_n\}$, a set of n feature vectors in the m-dimensional measurement space $\Re^m$. The ith object observed in the process has vector x_i as its numerical representation; x_{ij} is the jth ($j = 1, \ldots, m$) feature associated with the ith object. On the other hand, relational data are a set of n^2 numerical relationships, say r_{ij}, between pairs of objects. In other words, r_{ij} represents the extent to which objects x_i and x_j are related in the sense of some binary relationship ρ. If the objects that are pairwise related by ρ are called $O = \{o_1, \ldots, o_i, \ldots, o_n\}$, then $\rho : O \times O \to \Re$.

1.2.2 Feature Selection

Feature selection or extraction is a process of selecting a map by which a sample in an m-dimensional measurement space is transformed into a point in a d-dimensional feature space, where $d < m$ [1, 8]. Mathematically, it finds a mapping of the form $y = f(x)$, by which a sample $x = [x_1, \ldots, x_j, \ldots, x_m]$ in an m-dimensional measurement space $\mathcal{M}$ is transformed into an object $y = [y_1, \ldots, y_j, \ldots, y_d]$ in a d-dimensional feature space $\mathcal{F}$.

The main objective of this task is to retain or generate the optimum salient characteristics necessary for the recognition process and to reduce the dimensionality of the measurement space so that effective and easily computable algorithms can be devised for efficient classification. The problem of feature selection or extraction has two aspects, namely, formulating a suitable criterion to evaluate the goodness of a feature set and searching the optimal set in terms of the criterion. In general, those features are considered to have optimal saliencies for which interclass (respectively, intraclass) distances are maximized (respectively, minimized). The criterion for a good feature is that it should be unchanging with any other possible variation within a class, while emphasizing differences that are important in discriminating between patterns of different types.

The major mathematical measures so far devised for the estimation of feature quality are mostly statistical in nature, and can be broadly classified into two categories, namely, feature selection in the measurement space and feature selection in a transformed space. The techniques in the first category generally reduce the dimensionality of the measurement space by discarding redundant or least information-carrying features. On the other hand, those in the second category utilize all the information contained in the measurement space to obtain a new transformed space, thereby mapping a higher dimensional pattern to a lower dimensional one. This is referred to as *feature extraction* [1, 2, 8].

1.2.3   Classification and Clustering

The problem of classification and clustering is basically one of partitioning the feature space into regions, one region for each category of input. Hence, it attempts to assign every data object in the entire feature space to one of the possible classes or clusters. In real life, the complete description of the classes is not known. Instead, a finite and usually smaller number of samples are available, which often provide partial information for optimal design of feature selector or extractor or classification or clustering system. Under such circumstances, it is assumed that these samples are representative of the classes or clusters. Such a set of typical patterns is called a *training set*. On the basis of the information gathered from the samples in the training set, the pattern recognition systems are designed. That is, the values of the parameters of various pattern recognition methods are decided.

Design of a classification or clustering scheme can be made with labeled or unlabeled data. When the algorithm is given a set of objects with known classifications, that is, labels, and is asked to classify an unknown object based on the information acquired by it during training, the design scheme is called *supervised learning*; otherwise it is *unsupervised learning*. Supervised learning is used for classifying different objects, while clustering is performed through unsupervised learning. Through cluster analysis, a given data set is divided into a set of clusters in such a way that two objects from the same cluster are as similar as possible and the objects from different clusters are as dissimilar as possible. In effect, it tries to mimic the human ability to group similar objects into classes and categories. A number of clustering algorithms have been proposed to suit different requirements [2, 26, 27].

Pattern classification or clustering, by its nature, admits many approaches, sometimes complementary, sometimes competing, to provide the solution to a given problem. These include decision theoretic approach (both deterministic and probabilistic), syntactic approach, connectionist approach, fuzzy and rough set theoretic approaches and hybrid or soft computing approach. Let $\beta = \{\beta_1, \ldots, \beta_i, \ldots, \beta_c\}$ represent the c possible classes or clusters in a d-dimensional feature space $\mathcal{F}$, and $y = [y_1, \ldots, y_j, \ldots, y_d]$ be an unknown pattern vector whose class is to be identified. In deterministic classification or clustering approach, the object is assigned to only one unambiguous pattern class or cluster β_i if the decision function D_i associated with the class β_i satisfies the following relation:

$$D_i(y) > D_j(y), \qquad j = 1, \ldots, c, \text{ and } j \neq i. \qquad (1.2)$$

In the decision theoretic approach, once a pattern is transformed through feature evaluation to a vector in the feature space, its characteristics are expressed only by a set of numerical values. Classification can be done by using deterministic or probabilistic techniques [1, 2, 8]. The nearest neighbor classifier [2] is an example of deterministic classification approach, where it is assumed that there exists only one unambiguous pattern class corresponding to each of the unknown pattern vectors. In most of the practical problems, the features are often noisy and the classes in the feature space are overlapping. In order to model such systems, the features are considered as random variables in the probabilistic approach. The most commonly used classifier in such probabilistic systems is the Bayes maximum likelihood classifier [2].

When a pattern is rich in structural information such as picture recognition, character recognition, scene analysis, that is, the structural information plays an important role in describing and recognizing the patterns, it is convenient to use the syntactic approach [3]. It deals with the representation of structures via sentences, grammars, and automata. In the syntactic method [3], the ability of selecting and classifying the simple pattern primitives and their relationships represented by the composition operations is the vital criterion for making a system effective. Since the techniques of composition of primitives into patterns are usually governed by the formal language theory, the approach is often referred to as a *linguistic approach*. An introduction to a variety of approaches based on this idea can be found in Fu [3]. Other approaches to pattern recognition are discussed in Section 1.4 under soft computing methods.

1.3 DATA MINING

Data mining involves fitting models to or determining patterns from observed data. The fitted models play the role of inferred knowledge. Typically, a data mining algorithm constitutes some combination of three components, namely, model, preference criterion, and search algorithm [13].

The model represents its function (e.g., classification, clustering) and its representational form (e.g., linear discriminants, neural networks). A model contains parameters that are to be determined from the data. The preference criterion is a basis to decide the preference of one model or a set of parameters over another, depending on the given data. The criterion is usually some form of goodness of fit function of the model to the data, perhaps tempered by a smoothing term to avoid overfitting, or generating a model with too many degrees of freedom to be constrained by the given data. On the other hand, the search algorithm is the specification of an algorithm for finding particular models and parameters, given the data, models, and a preference criterion [5].

1.3.1 Tasks, Tools, and Applications

The common tasks or functions in current data mining practice include association rule discovery, clustering, classification, sequence analysis, regression, summarization, and dependency modeling.

The association rule discovery describes association relationship among different attributes. The origin of association rules is in market basket analysis. A market basket is a collection of items purchased by a customer in an individual customer transaction. One common analysis task in a transaction database is to find sets of items or itemsets that frequently appear together. Each pattern extracted through the analysis consists of an itemset and its support, that is, the number of transactions that contain it. Knowledge of these patterns can be used to improve placement of items in a store or for mail-order marketing. The huge size of transaction databases and the exponential increase in the number of potential frequent itemsets with increase in the number of attributes or items make the above problem a challenging one. The a priori algorithm [28] provides an early solution, which is improved by subsequent algorithms using partitioning, hashing, sampling, and dynamic itemset counting.

The clustering technique maps a data item into one of several clusters, where clusters are natural groupings of data items based on similarity metrics or probability density models. Clustering is used in several exploratory data analysis tasks, customer retention and management, and web mining. The clustering problem has been studied in many fields, including statistics, machine learning, and pattern recognition. However, large data considerations were absent in these approaches. To address those issues, several new algorithms with greater emphasis on scalability have been developed in the framework of data mining, including those based on summarized cluster representation called *cluster feature* [29], *sampling* [30], and *density joins* [31].

On the other hand, the classification algorithm classifies a data item into one of several predefined categorical classes. It is used for the purpose of predictive data mining in several fields such as scientific discovery, fraud detection, atmospheric data mining, and financial engineering. Several classification methodologies have been mentioned earlier in Section 1.2.3. Some typical algorithms suitable for large databases are based on Bayesian techniques [32] and decision trees [33, 34].

Sequence analysis [35] models sequential patterns such as time series data. The goal is to model the process of generating the sequence or to extract and report deviation and trends over time. The framework is increasingly gaining importance because of its application in bioinformatics and streaming data analysis. The regression [13, 36] technique maps a data item to a real-valued prediction variable. It is used in different prediction and modeling applications.

The summarization [13] procedure provides a compact description for a subset of data. A simple example would be mean and standard deviation for all fields. More sophisticated functions involve summary rules, multivariate visualization techniques, and functional relationship between variables. Summarization functions are often used in interactive data analysis, automated report generation, and text mining. On the other hand, the dependency modeling [37] describes significant dependencies among variables. Some other tasks required in some data mining applications are outlier or anomaly detection, link analysis, optimization, and planning.

A wide variety and number of data mining algorithms are described in the literature, from the fields of statistics, pattern recognition, machine learning, and databases. They represent a long list of seemingly unrelated and often highly specific algorithms. Some representative groups are statistical models [2, 14], probabilistic graphical dependency models [38], decision trees and rules [39], inductive-logic-programming-based models, example-based methods [40, 41], neural-network-based models [42, 43], fuzzy set theoretic models [12, 44, 45], rough set theory-based models [46–48], genetic-algorithm-based models [49], and hybrid and soft computing models [50].

Data mining algorithms determine both the flexibility of the model in representing the data and the interpretability of the model in human terms. Typically, the more complex models may fit the data better but may also be more difficult to understand and to fit reliably. Also, each representation suits some problems better than others. For example, decision tree classifiers can be very useful for finding structure in high dimensional spaces and are also useful in problems with mixed continuous and categorical data. However, they may not be suitable for problems where the true decision boundaries are nonlinear multivariate functions.

A wide range of organizations including business companies, scientific laboratories, and governmental departments have deployed successful applications of data mining. Although early adopters of this technology have tended to be in information-intensive industries such as financial services and direct mail marketing, the technology is applicable to any company looking to leverage a large data warehouse to better manage their operations. Two critical factors for success with data mining are a large, well-integrated data warehouse and a well-defined understanding of the process within which data mining is to be applied. Several domains where large volumes of data are stored in centralized or distributed databases include financial investment, hospital management systems, manufacturing and production, telecommunication network, astronomical object detection, genomic and biological data mining, and information retrieval [5].

1.3.2 Pattern Recognition Perspective

At present, pattern recognition and machine learning provide the most fruitful framework for data mining [5, 51, 52]. They provide a wide range of linear and nonlinear, comprehensible and complex, predictive and descriptive, instance and rule-based models for different data mining tasks such as clustering, classification, and rule discovery. Also, the methods for modeling probabilistic and fuzzy uncertainties in the discovered patterns form a part of pattern recognition research. Another aspect that makes pattern recognition algorithms attractive for data mining is their capability of learning or induction. As opposed to many statistical techniques that require the user to have a hypothesis in mind first, pattern recognition algorithms automatically analyze the data and identify relationships among attributes and entities in the data to build models that allow domain experts to understand the relationship between the attributes and the class. Several data preprocessing tasks such as instance selection, data cleaning, dimensionality reduction, and handling missing data are also extensively studied in pattern recognition framework. Besides these, other data mining issues addressed by pattern recognition methodologies include handling of relational, sequential, and symbolic data (syntactic pattern recognition; pattern recognition in arbitrary metric spaces); human interaction (knowledge encoding and extraction); knowledge evaluation (description length principle); and visualization.

Pattern recognition is at the core of data mining systems. However, pattern recognition and data mining are not equivalent considering their original definitions. There exists a gap between the requirements of a data mining system and the goals achieved by present-day pattern recognition algorithms. Development of new generation pattern recognition algorithms is expected to encompass more massive data sets involving diverse sources and types of data that will support mixed initiative data mining, where human experts collaborate with the computer to form hypotheses and test them.

1.4 RELEVANCE OF SOFT COMPUTING

A good pattern recognition system should possess several characteristics. These are online adaptation to cope with the changes in the environment, handling nonlinear class separability to tackle real-life problems, handling of overlapping classes or clusters for discriminating almost similar but different objects, real time processing for making a decision in a reasonable time, generation of soft and hard decisions to make the system flexible, verification and validation mechanisms for evaluating its performance, and minimizing the number of parameters in the system that have to be tuned for reducing the cost and complexity. Moreover, the system should be made artificially intelligent in order to emulate some aspects of the human processing system. Connectionist or artificial neural-network-based approaches to pattern recognition are attempts to achieve some of these goals because of their major characteristics such as adaptivity, robustness or ruggedness, speed, and optimality [53–57]. They are also suitable in data-rich environments and are typically used for extracting embedded knowledge in the form of rules, quantitative evaluation of these rules, clustering, self-organization, classification, and regression. They have an advantage, over other types of machine learning algorithms, for scaling [58, 59].

The fuzzy set theoretic classification approach is developed on the basis of the realization that a pattern may belong to more than one class, with varying degrees of class membership. Accordingly, fuzzy decision theoretic, fuzzy syntactic, fuzzy neural approaches are developed [4, 6, 60, 61]. These approaches can handle uncertainties, arising from vague, incomplete, linguistic, and overlapping patterns at various stages of pattern recognition systems [4, 15, 60, 62].

The theory of rough sets [16, 63, 64] has emerged as another major mathematical approach for managing uncertainty that arises from inexact, noisy, or incomplete information. It is turning out to be methodologically significant to the domains of artificial intelligence and cognitive sciences, especially in the representation of and reasoning with vague and/or imprecise knowledge, data classification, data analysis, machine learning, and knowledge discovery [48, 64–66]. This approach is relatively new when compared to connectionist and fuzzy set theoretic approaches.

Investigations have also been made in the area of pattern recognition using genetic algorithms [67, 68]. Similar to neural networks, genetic algorithms [69] are also based on powerful metaphors from the natural world. They mimic some of the processes observed in natural evolution, which include crossover, selection, and mutation, leading to a stepwise optimization of organisms.

There have been several attempts over the past two decades to evolve new approaches to pattern recognition and to derive their hybrids by judiciously combining the merits of several techniques [6, 70] involving mainly fuzzy logic, artificial neural networks, genetic algorithms, and rough set theory, for developing an efficient new paradigm called *soft computing* [71]. Here integration is done in a cooperative, rather than a competitive, manner. The result is a more intelligent and robust system providing a human interpretable, low cost, approximate solution, as compared to traditional techniques. Neuro-fuzzy approach is perhaps the most visible hybrid paradigm [6, 61, 72–75] realized so far in soft computing framework. Besides the generic advantages, neuro-fuzzy approach provides the corresponding application-specific merits [76–82]. Rough-fuzzy [63, 83] and neuro-rough [84–86] hybridizations are also proving to be fruitful frameworks for modeling human perceptions and providing means for computing with words. Rough-fuzzy computing provides a powerful mathematical framework to capture uncertainties associated with the data. Its relevance in modeling the fuzzy granulation (f-granulation) characteristics of the computational theory of perceptions may also be mentioned in this regard [87–89]. Other hybridized models for pattern recognition and data mining include neuro-genetic [90–94], rough-genetic [95–97], fuzzy-genetic [98–103], rough-neuro-genetic [104], rough-neuro-fuzzy [105–109], and neuro-fuzzy-genetic [110–115] approaches.

1.5 SCOPE AND ORGANIZATION OF THE BOOK

This book has nine chapters describing various theories, methodologies, and algorithms, along with extensive experimental results, addressing certain pattern recognition and mining tasks in rough-fuzzy computing paradigm with real-life applications. Various methodologies are described using soft computing approaches, judiciously integrating fuzzy logic and rough sets for clustering, classification, and feature selection. The emphasis is placed on the use of the methodologies for handling both object and relational data sets that are large both in size and dimension, and involve classes that are overlapping, intractable and/or having nonlinear boundaries. The effectiveness of the algorithms is demonstrated on different real-life data sets taken from varied domains such as remote sensing, medical imagery, speech recognition, protein sequence encoding and gene expression analysis with special emphasis on problems in medical imaging and mining patterns in bioinformatics. The superiority of the rough-fuzzy models presented in this book over several related ones is found to be statistically significant.

The basic notions and characteristics of two soft computing tools, namely, fuzzy sets and rough sets are briefly presented in Chapter 2. These are followed by the concept of information granules, f-granulations, emergence of rough-fuzzy computing paradigm, and their relevance to pattern recognition. It also provides a mathematical framework for generalized rough sets incorporating the concept of fuzziness in defining the granules as well as the set. Various roughness and uncertainty measures with properties are introduced. Different research issues related to rough granules are stated.

A generalized hybrid unsupervised learning algorithm, termed as *rough-fuzzy-possibilistic c-means*, is reported in Chapter 3. It comprises a judicious integration of the principles of rough sets and fuzzy sets. Although the concept of lower and upper approximations of rough sets deals with uncertainty, vagueness, and incompleteness in class definition, the membership function of fuzzy sets enables efficient handling of overlapping partitions. It incorporates both probabilistic and possibilistic memberships simultaneously to avoid the problems of noise sensitivity of fuzzy c-means and the coincident clusters of possibilistic c-means. The concept of crisp lower bound and fuzzy boundary of a class, introduced in rough-fuzzy-possibilistic c-means, enables efficient selection of cluster prototypes. The algorithm is generalized in the sense that all the existing variants of c-means algorithms can be derived from this algorithm as a special case. Several quantitative indices are described on the basis of rough sets for evaluating the performance of different c-means algorithms on real-life data sets.

A rough-fuzzy model for pattern classification based on granular computing is described in Chapter 4. In this model, the formulation of class-dependent granules in fuzzy environment is introduced. Fuzzy membership functions are used to represent the feature-wise belonging to different classes, thereby producing fuzzy granulation of the feature space. The fuzzy granules thus generated possess better class discriminatory information that is useful in pattern classification with overlapping classes. The neighborhood rough sets are used in the selection of a subset of granulated features that explore the local or contextual information from neighborhood granules. The model thus explores the mutual advantages of class-dependent fuzzy granulation and neighborhood rough sets. The superiority of this model over other similar methods is established with seven completely labeled data sets, including a synthetic remote sensing image, and two partially labeled real remote sensing images collected from satellites. Various performance measures, including a method of dispersion estimation, are used for comparative analysis. The dispersion score quantifies the nature of distribution of the classified patterns among different classes so that lower the dispersion, better the classifier. The rough-fuzzy granular space-based model is able to learn well even with a lower percentage of training set that makes the system faster. The model is seen to have the lowest dispersion measure (i.e., misclassified patterns are confined to minimum number of classes) compared to others, thereby reflecting well the overlapping characteristics of a class with others, and providing a strong clue for the class-wise performance improvement with available higher level information. The statistical significance of this model is also supported by the χ^2 test.

The selection of nonredundant and relevant features of real-valued data sets is a highly challenging problem. Chapter 5 deals with a feature selection method based on fuzzy-rough sets by maximizing the relevance and minimizing the redundancy of the selected features. By introducing the concept of fuzzy equivalence partition matrix, a new representation of Shannon's entropy for fuzzy approximation spaces is presented to measure the relevance and redundancy of features suitable for real-valued data sets. The fuzzy equivalence partition matrix is based on the theory of fuzzy-rough sets, where each row of the matrix represents a fuzzy equivalence partition that can be automatically derived from the given data set. The fuzzy equivalence partition matrix also offers an efficient way to calculate many more information measures, termed as *f-information measures*. Several *f*-information measures are shown to be effective for selecting nonredundant and relevant features of real-valued data sets. The experimental study also includes a comparison of the performance of different *f*-information measures for feature selection in fuzzy approximation spaces. Several quantitative indices are described on the basis of fuzzy-rough sets for evaluating the performance of different methods.

In pattern recognition, there are mainly two types of data: object and relational data. The former is the most common type of data and is in the form of the usual data set of feature vectors. On the other hand, the latter is less common and consists of the pairwise relations such as similarities or dissimilarities between each pair of implicit objects. Such a relation is usually stored in a relation matrix and no other knowledge is available about the objects being clustered. As the relational data is less common than object data, relational pattern recognition methods are not as well developed as their object counterparts, particularly in the area of robust clustering. However, relational methods are becoming a necessity as relational data becomes more and more common. For instance, information retrieval, data mining, web mining, and bioinformatics are all applications which could greatly benefit from pattern recognition methods that can deal with relational data. In this regard, the next chapter discusses a rough-fuzzy relational clustering algorithm, termed as *rough-fuzzy c-medoids algorithm*, and demonstrates its effectiveness in amino acid sequence analysis.

Although several experimental results on both artificial and real-life data sets, including speech and remotely sensed multispectral image data, are provided in Chapters 3, 4, and 5 to demonstrate the effectiveness of the respective rough-fuzzy methodologies, the next four chapters are concerned only with certain specific applications in bioinformatics and medical imaging. Problems considered include selection of a minimum set of basis strings with maximum information for amino acid sequence analysis (Chapter 6), grouping functionally similar genes from microarray gene expression data through clustering (Chapter 7), selection of relevant genes from high dimensional microarray gene expression data (Chapter 8), and segmentation of brain magnetic resonance (MR) images using clustering (Chapter 9).

In most pattern recognition algorithms, biological molecules such as amino acids cannot be used directly as inputs as they are nonnumerical variables. They, therefore, need encoding before being used as input. In this regard, bio-basis function maps a nonnumerical sequence space to a numerical feature space. It is designed using an amino acid mutation matrix. One of the important issues for the bio-basis function is how to select a minimum set of bio-basis strings with maximum information. In Chapter 6, the rough-fuzzy c-medoids algorithm is used to select most informative bio-basis strings. It comprises a judicious integration of the principles of rough sets, fuzzy sets, c-medoids algorithm, and amino acid mutation matrix. The concept of crisp lower bound and fuzzy boundary of a cluster, introduced in rough-fuzzy c-medoids, enables efficient selection of a minimum set of most informative bio-basis strings. Several indices are stated for evaluating quantitatively the quality of selected bio-basis strings.

Microarray technology is one of the important biotechnological means that allows recording the expression levels of thousands of genes during important biological processes and across collections of related samples. An important application of microarray data is to elucidate the patterns hidden in gene expression data for an enhanced understanding of functional genomics. However, the large number of genes and the complexity of biological networks greatly increase the challenges of comprehending and interpreting the resulting mass of data. A first step toward addressing this challenge is the use of clustering techniques. In this regard, different rough-fuzzy clustering algorithms are used in Chapter 7 to cluster functionally similar genes from microarray data sets. The effectiveness of these algorithms, along with a comparison with other related gene clustering algorithms, is demonstrated on a set of microarray gene expression data sets using some standard validity indices.

Several information measures such as entropy, mutual information, and f-information have been shown to be successful for selecting a set of relevant and nonredundant genes from high dimensional microarray data set. However, for continuous gene expression values, it is very difficult to find the true density functions and to perform the integrations required to compute different information measures. In this regard, the concept of fuzzy equivalence partition matrix, explained in Chapter 5, is used in Chapter 8 to approximate the true marginal and joint distributions of continuous gene expression values.

The performance of this methodology in selecting relevant and nonredundant continuous valued genes from microarray data is compared with that of existing ones using the class separability index and predictive accuracy of support vector machine.

Image segmentation is an indispensable process in the visualization of human tissues, particularly during clinical analysis of MR images. In Chapter 9, different rough-fuzzy clustering algorithms are used for the segmentation of brain MR images. One of the major issues of the rough-fuzzy clustering-algorithm-based brain MR image segmentation is how to select initial prototypes of different classes or categories. The concept of discriminant analysis, based on the maximization of class separability, is used to circumvent the initialization and local minima problems of the rough-fuzzy clustering algorithms. Some quantitative indices are described to extract local features of brain MR images, when applied on a set of synthetic and real brain MR images, for segmentation.

REFERENCES

1. P. A. Devijver and J. Kittler. *Pattern Recognition: A Statistical Approach*. Prentice Hall, Englewood Cliffs, NJ, 1982.

2. R. O. Duda, P. E. Hart, and D. G. Stork. *Pattern Classification and Scene Analysis*. John Wiley & Sons, New York, 1999.

3. K. S. Fu. *Syntactic Pattern Recognition and Applications*. Academic Press, London, 1982.

4. S. K. Pal and D. D. Majumder. *Fuzzy Mathematical Approach to Pattern Recognition*. John Wiley (Halsted Press), New York, 1986.

5. S. K. Pal and P. Mitra. *Pattern Recognition Algorithms for Data Mining*. CRC Press, Boca Raton, FL, 2004.

6. S. K. Pal and S. Mitra. *Neuro-Fuzzy Pattern Recognition: Methods in Soft Computing*. New York: John Wiley & Sons, 1999.

7. S. K. Pal and A. Pal, editors. *Pattern Recognition: from Classical to Modern Approaches*. World Scientific, Singapore, 2001.

8. J. T. Tou and R. C. Gonzalez. *Pattern Recognition Principles*. Addison-Wesley, Reading, MA, 1974.

9. L. Kanal. Patterns in Pattern Recognition. *IEEE Transactions on Information Theory*, 20:697–722, 1974.

10. A. K. Jain, R. P. W. Duin, and J. Mao. Statistical Pattern Recognition: A Review. *IEEE Transactions on Pattern Analysis and Machine Intelligence*, 22:4–37, 2000.

11. A. Pal and S. K. Pal. Pattern Recognition: Evolution of Methodologies and Data Mining. In S. K. Pal and A. Pal, editors, *Pattern Recognition: from Classical to Modern Approaches*, pages 1–23. World Scientific, Singapore, 2001.

12. K. Cios, W. Pedrycz, and R. Swiniarski. *Data Mining Methods for Knowledge Discovery*. Kluwer Academic Publishers, Boston, MA, 1998.

13. U. M. Fayyad, G. Piatetsky-Shapiro, P. Smyth, and R. Uthurusamy, editors. *Advances in Knowledge Discovery and Data Mining*. MIT Press, Cambridge, MA, 1996.

14. T. Hastie, R. Tibshirani, and J. Friedman. *The Elements of Statistical Learning: Data Mining, Inference, and Prediction*. Springer-Verlag, New York, 2001.

15. L. A. Zadeh. Fuzzy Sets. *Information and Control*, 8:338–353, 1965.

16. Z. Pawlak. *Rough Sets: Theoretical Aspects of Reasoning about Data*. Kluwer, Dordrecht, The Netherlands, 1991.

17. D. Dubois and H. Prade. Rough Fuzzy Sets and Fuzzy Rough Sets. *International Journal of General Systems*, 17:191–209, 1990.

18. R. Jensen and Q. Shen. Fuzzy-Rough Attribute Reduction with Application to Web Categorization. *Fuzzy Sets and Systems*, 141:469–485, 2004.

19. R. Jensen and Q. Shen. Semantics-Preserving Dimensionality Reduction: Rough and Fuzzy-Rough-Based Approach. *IEEE Transactions on Knowledge and Data Engineering*, 16(12): 1457–1471, 2004.

20. S. K. Pal, S. Meher, and S. Dutta. Class-Dependent Rough-Fuzzy Granular Space, Dispersion Index and Classification. *Pattern Recognition*, under revision.

21. D. Sen and S. K. Pal. Generalized Rough Sets, Entropy and Image Ambiguity Measures. *IEEE Transactions on Systems Man and Cybernetics Part B-Cybernetics*, 39(1): 117–128, 2009.

22. Y. F. Wang. Mining Stock Price Using Fuzzy Rough Set System. *Expert Systems with Applications*, 24(1): 13–23, 2003.

23. P. Srinivasan, M. E. Ruiz, D. H. Kraft, and J. Chen. Vocabulary Mining for Information Retrieval: Rough Sets and Fuzzy Sets. *Information Processing and Management*, 37(1): 15–38, 1998.

24. P. Maji and S. K. Pal. Rough-Fuzzy C-Medoids Algorithm and Selection of Bio-Basis for Amino Acid Sequence Analysis. *IEEE Transactions on Knowledge and Data Engineering*, 19(6): 859–872, 2007.

25. P. Maji and S. K. Pal. Rough Set Based Generalized Fuzzy C-Means Algorithm and Quantitative Indices. *IEEE Transactions on Systems Man and Cybernetics Part B-Cybernetics*, 37(6): 1529–1540, 2007.

26. A. K. Jain and R. C. Dubes. *Algorithms for Clustering Data*. Prentice Hall, Englewood Cliffs, NJ, 1988.

27. A. K. Jain, M. N. Murty, and P. J. Flynn. Data Clustering: A Review. *ACM Computing Surveys*, 31(3): 264–323, 1999.

28. R. Agrawal, H. Mannila, R. Srikant, H. Toivonen, and I. Verkamo. Fast Discovery of Association Rules. In U. M. Fayyad, G. Piatetsky-Shapiro, P. Smyth, and R. Uthuruswamy, editors, *Advances in Knowledge Discovery and Data Mining*, pages 307–328. MIT Press, Cambridge, MA, 1996.

29. P. Bradley, U. M. Fayyad, and C. Reina. Scaling Clustering Algorithms to Large Databases. In *Proceedings of the 4th International Conference on Knowledge Discovery and Data Mining, New York*, pages 9–15, AAAI Press, Menlo Park, CA, 1998.

30. S. Guha, R. Rastogi, and K. Shim. CURE: An Efficient Clustering Algorithm for Large Databases. In *Proceedings of the ACM SIGMOD International Conference on Management of Data*, pages 73–84, ACM Press, New York, 1998.

31. J. Ester, H.-P. Kriegel, J. Sander, and X. Xu. A Density-Based Algorithm for Discovering Clusters in Large Spatial Databases with Noise. In *Proceedings of the 2nd International Conference on Knowledge Discovery and Data Mining, Portland, OR*, pages 226–231, AAAI Press, Menlo Park, CA, 1996.

32. P. Cheeseman, J. Kelly, M. Self, and J. Stutz. Autoclass: A Bayesian Classification System. In *Proceedings of the 5th International Conference on Machine Learning, Ann Arbor, MI*, Morgan Kaufmann, San Mateo, CA, 1988.

33. J. Gehrke, R. Ramakrishnan, and V. Ganti. RainForest: A Framework for Large Decision Tree Construction for Large Datasets. In *Proceedings of the 24th International Conference on Very Large Databases, San Francisco, CA*, pages 416–427, Morgan Kaufmann, San Mateo, CA, 1998.

34. J. Shafer, R. Agrawal, and M. Mehta. SPRINT: A Scalable Parallel Classifier for Data Mining. In *Proceedings of the 22nd International Conference on Very Large Databases*, San Francisco, pages 544–555. Morgan Kaufmann, San Mateo, CA, 1996.

35. D. Gusfield. *Algorithms on Strings, Trees, and Sequences: Computer Science and Computational Biology*. Cambridge University Press, Cambridge, 1997.

36. P. Maji and S. Paul. Rough Sets for Selection of Molecular Descriptors to Predict Biological Activity of Molecules. *IEEE Transactions on Systems Man and Cybernetics Part C-Applications and Reviews*, 40(6): 639–648, 2010.

37. J. Hale and S. Shenoi. Analyzing FD Inference in Relational Databases. *Data and Knowledge Engineering*, 18:167–183, 1996.

38. F. V. Jensen. *Bayesian Networks and Decision Diagrams*. Springer-Verlag, New York, 2001.

39. L. Breiman, J. H. Friedman, R. A. Olshen, and C. J. Stone. *Classification and Regression Trees*. Wadsworth and Brooks/Cole, Monterey, CA, 1984.

40. D. W. Aha, D. Kibler, and M. K. Albert. Instance-Based Learning Algorithms. *Machine Learning*, 6:37–66, 1991.

41. S. K. Pal and S. C. K. Shiu. *Foundations of Soft Case Based Reasoning*. John Wiley & Sons, New York, 2004.

42. M. Craven and J. Shavlik. Using Neural Networks for Data Mining. *Future Generation Computer Systems*, 13:211–219, 1997.

43. H. Lu, R. Setiono, and H. Liu. Effective Data Mining Using Neural Networks. *IEEE Transactions on Knowledge and Data Engineering*, 8(6): 957–961, 1996.

44. J. F. Baldwin. Knowledge from Data Using Fuzzy Methods. *Pattern Recognition Letters*, 17:593–600, 1996.

45. W. Pedrycz. Fuzzy Set Technology in Knowledge Discovery. *Fuzzy Sets and Systems*, 98:279–290, 1998.

46. J. Komorowski, Z. Pawlak, L. Polkowski, and A. Skowron. A Rough Set Perspective on Data and Knowledge. In W. Klosgen and J. Zytkow, editors, *The Handbook of Data Mining and Knowledge Discovery*. Oxford University Press, Oxford, 1999.

47. T. Y. Lin and N. Cercone, editors. *Rough Sets and Data Mining: Analysis of Imprecise Data*. Kluwer Academic Publications, Boston, MA, 1997.

48. L. Polkowski and A. Skowron, editors. *Rough Sets in Knowledge Discovery*, volumes 1 and 2. Physica-Verlag, Heidelberg, 1998.

49. I. W. Flockhart and N. J. Radcliffe. A Genetic Algorithm-Based Approach to Data Mining. In *Proceedings of the 2nd International Conference on Knowledge Discovery and Data Mining, Portland, OR*, page 299, AAAI Press, Menlo Park, CA, 1996.

50. S. Mitra, S. K. Pal, and P. Mitra. Data Mining in Soft Computing Framework: A Survey. *IEEE Transactions on Neural Networks*, 13(1): 3–14, 2002.

51. R. L. Kennedy, Y. Lee, B. van Roy, C. D. Reed, and R. P. Lippman. *Solving Data Mining Problems Through Pattern Recognition*. Prentice Hall, NJ, 1998.

52. T. Mitchell. Machine Learning and Data Mining. *Communications of the ACM*, 42(11): 30–36, 1999.

53. W. Bian and X. Xue. Subgradient-Based Neural Networks for Nonsmooth Nonconvex Optimization Problems. *IEEE Transactions on Neural Networks*, 20(6): 1024–1038, 2009.

54. H. Chen and X. Yao. Regularized Negative Correlation Learning for Neural Network Ensembles. *IEEE Transactions on Neural Networks*, 20(12): 1962–1979, 2009.

55. S. Haykin. *Neural Networks: A Comprehensive Foundation*, 2nd edition. Prentice Hall, New Jersey, 1998.

56. R. Lippmann. An Introduction to Computing with Neural Nets. *IEEE ASSP Magazine*, 4(22), 1987.

57. S. L. Phung and A. Bouzerdoum. A Pyramidal Neural Network for Visual Pattern Recognition. *IEEE Transactions on Neural Networks*, 18(2): 329–343, 2007.

58. Y. Bengio, J. M. Buhmann, M. Embrechts, and J. M. Zurada. Introduction to the Special Issue on Neural Networks for Data Mining and Knowledge Discovery. *IEEE Transactions on Neural Networks*, 11:545–549, 2000.

59. A. A. Frolov, D. Husek, and P. Y. Polyakov. Recurrent-Neural-Network-Based Boolean Factor Analysis and Its Application to Word Clustering. *IEEE Transactions on Neural Networks*, 20(7): 1073–1086, 2009.

60. J. C. Bezdek and S. K. Pal, editors. *Fuzzy Models for Pattern Recognition: Methods that Search for Structures in Data*. IEEE Press, New York, 1992.

61. H. Bunke and A. Kandel, editors. *Neuro-Fuzzy Pattern Recognition*. World Scientific, Singapore, 2001.

62. A. Kandel. *Fuzzy Techniques in Pattern Recognition*. Wiley Interscience, New York, 1982.

63. S. K. Pal and A. Skowron, editors. *Rough-Fuzzy Hybridization: A New Trend in Decision Making*. Springer-Verlag, Singapore, 1999.

64. A. Skowron and R. Swiniarski. Rough Sets in Pattern Recognition. In S. K. Pal and A. Pal, editors, *Pattern Recognition: From Classical to Modern Approaches*, pages 385–428. World Scientific, Singapore, 2001.

65. E. Orlowska, editor. *Incomplete Information: Rough Set Analysis*. Physica-Verlag, Heidelberg, 2010.

66. L. Polkowski. *Rough Sets*. Physica-Verlag, Heidelberg, 2002.

67. S. Bandyopadhyay and S. K. Pal. *Classification and Learning Using Genetic Algorithms: Applications in Bioinformatics and Web Intelligence*. Springer-Verlag, Hiedelberg, Germany, 2007.

68. S. K. Pal and P. P. Wang, editors. *Genetic Algorithms for Pattern Recognition*. CRC Press, Boca Raton, FL, 1996.

69. D. E. Goldberg. *Genetic Algorithms in Search, Optimization and Machine Learning*. Addison-Wesley, Reading, MA, 1989.

70. S. K. Pal. Soft Computing Pattern Recognition: Principles, Integrations and Data Mining. In T. Terano, T. Nishida, A. Namatame, S. Tsumoto, Y. Ohswa, and T. Washio, editors, *Advances in Artificial Intelligence , Lecture Notes in Artificial Intelligence*, volume 2253, pages 261–268. Springer-Verlag, Berlin, 2002.

71. L. A. Zadeh. Fuzzy logic, Neural Networks, and Soft Computing. *Communications of the ACM*, 37:77–84, 1994.

72. S. Mitra, R. K. De, and S. K. Pal. Knowledge-Based Fuzzy MLP for Classification and Rule Generation. *IEEE Transactions on Neural Networks*, 8:1338–1350, 1997.

73. S. Mitra and S. K. Pal. Fuzzy Multi-Layer Perceptron, Inferencing and Rule Generation. *IEEE Transactions on Neural Networks*, 6:51–63, 1995.

74. S. Mitra and S. K. Pal. Fuzzy Self Organization, Inferencing and Rule Generation. *IEEE Transactions on Systems Man and Cybernetics Part A-Systems and Humans*, 26:608–620, 1996.

75. S. K. Pal and A. Ghosh. Neuro-Fuzzy Computing for Image Processing and Pattern Recognition. *International Journal of System Science*, 27(12): 1179–1193, 1996.

76. K. Cpalka. A New Method for Design and Reduction of Neuro-Fuzzy Classification Systems. *IEEE Transactions on Neural Networks*, 20(4): 701–714, 2009.

77. A. Gajate, R. E. Haber, P. I. Vega, and J. R. Alique. A Transductive Neuro-Fuzzy Controller: Application to a Drilling Process. *IEEE Transactions on Neural Networks*, 21(7): 1158–1167, 2010.

78. L. B. Goncalves, M. M. B. R. Vellasco, M. A. C. Pacheco, and F. J. de Souza. Inverted Hierarchical Neuro-Fuzzy BSP System: A Novel Neuro-Fuzzy Model for Pattern Classification and Rule Extraction in Databases. *IEEE Transactions on Systems Man and Cybernetics Part C-Applications and Reviews*, 36(2): 236–248, 2006.

79. Z.-L. Sun, K.-F. Au, and T.-M. Choi. A Neuro-Fuzzy Inference System Through Integration of Fuzzy Logic and Extreme Learning Machines. *IEEE Transactions on Systems Man and Cybernetics Part B-Cybernetics*, 37(5): 1321–1331, 2007.

80. W.-C. Wong, S.-Y. Cho, and C. Quek. R-POPTVR: A Novel Reinforcement-Based POPTVR Fuzzy Neural Network for Pattern Classification. *IEEE Transactions on Neural Networks*, 20(11): 1740–1755, 2009.

81. J. Zhang. Modeling and Optimal Control of Batch Processes Using Recurrent Neuro-Fuzzy Networks. *IEEE Transactions on Fuzzy Systems*, 13(4): 417–427, 2005.

82. L. Zhou and A. Zenebe. Representation and Reasoning Under Uncertainty in Deception Detection: A Neuro-Fuzzy Approach. *IEEE Transactions on Fuzzy Systems*, 16(2): 442–454, 2008.

83. S. K. Pal and A. Skowron, editors. Special Issue on Rough Sets, Pattern Recognition and Data Mining. *Pattern Recognition Letters*, 24(6), 2003.

84. J.-H. Chiang and S.-H. Ho. A Combination of Rough-Based Feature Selection and RBF Neural Network for Classification Using Gene Expression Data. *IEEE Transactions on NanoBioscience*, 7(1): 91–99, 2008.

85. J. Jiang, D. Yang, and H. Wei. Image Segmentation Based on Rough Set Theory and Neural Networks. In *Proceedings of the 5th International Conference on Visual Information Engineering*, pages 361–365. IET, UK, 2008.

86. S. K. Pal, L. Polkowski, and A. Skowron, editors. *Rough-Neuro Computing: Techniques for Computing with Words*. Springer, Heidelberg, 2003.

87. S. K. Pal. Soft Data Mining, Computational Theory of Perceptions, and Rough-Fuzzy Approach. *Information Sciences*, 163(1–3): 5–12, 2004.

88. S. K. Pal. Computational Theory of Perception (CTP), Rough-Fuzzy Uncertainty Analysis and Mining in Bioinformatics and Web Intelligence: A Unified Framework. *LNCS Transactions on Rough Sets*, 5946:106–129, 2009.

89. L. A. Zadeh. A New Direction in AI: Toward a Computational Theory of Perceptions. *AI Magazine*, 22:73–84, 2001.

90. S. Bornholdt and D. Graudenz. General Asymmetric Neural Networks and Structure Design by Genetic Algorithms. *Neural Networks*, 5:327–334, 1992.

91. V. Maniezzo. Genetic Evolution of the Topology and Weight Distribution of Neural Networks. *IEEE Transactions on Neural Networks*, 5:39–53, 1994.

92. S. K. Pal and D. Bhandari. Selection of Optimal Set of Weights in a Layered Network Using Genetic Algorithms. *Information Sciences*, 80:213–234, 1994.

93. S. Saha and J. P. Christensen. Genetic Design of Sparse Feedforward Neural Networks. *Information Sciences*, 79:191–200, 1994.

94. D. Whitley, T. Starkweather, and C. Bogart. Genetic Algorithms and Neural Networks: Optimizing Connections and Connectivity. *Parallel Computing*, 14:347–361, 1990.

95. A. T. Bjorvand and J. Komorowski. Practical Applications of Genetic Algorithms for Efficient Reduct Computation. In *Proceedings of the 15th IMACS World Congress on Scientific Computation, Modeling and Applied Mathematics*, Berlin, volume 4, pages 601–606, 1997.

96. D. Slezak. Approximate Reducts in Decision Tables. In *Proceedings of the 6th International Conference on Information Processing and Management of Uncertainty in Knowledge-Based Systems*, Granada, pages 1159–1164, 1996.

97. J. Wroblewski. Finding Minimal Reducts Using Genetic Algorithms. In *Proceedings of the 2nd Annual Joint Conference on Information Sciences*, North Carolina, pages 186–189, 1995.

98. G. Ascia, V. Catania, and D. Panno. An Integrated Fuzzy-GA Approach for Buffer Management. *IEEE Transactions on Fuzzy Systems*, 14(4): 528–541, 2006.

99. J. Casillas, B. Carse, and L. Bull. Fuzzy-XCS: A Michigan Genetic Fuzzy System. *IEEE Transactions on Fuzzy Systems*, 15(4): 536–550, 2007.

100. C.-H. Chen, V. S. Tseng, and T.-P. Hong. Cluster-Based Evaluation in Fuzzy-Genetic Data Mining. *IEEE Transactions on Fuzzy Systems*, 16(1): 249–262, 2008.

101. V. Giordano, D. Naso, and B. Turchiano. Combining Genetic Algorithms and Lyapunov-Based Adaptation for Online Design of Fuzzy Controllers. *IEEE Transactions on Systems Man and Cybernetics Part B-Cybernetics*, 36(5): 1118–1127, 2006.

102. C.-S. Lee, S.-M. Guo, and C.-Y. Hsu. Genetic-Based Fuzzy Image Filter and Its Application to Image Processing. *IEEE Transactions on Systems Man and Cybernetics Part B-Cybernetics*, 35(4): 694–711, 2005.

103. A. Mukhopadhyay, U. Maulik, and S. Bandyopadhyay. Multiobjective Genetic Algorithm-Based Fuzzy Clustering of Categorical Attributes. *IEEE Transactions on Evolutionary Computation*, 13(5): 991–1005, 2009.

104. H. Kiem and D. Phuc. Using Rough Genetic and Kohonen's Neural Network for Conceptual Cluster Discovery in Data Mining. In *Proceedings of the 7th International Conference on Rough Sets, Fuzzy Sets, Data Mining, and Granular Computing*, pages 448–452, Yamaguchi, Japan, 1999.

105. K. K. Ang and C. Quek. Stock Trading Using RSPOP: A Novel Rough Set-Based Neuro-Fuzzy Approach. *IEEE Transactions on Neural Networks*, 17(5): 1301–1315, 2006.

106. M. Banerjee, S. Mitra, and S. K. Pal. Rough-Fuzzy MLP: Knowledge Encoding and Classification. *IEEE Transactions on Neural Networks*, 9(6): 1203–1216, 1998.

107. R. Nowicki. On Combining Neuro-Fuzzy Architectures with the Rough Set Theory to Solve Classification Problems with Incomplete Data. *IEEE Transactions on Knowledge and Data Engineering*, 20(9): 1239–1253, 2008.

108. R. Nowicki. Rough Neuro-Fuzzy Structures for Classification with Missing Data. *IEEE Transactions on Systems Man and Cybernetics Part B-Cybernetics*, 39(6): 1334–1347, 2009.

109. S. K. Pal, S. Mitra, and P. Mitra. Rough-Fuzzy MLP: Modular Evolution, Rule Generation, and Evaluation. *IEEE Transactions on Knowledge and Data Engineering*, 15(1): 14–25, 2003.

110. G. Leng, T. M. McGinnity, and G. Prasad. Design for Self-Organizing Fuzzy Neural Networks Based on Genetic Algorithms. *IEEE Transactions on Fuzzy Systems*, 14(6): 755–766, 2006.

111. G.-C. Liao and T.-P. Tsao. Application of a Fuzzy Neural Network Combined with a Chaos Genetic Algorithm and Simulated Annealing to Short-Term Load Forecasting. *IEEE Transactions on Evolutionary Computation*, 10(3): 330–340, 2006.

112. A. Quteishat, C. P. Lim, and K. S. Tan. A Modified Fuzzy Min-Max Neural Network with a Genetic-Algorithm-Based Rule Extractor for Pattern Classification. *IEEE Transactions on Systems Man and Cybernetics Part A-Systems and Humans*, 40(3): 641–650, 2010.

113. M. Russo. FuGeNeSys: A Fuzzy Genetic Neural System for Fuzzy Modeling. *IEEE Transactions on Fuzzy Systems*, 6(3): 373–388, 1998.

114. T. L. Seng, M. Bin Khalid, and R. Yusof. Tuning of a Neuro-Fuzzy Controller by Genetic Algorithm. *IEEE Transactions on Systems Man and Cybernetics Part B-Cybernetics*, 29(2): 226–236, 1999.

115. W.-Y. Wang and Y.-H. Li. Evolutionary Learning of BMF Fuzzy-Neural Networks Using a Reduced-Form Genetic Algorithm. *IEEE Transactions on Systems Man and Cybernetics Part B-Cybernetics*, 33(6): 966–976, 2003.

2

ROUGH-FUZZY HYBRIDIZATION AND GRANULAR COMPUTING

2.1 INTRODUCTION

Soft computing denotes a consortium of methodologies that works synergistically and provides in one form or another flexible information processing capability for handling real life ambiguous situations. Its aim is to exploit the tolerance for imprecision, uncertainty, approximate reasoning, and partial truth in order to achieve tractability, robustness, and low cost solutions. The guiding principle is to devise methods of computation that lead to an acceptable solution at low cost by seeking an approximate solution to an intractable problem. Chapter 1 described the relevance of different essential components of the soft computing paradigm such as artificial neural networks, genetic algorithms, fuzzy logic, and rough sets to pattern recognition and data mining problems. Various integrations of these tools to complement each other have also been mentioned.

During the past decade, there have been several attempts to derive hybrid methods by judiciously combining the merits of fuzzy logic and rough sets under the name rough-fuzzy or fuzzy-rough computing. One of the challenges of this hybridization is how to integrate these two tools synergistically to achieve both generic and application-specific merits. This is done in a cooperative, rather than a competitive, manner. The result is a more intelligent and robust system providing a human interpretable, low cost, approximate solution, as compared to traditional techniques. This chapter discusses some of the theoretical developments relevant to pattern recognition.

Rough-Fuzzy Pattern Recognition: Applications in Bioinformatics and Medical Imaging,
First Edition. Pradipta Maji and Sankar K. Pal.
© 2012 John Wiley & Sons, Inc. Published 2012 by John Wiley & Sons, Inc.

The structure of the rest of this chapter is as follows: Sections 2.2 and 2.3 briefly introduce the necessary notions of fuzzy sets and rough sets. In Section 2.4, the concepts of granular computing and fuzzy granulation and emergence of rough-fuzzy computing are discussed. Section 2.5 presents a mathematical framework of generalized rough sets for uncertainty handling and defining rough entropy. Various roughness and entropy measures with properties are reported in Section 2.6. Concluding remarks are given in Section 2.7.

2.2 FUZZY SETS

The development of fuzzy logic has primarily led to the emergence of soft computing. It is the earliest and most widely reported constituent of soft computing. This section provides a glimpse of the available literature pertaining to the use of fuzzy sets in pattern recognition and data mining [1].

Fuzzy set was introduced by Zadeh [2] as a generalization of the classical set theory. To a reasonable extent, fuzzy logic is capable of supporting human type reasoning in natural form. The uncertainty can arise either implicitly or explicitly in each and every phase of a pattern recognition system. It results from the incomplete, imprecise, or ambiguous input information, the ill-defined and/or overlapping boundaries among the classes, and the indefiniteness in defining or extracting features and relations among them. Any decision taken at a particular level may have an impact on all other higher level activities. The modeling of imprecise and qualitative knowledge, as well as the transmission and handling of uncertainty at various stages, is possible through the use of fuzzy sets [3, 4].

Data mining from pattern recognition perspective is mainly concerned with identifying interesting patterns and describing them in a concise and meaningful manner. Fuzzy models can be said to represent a prudent and user-oriented sifting of data, qualitative observations, and calibration of common sense rules in an attempt to establish meaningful and useful relationships between system variables [5]. Despite a growing versatility of knowledge discovery systems, there is an important component of human interaction that is inherent to any process of knowledge representation, manipulation, and processing. Fuzzy sets are inherently inclined toward coping with linguistic domain knowledge and producing more interpretable solutions.

A fuzzy set A in a space of objects $U = \{x_i\}$ is a class of events with a continuum of grades of membership and is characterized by a membership function $\mu_A(x_i)$ that associates with each element in U a real number in the interval [0, 1] with the value of $\mu_A(x_i)$ at x_i representing the grade of membership of x_i in A. Formally, a fuzzy set A with its finite number of supports $x_1, \ldots, x_i, \ldots, x_n$ is defined as a collection of ordered pairs $A = \{\mu_A(x_i)/x_i, i = 1, \ldots, n\}$, where the support of A is an ordinary subset of U and is defined as

$$S(A) = \{x_i | x_i \in U \text{ and } \mu_A(x_i) > 0\}. \tag{2.1}$$

Here, $\mu_A(x_i)$ represents the degree to which an object x_i may be a member of A or belong to A. If the support of a fuzzy set is only a single object $x_1 \in U$, then $A = \mu_A(x_1)/x_1$ is called a *fuzzy singleton*. Hence, if $\mu_A(x_1) = 1$, $A = 1/x_1$ denotes a nonfuzzy singleton. In terms of the constituent singletons, the fuzzy set A with its finite number of supports $x_1, \ldots, x_i, \ldots, x_n$ can also be expressed in union form as

$$A = \{\mu_A(x_1)/x_1 + \cdots + \mu_A(x_i)/x_i + \cdots + \mu_A(x_n)/x_n\} \qquad (2.2)$$

where the sign $+$ denotes the union [4]. Assignment of membership functions of a fuzzy subset is subjective in nature and reflects the context in which the problem is viewed.

The fuzzy set theory has greater flexibility to capture various aspects of incompleteness, impreciseness, or imperfection in information about a situation as it is a generalization of the classical set theory [2]. The relevance of fuzzy set theory in the realm of pattern recognition and data mining is adequately justified in [3, 6–9]. Fuzzy sets have been successfully applied in pattern classification [10], clustering [3, 6–8, 11–14], and image processing [15–24]. In addition to pattern recognition, fuzzy sets find widespread applications in solving different problems in association rule mining [25–27], fuzzy information storage and retrieval [28], functional dependency [29], data summarization [30, 31], web mining [32, 33], granular computing [34, 35], microarray data analysis [36, 37], case-based reasoning [38], and so forth.

Pattern recognition and data mining aim at sifting through large volumes of data in order to reveal useful information in the form of new relationships, patterns, or clusters for decision making. Fuzzy sets support a focused search, specified in linguistic terms, through data. They also help discover dependencies between the data in qualitative format. In pattern recognition and data mining, one is typically interested in a focused discovery of structure and an eventual quantification of functional dependencies existing therein. This helps prevent searching for meaningless or trivial patterns in a database. Researchers have developed fuzzy clustering algorithms for this purpose [39, 40]. In this context, it may be mentioned that granular computing [41–43] is useful in finding meaningful patterns in data by expressing and processing chunks of information or granules. Coarse granulation reduces attribute distinctiveness, resulting in loss of useful information, while finer granulation leads to partitioning difficulty. Soft granules can be defined in terms of fuzzy membership functions.

2.3 ROUGH SETS

The theory of rough sets [44, 45] has emerged as a major mathematical tool for managing uncertainty that arises from granularity in the domain of discourse, that is, from the indiscernibility between objects in a set. It has proved to be useful in a variety of pattern recognition and data mining problems [1].

It offers mathematical tools to discover hidden patterns in data, and therefore its importance, as far as pattern recognition and data mining are concerned, can in no way be overlooked. A fundamental principle of a rough set-based learning system is to discover redundancy and dependency between the given features of a problem to be classified. It approximates a given concept using lower and upper approximations of that concept [44, 45].

The rough sets begin with the notion of an approximation space, which is a pair $<U, R>$, where U, the universe of discourse, is a nonempty set and R an equivalence relation on U, that is, R is reflexive, symmetric, and transitive. The relation R decomposes the set U into disjoint classes in such a way that two elements x and y are in the same class iff $(x, y) \in R$. Let U/R denote the quotient set of U by the relation R, and

$$U/R = \{X_1, \ldots, X_i, \ldots, X_p\} \tag{2.3}$$

where X_i is an equivalence class or information granule of R, $i = 1, 2, \ldots, p$. If two elements x and y in U belong to the same equivalence class $X_i \in U/R$, we say that x and y are indistinguishable. The equivalence classes of R and the empty set $\emptyset$ are the elementary sets in the approximation space $<U, R>$. Given an arbitrary set $X \in 2^U$, in general, it may not be possible to describe X precisely in $<U, R>$. One may characterize X by a pair of lower and upper approximations defined as follows [45, 46]:

$$\underline{R}X = \bigcup_{X_i \subseteq X} X_i; \tag{2.4}$$

$$\overline{R}X = \bigcup_{X_i \cap X \neq \emptyset} X_i. \tag{2.5}$$

Hence, the lower approximation $\underline{R}X$ is the union of all the elementary sets that are subsets of X, and the upper approximation $\overline{R}X$ is the union of all the elementary sets that have a nonempty intersection with X. The interval $<\underline{R}X, \overline{R}X>$ is the representation of an ordinary set X in the approximation space $<U, R>$ or simply called the *rough sets of* X. The lower (respectively, upper) approximation $\underline{R}X$ (respectively, $\overline{R}X$) is interpreted as the collection of those elements of U that definitely (respectively, possibly) belong to X. Further,

- a set $X \in 2^U$ is said to be definable or exact in $<U, R>$ iff $\underline{R}X = \overline{R}X$.
- for any $X, Y \in 2^U$, X is said to be roughly included in Y, denoted by $X \tilde{\subset} Y$, iff $\underline{R}X \subseteq \underline{R}Y$ and $\overline{R}X \subseteq \overline{R}Y$.
- X and Y are said to be roughly equal, denoted by $X \simeq_R Y$, in $<U, R>$ iff $\underline{R}X = \underline{R}Y$ and $\overline{R}X = \overline{R}Y$.

Pawlak [45] discusses two numerical characterizations of imprecision of a subset X in the approximation space $<U, R>$: accuracy and roughness. Accuracy of X, denoted by $\alpha_R(X)$, is the ratio of the number of objects in its lower approximation to that in its upper approximation, namely

$$\alpha_R(X) = \frac{|\underline{R}X|}{|\overline{R}X|}. \tag{2.6}$$

The roughness of X, denoted by $\rho_R(X)$, is defined by subtracting the accuracy from 1:

$$\rho_R(X) = 1 - \alpha_R(X) = 1 - \frac{|\underline{R}X|}{|\overline{R}X|}. \tag{2.7}$$

Note that the lower the roughness of a subset, the better its approximation. Further, the following observations are easily obtained:

1. As $\underline{R}X \subseteq X \subseteq \overline{R}X$, $0 \le \rho_R(X) \le 1$.
2. By convention, when $X = \emptyset$, $\underline{R}X = \overline{R}X = \emptyset$, and $\rho_R(X) = 0$.
3. $\rho_R(X) = 0$ if and only if X is definable in $<U, R>$.

The rough set method provides an effective tool for extracting knowledge from databases. First, a knowledge base is created by classifying objects and attributes within the created decision tables. Then, a knowledge discovery process is initiated to remove some undesirable attributes. Finally, the data dependency is analyzed, in the reduced database, to find the minimal subset of attributes, called *reduct*. A rough set learning algorithm can be used to obtain a set of rules from a decision table in IF-THEN form [45].

The applications of rough sets to pattern recognition and data mining mainly proceed along the following two directions, namely, decision rule induction from attribute value table and data filtration by template generation. Most of the methods of the former category are based on generation of discernibility matrices and reducts [47–53]. On the other hand, the latter category mainly involves extracting elementary blocks from data based on equivalence relation [54]. Besides these, reduction of memory and computational requirements for rule generation, and working on dynamic databases [51] are also considered.

The rough set theory has been applied successfully to fuzzy rule extraction [55], reasoning with uncertainty [45, 56], fuzzy modeling, feature selection [57–60], microarray data analysis [61–66], prediction of biological activity of molecules [67], image processing [68–73], and web mining [74]. It is proposed for indiscernibility in classification according to some similarity [45, 75]. The variable precision rough set model [76–79], multigranulation rough sets [80], covering-based rough sets [81], tolerance rough sets [82, 83], fuzzy-rough sets [84–88], neighborhood rough sets [89], and probabilistic rough sets [90–92] are the extensions of the original rough-set-based knowledge representation.

2.4   EMERGENCE OF ROUGH-FUZZY COMPUTING

This section provides the basic notions of granular computing, fuzzy granulation, and emergence of rough-fuzzy computing.

2.4.1   Granular Computing

Granulation is a computing paradigm, among others such as self-reproduction, self-organization, functioning of brain, Darwinian evolution, group behavior, cell membranes, and morphogenesis, which are abstracted from natural phenomena. Granulation is inherent in human thinking and reasoning processes. Granular computing provides an information processing framework where computation and operations are performed on information granules and is based on the realization that precision is sometimes expensive and not much meaningful in modeling and controlling complex systems.

Granular computing may be regarded as a unified framework for theories, methodologies, and techniques that make use of granules, that is, groups, classes, or clusters of objects in a universe, in the process of problem solving. In many situations, when a problem involves incomplete, uncertain and vague information, it may be difficult to differentiate distinct elements and one may find it convenient to consider granules for its handling [41, 93]. On the other hand, in some situations although detailed information is available, it may be sufficient to use granules in order to have an efficient and practical solution. Hence, granulation is an important step in the human cognition process [35]. From a more practical point of view, the simplicity derived from granular computing is useful for designing scalable pattern recognition and data mining algorithms [34, 42, 94–98]. There are two aspects of granular computing, namely, algorithmic aspect and semantic aspect. In algorithmic aspect, one deals with formation, representation, and interpretation of granules, while the semantic aspect deals with utilization of granules for problem solving. Since, in granular computing, computations or operations are performed on granules, that is, clump of similar objects or points, rather than on the individual data points, computation time is greatly reduced. Several approaches for granular computing have been suggested in the literature including fuzzy set theory [43, 99], rough set theory [45, 100, 101], power algebras, and interval analysis.

2.4.2   Computational Theory of Perception and f-Granulation

Any discussion about pattern recognition and decision making in the twenty first century will be incomplete without the mention of computational theory of perceptions (CTP) explained by Zadeh [102], which is governed by perception-based computation. CTP is inspired by the remarkable human capability to perform a wide variety of physical and mental tasks, which include recognition tasks without any measurements and any computations. This capability is due to the crucial ability of the human brain to manipulate perceptions of time, distance, force, direction, shape, color, taste, number, intent, likelihood, and truth, among others.

Recognition and perception are closely related. In a fundamental way, a recognition process may be viewed as a sequence of decisions. Decisions are based on information. In most realistic settings, decision-relevant information is a mixture of measurements and perceptions. An essential difference between measurement and perception is that, in general, measurements are crisp while perceptions are fuzzy. In existing theories, perceptions are converted into measurements, but such conversions in many cases, are infeasible, unrealistic, or counterproductive. An alternative suggested by the CTP is to convert perceptions into propositions expressed in a natural language.

Perceptions are intrinsically imprecise. More specifically, perceptions are fuzzy-granular (f-granular), that is, both fuzzy and granular, with a granule being a clump of elements of a class that are drawn together by indistinguishability, similarity, proximity, or functionality. The f-granularity of perceptions reflects the finite ability of sensory organs and, ultimately, the brain, to resolve detail and store information. In effect, f-granulation is a human way of achieving data compression. It may be mentioned here that although information granulation in which the granules are crisp, that is, c-granular, plays key roles in both human and machine intelligence, it fails to reflect the fact that in much, perhaps most, of human reasoning and concept formation the granules are fuzzy (f-granular) rather than crisp. In this respect, generality increases as the information ranges from singular, c-granular to f-granular. It means that the CTP has, in principle,a higher degree of generality than qualitative reasoning and qualitative process theory in artificial intelligence [103]. The types of problems that fall under the scope of the CTP typically include perception-based function modeling, perception-based system modeling, perception-based time series analysis, solution of perception-based equations, and computation with perception-based probabilities where perceptions are described as a collection of different linguistic IF-THEN rules.

The f-granularity of perceptions puts them well beyond the meaning representation capabilities of predicate logic and other available meaning representation methods [102]. In the CTP, meaning representation is based on the use of the so-called constraint-centered semantics, and reasoning with perceptions is carried out by goal-directed propagation of generalized constraints. In this way, the CTP adds to the existing theories the capability to operate on and reason with perception-based information. This capability is already provided, to an extent, by fuzzy logic and, in particular, by the concept of a linguistic variable and the calculus of fuzzy IF-THEN rules. The CTP extends this capability much further and in new directions. In application to pattern recognition and data mining, the CTP opens the door to a much wider and more systematic use of natural languages in the description of patterns, classes, perceptions and methods of recognition, organization, and knowledge discovery. Upgrading a search engine to a question-answering system is another prospective candidate in web mining for the CTP application. However, one may note that dealing with perception-based information is more complex and more effort intensive than dealing with measurement-based information, and this complexity is the price that has to be paid to achieve superiority.

2.4.3   Rough-Fuzzy Computing

Rough set theory [45, 56] provides an effective means for the analysis of data by synthesizing or constructing approximations (upper and lower) of set concepts from the acquired data. The key notions here are those of information granule and reducts. Information granule formalizes the concept of finite precision representation of objects in real life situation, and reducts represent the core of an information system, both in terms of objects and features, in a granular universe. Granular computing refers to that domain where computation and operations are performed on information granules, that is, clumps of similar objects or points. Therefore, it leads to both data compression and gain in computation time, and finds wide applications [104]. An important use of rough set theory and granular computing in pattern recognition and data mining has been in generating logical rules for classification and association [45]. These logical rules correspond to different important regions of a feature space, which represent data clusters roughly.

For the past few years, rough set theory and granular computation have proved to be another soft computing tool which, in various synergistic combinations with fuzzy logic, artificial neural networks, and genetic algorithms, provides a stronger framework to achieve tractability, robustness, and low cost solution and closely resembles human decision making [104]. There are usually real valued data and fuzzy information in real world applications. Combining fuzzy sets and rough sets provides an important direction in reasoning with uncertainty for real valued data [58, 105–108]. Both fuzzy sets and rough sets provide a mathematical framework to capture uncertainties associated with the data [106, 107]. They are complementary in some aspects. The significance of roughness of a fuzzy set [109] and fuzziness in rough sets [110] may also be mentioned in this regard.

It may be noted that fuzzy set theory hinges on the notion of a membership function on the domain of discourse, assigning to each object a grade of belongingness in order to represent an imprecise or overlapping concept. The form of rough set theory is on the ambiguity caused by limited discernibility of objects in the domain of discourse. The idea is to approximate any concept (a crisp subset in the domain) by a pair of exact sets, called the *lower and upper approximations*. But, the concepts, in such a granular universe, may well be imprecise in the sense that these may not be represented by crisp subsets. This leads to a direction, among others, in which the notions of rough sets and fuzzy sets can be integrated, the aim being to develop a model of uncertainty stronger than either under the umbrella called *rough-fuzzy computing*.

Recently, rough-fuzzy computing has drawn the attention of researchers in the pattern recognition and machine learning community [85, 106, 107, 111–125]. Rough-fuzzy techniques are efficient hybrid techniques based on judicious integration of the principles of rough sets and fuzzy sets. In pattern recognition, for example, while the membership functions of fuzzy sets enable efficient handling of overlapping classes, the concept of lower and upper approximations of rough sets deals with uncertainty, vagueness, and incompleteness in class definitions. Since the rough-fuzzy approach has the capability of providing a stronger paradigm for uncertainty handling, it has greater promise in application domains of pattern recognition and data mining, where fuzzy sets and/or rough sets are being effectively used and proved to be successful. Its effectiveness in handling large data sets, both in size and dimension, is also evident because of its fuzzy granulation (f-granulation) characteristics. Some of the challenges arising out of those posed by massive data and high dimensionality, nonstandard and incomplete data, and knowledge discovery using linguistic rules and over-fitting problems can be dealt well using soft computing and rough-fuzzy approaches. The generalized theories of rough-fuzzy techniques have been applied successfully to feature selection of real valued data [58, 84, 86, 126–128], clustering [114, 115, 129, 130], classification of incomplete data [131, 132], mining stock price [122], vocabulary mining for information retrieval [120], fuzzy decision rule extraction, image processing [133–136], case-based reasoning [137–139], microarray data analysis [117, 129], and web mining [127, 140]. Rough-fuzzy integration can also be considered as a way of emulating the basis for f-granulation in the CTP, where perceptions have fuzzy boundaries and granular attribute values [103].

The next section presents a mathematical framework of generalized rough sets, proposed by Sen and Pal [136], for uncertainty handling and defining rough entropy based on four criteria, namely, (i) set is crisp and granules are crisp, (ii) set is fuzzy and granules are crisp, (iii) set is crisp and granules are fuzzy, and (iv) set is fuzzy and granules are fuzzy. The f-granulation property of the CTP can therefore be modeled using the rough-fuzzy computing framework with one or more of the aforesaid criteria. Various roughness and entropy measures with properties are then introduced in Section 2.6.

2.5 GENERALIZED ROUGH SETS

The expressions in Equations (2.4) and (2.5) for the lower and upper approximations of the set X, respectively, depend on the type of relation R and whether X is a crisp or a fuzzy set [4]. In the following discussion, the upper and lower approximations of the set X are considered when R denotes an equivalence, a fuzzy equivalence, a tolerance, or a fuzzy tolerance relation and X is a crisp or a fuzzy set.

When X is a crisp or a fuzzy set and the relation R is a crisp or a fuzzy equivalence relation, the expressions for the lower and the upper approximations of the set X are given as

$$\underline{R}X = \{(u, \underline{M}(u))\,|\,u \in U\}$$
$$\overline{R}X = \{(u, \overline{M}(u))\,|\,u \in U\} \tag{2.8}$$

where

$$\underline{M}(u) = \sum_{Y \in U/R} m_Y(u) \times \inf_{\varphi \in U} \max(1 - m_Y(\varphi), \mu_X(\varphi))$$

$$\overline{M}(u) = \sum_{Y \in U/R} m_Y(u) \times \sup_{\varphi \in U} \min(m_Y(\varphi), \mu_X(\varphi)) \tag{2.9}$$

where the membership function m_Y represents the belongingness of every element u in the universe U to a granule $Y \in U/R$ and it takes values in the interval $[0, 1]$ such that $\sum_Y m_Y(u) = 1$, and μ_X, which takes values in the interval $[0, 1]$, is the membership function associated with X. When X is a crisp set, μ_X would take values only from the set $\{0, 1\}$. Similarly, when R is a crisp equivalence relation m_Y would take values only from the set $\{0, 1\}$. In the above, the symbols $\sum$ (sum) and $\times$ (product) represent specific fuzzy union and intersection operations, respectively [4], which are chosen on the basis of their suitability with respect to the underlying application of measuring ambiguity.

In the above, the isndiscernibility relation $R \subseteq U \times U$ is considered to be an equivalence relation, that is, R satisfies crisp or fuzzy reflexivity, symmetry, and transitivity properties [4]. However, if R does not satisfy any one of these three properties, the expressions in Equation (2.8) can no longer be used. Next, the case where the transitivity property is not satisfied is considered. Such a relation R is said to be a tolerance relation and the space $<U, R>$ obtained is referred to as a *tolerance approximation space* [141]. When R is a tolerance relation, the expressions for the membership values corresponding to the lower and upper approximations of an arbitrary set X in U are given as

$$\underline{M}(u) = \inf_{\varphi \in U} \max(1 - S_R(u, \varphi), \mu_X(\varphi))$$

$$\overline{M}(u) = \sup_{\varphi \in U} \min(S_R(u, \varphi), \mu_X(\varphi)) \tag{2.10}$$

where $S_R(u, \varphi)$ is a value representing the tolerance relation R between u and φ. The pair of sets $<\underline{R}X, \overline{R}X>$ and the approximation space $<U, R>$ are referred to differently, depending on whether X is a crisp or a fuzzy set; the relation R is a crisp or a fuzzy equivalence, or a crisp or a fuzzy tolerance relation. The different forms of the lower and upper approximations of the set X are shown graphically in Fig. 2.1, while the different names are listed in Table 2.1.

2.6 ENTROPY MEASURES

Defining entropy measures on the basis of rough set theory has been considered by researchers in the past decade. Probably, the first such work was reported in Beaubouef et al. [142], where a rough entropy of a set in a universe has been

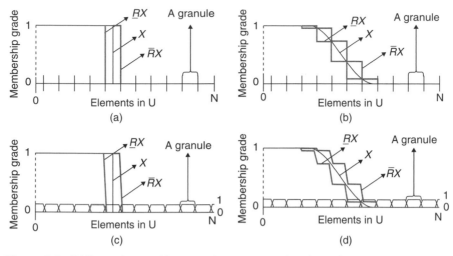

Figure 2.1 Different forms of lower and upper approximations of set X. (a) Rough sets, (b) rough-fuzzy sets, (c) fuzzy-rough sets, and (d) fuzzy rough-fuzzy sets.

TABLE 2.1 Different Types of Rough Sets and Approximation Space $<U, R>$

Relation R	X Is Crisp $<\underline{R}X, \overline{R}X>$	Approximation Space $<U, R>$
$m_Y(u) \in \{0, 1\}$ (crisp equivalence relation)	Rough set of X	Crisp equivalence
$m_Y(u) \in [0, 1]$ (fuzzy equivalence relation)	Fuzzy-rough set of X	Fuzzy equivalence
$S_R : U \times U \to \{0, 1\}$ (crisp tolerance relation)	Tolerance rough set of X	Crisp tolerance
$S_R : U \times U \to [0, 1]$ (fuzzy tolerance relation)	Tolerance fuzzy-rough set of X	Fuzzy tolerance

Relation R	X Is Fuzzy $<\underline{R}X, \overline{R}X>$	Approximation Space $<U, R>$
$m_Y(u) \in \{0, 1\}$ (crisp equivalence relation)	Rough-fuzzy set of X	Crisp equivalence
$m_Y(u) \in [0, 1]$ (fuzzy equivalence relation)	Fuzzy rough-fuzzy set of X	Fuzzy equivalence
$S_R : U \times U \to \{0, 1\}$ (crisp tolerance relation)	Tolerance rough-fuzzy set of X	Crisp tolerance
$S_R : U \times U \to [0, 1]$ (fuzzy tolerance relation)	Tolerance fuzzy rough-fuzzy set of X	Fuzzy tolerance

proposed. This rough entropy measure is defined on the basis of the uncertainty in granulation obtained using a relation defined over universe [45] and the definability of the set. Other entropy measures that quantify the uncertainty in crisp or fuzzy granulation alone have been reported in the literature [143–145]. An entropy measure is presented in Yager [146], which quantifies information with the underlying elements with limited discernibility between them, although it is not based on rough set theory.

Incompleteness of knowledge about a universe leads to granulation [45] and hence a measure of the uncertainty in granulation quantifies this incompleteness of knowledge. Therefore, apart from the rough entropy discussed in Beaubouef et al. [142], which quantifies the incompleteness of knowledge about a set in a universe, the other aforesaid entropy measures quantify the incompleteness of knowledge about a universe. The effect of incompleteness of knowledge about a universe becomes evident only when an attempt is made to define a set in it. Note that the definability of a set in a universe is not always affected by a change in the uncertainty in granulation. This is evident in a few examples given in Beaubouef et al. [142]. Hence, a measure of incompleteness of knowledge about a universe with respect to only the definability of a set is required.

First attempt of formulating an entropy measure with respect to the definability of a set was made by Pal et al. [94], which has been successfully used for image segmentation. However, as pointed out in Sen and Pal [147], this measure does not satisfy the necessary property that the entropy value is maximum or optimum when the uncertainty is maximum. In this case, the uncertainty is the incompleteness of knowledge. Next, some recently proposed entropy measures [136] are reported, which quantify the incompleteness of knowledge about a universe with respect to the definability of a set of elements in the universe holding a particular property or representing a category. The lower and upper approximations of a vaguely definable set X in a universe U can be used in the expression given in Equation (2.7) in order to get an inexactness measure of the set X called the *roughness measure* $\rho_R(X)$. The vague definition of X in U signifies incompleteness of knowledge about U. The entropy measures [136] are presented based on the roughness measures of a set and its complement in order to quantify the incompleteness of knowledge about a universe.

One of the two classes of entropy measures is obtained by measuring the gain in information or in present case the gain in incompleteness using a logarithmic function as suggested in the Shannon's theory. This class of entropy measures for quantifying the incompleteness of knowledge about U with respect to the definability of a set $X \subseteq U$ is given as [136]

$$H_R^L(X) = -\frac{1}{2}\left[\varkappa(X) + \varkappa(X^{\complement})\right] \qquad (2.11)$$

where $\varkappa(X) = \rho_R(X)\log_\beta \rho_R(X)/\beta$ for any set $X \subseteq U$, β denotes the base of the logarithmic function used and $X^{\complement} \subseteq U$ stands for the complement of the set X in the universe. The various entropy measures of this class are obtained by calculating the roughness values $\rho_R(X)$ and $\rho_R(X^C)$ considering the different ways of obtaining the lower and upper approximations of the vaguely definable set X. Note that, the gain in incompleteness term is taken as $-\log_\beta(\rho_R/\beta)$ in Equation (2.11) and for $\beta > 1$ it takes a value in the interval $[1, \infty]$.

The other class of entropy measures is obtained by considering an exponential function [148] to measure the gain in incompleteness. This second class of entropy measures for quantifying the incompleteness of knowledge about U with respect to the definability of a set $X \subseteq U$ is given as [136]

$$H_R^E(X) = \frac{1}{2}\left[\rho_R(X)\beta^{\left(\overline{\rho}_R(X)\right)} + \rho_R(X^C)\beta^{\left(\overline{\rho}_R(X^C)\right)}\right] \tag{2.12}$$

where $\overline{\rho}_R(X) = 1 - \rho_R(X)$ for any set $X \subseteq U$, β denotes the base of the exponential function used. N.R. Pal and S.K. Pal [148] had considered only the case when β equaled $e(\approx 2.7183)$. Similar to the class of entropy measures H_R^L, the various entropy measures of this class are obtained by using the different ways of obtaining the lower and upper approximations of X in order to calculate $\rho_R(X)$ and $\rho_R(X^C)$. The gain in incompleteness term is taken as $\beta^{\left(1-\rho_R\right)}$ in Equation (2.12) and for $\beta > 1$ it takes a value in the finite interval $[1, \beta]$.

In the following discussion, an entropy measure is named using attributes that represent the class (logarithmic or exponential) it belongs to, and the type of the pair of sets $<\underline{R}X, \overline{R}X>$ considered. If $<\underline{R}X, \overline{R}X>$ represents a tolerance rough-fuzzy set and the expression of the entropy in Equation (2.12) is considered, then such an entropy is called as the *exponential tolerance rough-fuzzy entropy*. Some other examples of names for the entropy measures are the logarithmic rough entropy, the exponential fuzzy-rough entropy, and the logarithmic tolerance fuzzy rough-fuzzy entropy.

Let a set A is fuzzy in nature and it is associated with a membership function μ_A. As mentioned in Pal and Bezdek [149], most of the appropriate fuzzy-set-theory-based uncertainty measures can be grouped into two classes, namely, the multiplicative class and the additive class. It should be noted from Pal and Bezdek [149] that the measures belonging to these classes are functions of μ_A and μ_{A^C} where $\mu_A = 1 - \mu_{A^C}$. The existence of an exact relation between μ_A and μ_{A^C}, as mentioned in Pal and Bezdek [149], suggests that they theoretically convey the same. However, sometimes such unnecessary terms should to be retained as dropping them would cause the corresponding measures to fail certain important properties.

The following analysis establishes the relation between $\rho_R(X)$ and $\rho_R(X^C)$ and shows that there exist no unnecessary terms in the classes of entropy measures (Eqs. (2.11) and (2.12)) proposed using rough set theory and its certain generalizations. Let us consider

$$\rho_R(X) = \frac{1}{C}, \ 1 \le C \le \infty \tag{2.13}$$

as $\rho_R(X)$ takes a value in the interval $[0, 1]$ and let the total number of elements in the universe U under consideration be n. As we have $X \cup X^C = U$, it can be easily deduced that $\overline{R}X \cup \underline{R}X^{\complement} = U$ and $\underline{R}X \cup \overline{R}X^{\complement} = U$. Hence, from Equation (2.7), we get

$$\rho_R(X) = 1 - \frac{|\underline{R}X|}{|\overline{R}X|}; \tag{2.14}$$

$$\rho_R(X^{\complement}) = 1 - \frac{|\underline{R}X^{\complement}|}{|\overline{R}X^{\complement}|} = 1 - \frac{n - |\overline{R}X|}{n - |\underline{R}X|}. \tag{2.15}$$

From Equations (2.13), (2.14), and (2.15), the following relation between $\rho_R(X)$ and $\rho_R(X^{\complement})$ can be established:

$$\rho_R(X^{\complement}) = \rho_R(X)\frac{|\overline{R}X|}{n - |\underline{R}X|} = \frac{1}{C}\left(\frac{|\overline{R}X|}{n - |\underline{R}X|}\right). \tag{2.16}$$

C1: When $1 < C < \infty$, according to Equation (2.13), $|\underline{R}X|/|\overline{R}X| = (C - 1)/C$. Using this relation in Equation (2.16), the following relation can be obtained:

$$\rho_R(X^{\complement}) = \frac{1}{C}\left(\frac{|\underline{R}X|\left(\dfrac{C}{C-1}\right)}{n - |\underline{R}X|}\right), \tag{2.17}$$

which reduces to

$$\rho_R(X^{\complement}) = \frac{1}{C-1}\left(\frac{1}{\dfrac{n}{|\underline{R}X|} - 1}\right). \tag{2.18}$$

Note that $\rho_R(X)$ takes value in the interval $(0, 1)$ when $1 < C < \infty$. In this case, the value of $|\overline{R}X|$ could range from a positive infinitesimal quantity, say ϵ, to a maximum value of n. Hence,

$$\epsilon\frac{C-1}{C} \le |\underline{R}X| \le n\frac{C-1}{C}. \tag{2.19}$$

Using Equation (2.19) in Equation (2.18), we get

$$\frac{\epsilon}{nC - \epsilon(C-1)} \le \rho_R\left(X^{\complement}\right) \le 1. \tag{2.20}$$

As $1 < C < \infty$, $\epsilon \ll 1$ and usually $n \gg 1$, we may write Equation (2.20) as

$$0 < \rho_R(X^{\complement}) \le 1. \tag{2.21}$$

Hence, it may be concluded that for a given nonzero and nonunity value of $\rho_R(X)$, $\rho_R(X^{\complement})$ may take any value in the interval $(0, 1]$.

C2: When $C = 1$, $\rho_R(X)$ takes a unity value. In this case, $|\underline{R}X| = 0$ and the value of $|\overline{R}X|$ could range from ϵ to a maximum value of n. Hence, it is easily evident from Equation (2.16) that $\rho_R(X^C)$ may take any value in the interval $(0, 1]$ when $\rho_R(X) = 1$.

C3: When $C = \infty$, the $\rho_R(X) = 0$. In such a case, the value of $|\overline{R}X|$ could range from zero to a maximum value of n and $|\underline{R}X| = |\overline{R}X|$. As evident from Equation (2.16), when $C = \infty$, irrespective of any other term, we get $\rho_R(X^C) = 0$. This is obvious as a exactly definable set X should imply an exactly definable set X^C.

Hence, the relation between $\rho_R(X)$ and $\rho_R(X^C)$ is such that, if one of them is considered to take a nonzero value, that is, the underlying set is vaguely definable or inexact, the value of the other, which would also be a nonzero quantity, cannot be uniquely specified. Therefore, there exist no unnecessary terms in the classes of entropy measures given in Equations (2.11) and (2.12). However, from Equations (2.14) and (2.15), it is easily evident that $\rho_R(X)$ and $\rho_R(X^C)$ are positively correlated.

The base parameter β of the two classes of entropy measures (Eqs. (2.11) and (2.12)) incurs certain restrictions, so that the entropies satisfy some important properties. In this section, the restrictions regarding the base parameters are discussed, along with a few properties of the entropies.

Range of Values for the Base β   The classes of entropy measures H_R^L and H_R^E, given in Equations (2.11) and (2.12), respectively, must be consistent with the fact that maximum information (entropy) is available when the uncertainty is maximum and the entropy is zero when there is no uncertainty. Note that, in this case, maximum uncertainty represents maximum possible incompleteness of knowledge about the universe. Hence, maximum uncertainty occurs when both the roughness values used in H_R^L and H_R^E equal unity and uncertainty is zero when both of them are zero. It can be easily shown that in order to satisfy the aforesaid condition, the base β in H_R^L must take a finite value greater than or equal to $e^{\sim}(\approx 2.7183)$ and the base β in H_R^E must take a value in the interval $(1, e]$. When $\beta \geq e$ in H_R^L and $1 < \beta \leq e$ in H_R^E, the values taken by both H_R^L and H_R^E lie in the range $[0, 1]$. Note that, for an appropriate β value, the entropy measures attain the minimum value of zero only when $\rho_R(X) = \rho_R(X^C) = 0$ and the maximum value of unity only when $\rho_R(X) = \rho_R(X^C) = 1$.

Fundamental Properties   A few properties of the logarithmic and exponential classes of entropy measures expressing H_R^L and H_R^E as functions of two parameters representing roughness measures are presented next. The expressions given in Equations (2.11) and (2.12) can be rewritten in parametric form as follows:

$$H_R^L(A, B) = -\frac{1}{2}\left[A\log_\beta\left(\frac{A}{\beta}\right) + B\log_\beta\left(\frac{B}{\beta}\right)\right] \qquad (2.22)$$

$$H_R^E(A, B) = \frac{1}{2}\left[A\beta^{(1-A)} + B\beta^{(1-B)}\right] \qquad (2.23)$$

where the parameters A ($\in [0, 1]$) and B ($\in [0, 1]$) represent the roughness values $\rho_R(X)$ and $\rho_R(X^\complement)$, respectively. Considering the convention $0 \log_\beta 0 = 0$, the properties of $H_R^L(A, B)$ and $H_R^E(A, B)$ are discussed next along the lines of Ebanks [150].

P1. Nonnegativity: $H_R^L(A, B) \geq 0$ and $H_R^E(A, B) \geq 0$ with equality in both the cases if and only if $A = 0$ and $B = 0$.

P2. Continuity: Both $H_R^E(A, B)$ and $H_R^L(A, B)$ are continuous functions of A and B, where $A, B \in [0, 1]$.

P3. Sharpness: Both $H_R^L(A, B)$ and $H_R^E(A, B)$ equal zero if and only if the roughness values A and B equal zero, that is, A and B are sharp.

P4. Maximality and normality: Both $H_R^L(A, B)$ and $H_R^E(A, B)$ attain their maximum value of unity if and only if the roughness values A and B are unity. That is, $H_R^L(A, B) \leq H_R^L(1, 1) = 1$ and $H_R^E(A, B) \leq H_R^E(1, 1) = 1$, where $A, B \in [0, 1]$.

P5. Resolution: $H_R^L(A^*, B^*) \leq H_R^L(A, B)$ and $H_R^E(A^*, B^*) \leq H_R^E(A, B)$, where A^* and B^* are the sharpened version of A and B, respectively, that is, $A^* \leq A$ and $B^* \leq B$.

P6. Symmetry: Both $H_R^L(A, B)$ and $H_R^E(A, B)$ are symmetric about the line $A = B$.

P7. Monotonicity: Both $H_R^L(A, B)$ and $H_R^E(A, B)$ are monotonically nondecreasing functions of A and B.

P8. Concavity: Both $H_R^L(A, B)$ and $H_R^E(A, B)$ are concave functions of A and B.

2.7   CONCLUSION AND DISCUSSION

This chapter presents the basic notions and characteristics of fuzzy sets and rough sets. It also provides a glimpse of the available literature pertaining to the use of fuzzy sets and rough sets in pattern recognition and data mining. The concept of information granules is introduced, along with the emergence of rough-fuzzy computing paradigm and their relevance to pattern recognition and data mining. This chapter also provides a mathematical framework for generalized rough sets incorporating the concept of fuzziness in defining the granules as well as the set. Various roughness and uncertainty measures with properties are stated.

The new entropies have been used in Sen and Pal [136] to quantify ambiguities in images, and it has been shown that some of the entropies can be used to quantify ambiguities because of both fuzzy boundaries and rough resemblance. The utility and effectiveness of these entropy measures have been demonstrated by considering some elementary image processing applications and comparisons with the use of certain fuzziness measures. The classes of entropy measures based on rough set theory and its certain generalizations are not restricted to the few applications discussed [136]. They are, in general, applicable to all tasks where ambiguity measure-based techniques have been found suitable, provided that the rough resemblance aspect of ambiguities exists. It would be interesting to carry out such investigations as the new measures possess certain advantages over most fuzzy set theory-based uncertainty measures, which have been the prime tool for measuring ambiguities. Chapter 3 presents a generalized hybrid unsupervised learning algorithm, integrating the merits of rough sets and fuzzy sets.

REFERENCES

1. S. Mitra, S. K. Pal, and P. Mitra. Data Mining in Soft Computing Framework: A Survey. *IEEE Transactions on Neural Networks*, 13(1):3–14, 2002.

2. L. A. Zadeh. Fuzzy Sets. *Information and Control*, 8:338–353, 1965.

3. J. C. Bezdek and S. K. Pal, editors. *Fuzzy Models for Pattern Recognition: Methods that Search for Structures in Data*. IEEE Press, New York, 1992.

4. G. Klir and B. Yuan. *Fuzzy Sets and Fuzzy Logic: Theory and Applications*. Prentice Hall, New Delhi, India, 2005.

5. W. Pedrycz. Fuzzy Set Technology in Knowledge Discovery. *Fuzzy Sets and Systems*, 98:279–290, 1998.

6. J. C. Bezdek. *Pattern Recognition with Fuzzy Objective Function Algorithm*. Plenum, New York, 1981.

7. A. Kandel. *Fuzzy Techniques in Pattern Recognition*. Wiley Interscience, New York, 1982.

8. S. K. Pal and D. D. Majumder. *Fuzzy Mathematical Approach to Pattern Recognition*. John Wiley (Halsted Press), New York, 1986.

9. L. A. Zadeh. Fuzzy Sets and Information Granularity. In M. Gupta, R. Ragade, and R. Yager, editors, *Advances in Fuzzy Set Theory and Applications*, pages 3–18. North-Holland Publishing Co., Amsterdam, 1979.

10. L. I. Kuncheva. *Fuzzy Classifier Design*. Physica-Verlag, Heidelberg, 2010.

11. J. C. Dunn. A Fuzzy Relative of the ISODATA Process and Its Use in Detecting Compact, Well-Separated Clusters. *Journal of Cybernetics*, 3:32–57, 1974.

12. R. Krishnapuram and J. M. Keller. A Possibilistic Approach to Clustering. *IEEE Transactions on Fuzzy Systems*, 1(2):98–110, 1993.

13. F. Masulli and S. Rovetta. Soft Transition from Probabilistic to Possibilistic Fuzzy Clustering. *IEEE Transactions on Fuzzy Systems*, 14(4):516–527, 2006.

14. N. R. Pal, K. Pal, J. M. Keller, and J. C. Bezdek. A Possibilistic Fuzzy C-Means Clustering Algorithm. *IEEE Transactions on Fuzzy Systems*, 13(4):517–530, 2005.

15. M. Banerjee, M. K. Kundu, and P. Maji. Content-Based Image Retrieval Using Visually Significant Point Features. *Fuzzy Sets and Systems*, 160(23):3323–3341, 2009.

16. H. Frigui. Adaptive Image Retrieval Using the Fuzzy Integral. In *Proceedings of the 18th International Conference of the North American Fuzzy Information Processing Society*, pages 575–579. IEEE Press, Piscataway, NJ, 1999.

17. E. E. Kerre and M. Nachtegael, editors. *Fuzzy Techniques in Image Processing*. Physica-Verlag, Heidelberg, 2010.

18. P. Maji, M. K. Kundu, and B. Chanda. Second Order Fuzzy Measure and Weighted Co-Occurrence Matrix for Segmentation of Brain MR Images. *Fundamenta Informaticae*, 88(1–2):161–176, 2008.

19. S. K. Pal and A. Ghosh. Index of Area Coverage of Fuzzy Image Subsets and Object Extraction. *Pattern Recognition Letters*, 11(12):831–841, 1990.

20. S. K. Pal and A. Ghosh. Image Segmentation Using Fuzzy Correlation. *Information Sciences*, 62(3):223–250, 1992.

21. S. K. Pal, A. Ghosh, and B. U. Shankar. Segmentation of Remotely Sensed Images with Fuzzy Thresholding and Quantitative Evaluation. *International Journal of Remote Sensing*, 21(11):2269–2300, 2000.

22. S. K. Pal, R. A. King, and A. A. Hashim. Automatic Gray Level Thresholding through Index of Fuzziness and Entropy. *Pattern Recognition Letters*, (1):141–146, 1983.

23. O. J. Tobias and R. Seara. Image Segmentation by Histogram Thresholding Using Fuzzy Sets. *IEEE Transactions on Image Processing*, 11(12):1457–1465, 2002.

24. K. Xiao, S. H. Ho, and A. E. Hassanien. Automatic Unsupervised Segmentation Methods for MRI Based on Modified Fuzzy C-Means. *Fundamenta Informaticae*, 87(3–4):465–481, 2008.

25. W. H. Au and K. C. C. Chan. An Effective Algorithm for Discovering Fuzzy Rules in Relational Databases. In *Proceedings of the IEEE International Conference on Fuzzy Systems*, pages 1314–1319, Alaska, 1998.

26. A. Maeda, H. Ashida, Y. Taniguchi, and Y. Takahashi. Data Mining System Using Fuzzy Rule Induction. In *Proceedings of the IEEE International Conference on Fuzzy Systems*, pages 45–46, Yokohama, Japan, 1995.

27. Q. Wei and G. Chen. Mining Generalized Association Rules with Fuzzy Taxonomic Structures. In *Proceedings of the 18th International Conference of the North American Fuzzy Information Processing Society*, pages 477–481, IEEE Press, Piscataway, NJ, 1999.

28. J. Hale and S. Shenoi. Analyzing FD Inference in Relational Databases. *Data and Knowledge Engineering*, 18:167–183, 1996.

29. P. Bosc, O. Pivert, and L. Ughetto. Database Mining for the Discovery of Extended Functional Dependencies. In *Proceedings of the 18th International Conference of the North American Fuzzy Information Processing Society*, pages 580–584. IEEE Press, Piscataway, NJ, 1999.

30. D.-A. Chiang, L. R. Chow, and Y.-F. Wang. Mining Time Series Data by a Fuzzy Linguistic Summary System. *Fuzzy Sets and Systems*, 112:419–432, 2000.

31. D. H. Lee and M. H. Kim. Database Summarization Using Fuzzy ISA Hierarchies. *IEEE Transactions on Systems Man and Cybernetics Part B-Cybernetics*, 27:68–78, 1997.

32. R. Krishnapuram, A. Joshi, O. Nasraoui, and L. Yi. Low Complexity Fuzzy Relational Clustering Algorithms for Web Mining. *IEEE Transactions on Fuzzy System*, 9:595–607, 2001.

33. O. Nasraoui, R. Krishnapuram, and A. Joshi. Relational Clustering Based on a New Robust Estimator with Application to Web Mining. In *Proceedings of the 18th International Conference of the North American Fuzzy Information Processing Society*, pages 705–709, IEEE Press, Piscataway, NJ, 1999.

34. A. Bargiela and W. Pedrycz. *Granular Computing: An Introduction*. Kluwer Academic Publishers, Boston, MA, 2003.

35. A. Bargiela and W. Pedrycz. Toward a Theory of Granular Computing for Human-Centered Information Processing. *IEEE Transactions on Fuzzy Systems*, 16(2):320–330, 2008.

36. N. Belacel, M. Cuperlovic-Culf, M. Laflamme, and R. Ouellette. Fuzzy J-Means and VNS Methods for Clustering Genes from Microarray Data. *Bioinformatics*, 20(11):1690–1701, 2004.

37. D. Dembele and P. Kastner. Fuzzy C-Means Method for Clustering Microarray Data. *Bioinformatics*, 19(8):973–980, 2003.

38. S. K. Pal, T. S. Dillon, and D. S. Yeung, editors. *Soft Computing in Case Based Reasoning*. Springer-Verlag, London, 2000.

39. W. Pedrycz. Conditional Fuzzy C-Means. *Pattern Recognition Letters*, 17:625–632, 1996.

40. I. B. Turksen. Fuzzy Data Mining and Expert System Development. In *Proceedings of the IEEE International Conference on Systems Man and Cybernetics*, pages 2057–2061, San Diego, CA, 1998.

41. A. Bargiela, W. Pedrycz, and K. Hirota. Granular Prototyping in Fuzzy Clustering. *IEEE Transactions on Fuzzy Systems*, 12(5):697–709, 2004.

42. W. Pedrycz, editor. *Granular Computing: An Emerging Paradigm*. Physica-Verlag, Heidelberg, 2010.

43. L. A. Zadeh. Toward a Theory of Fuzzy Information Granulation and Its Centrality in Human Reasoning and Fuzzy Logic. *Fuzzy Sets and Systems*, 90:111–127, 1997.

44. Z. Pawlak. Rough Sets. *International Journal of Computer and Information Science*, 11:341–356, 1982.

45. Z. Pawlak. *Rough Sets: Theoretical Aspects of Reasoning about Data*. Kluwer, Dordrecht, The Netherlands, 1991.

46. L. Polkowski. *Rough Sets*. Physica-Verlag, Heidelberg, 2002.

47. J. Bazan, A. Skowron, and P. Synak. Dynamic Reducts as a Tool for Extracting Laws from Decision Tables. In Z. W. Ras and M. Zemankova, editors, *Proceedings of the 8th Symposium on Methodologies for Intelligent Systems, Lecture Notes in Artificial Intelligence*, volume 869, pages 346–355. Springer-Verlag, Charlotte, North Carolina, 1994.

48. X. Hu and N. Cercone. Mining Knowledge Rules from Databases: A Rough Set Approach. In *Proceedings of the 12th International Conference on Data Engineering, Washington*, pages 96–105, IEEE Computer Society Press, New York, 1996.

49. J. Komorowski, Z. Pawlak, L. Polkowski, and A. Skowron. Rough Sets: A Tutorial. In S. K. Pal and A. Skowron, editors, *Rough-Fuzzy Hybridization: A New Trend in Decision Making*, pages 3–98. Springer-Verlag, Singapore, 1999.

50. T. Mollestad and A. Skowron. A Rough Set Framework for Data Mining of Propositional Default Rules. In Z. W. Ras and M. Michalewicz, editors, *Foundations of Intelligent Systems, Lecture Notes in Computer Science*, volume 1079, pages 448–457. Springer-Verlag, Berlin, 1996.

51. N. Shan and W. Ziarko. Data-Based Acquisition and Incremental Modification of Classification Rules. *Computational Intelligence*, 11:357–370, 1995.

52. A. Skowron. Extracting Laws from Decision Tables: A Rough Set Approach. *Computational Intelligence*, 11:371–388, 1995.

53. A. Skowron and C. Rauszer. The Discernibility Matrices and Functions in Information Systems. In R. Slowinski, editor, *Intelligent Decision Support*, pages 331–362. Kluwer Academic Publishers, Dordrecht, 1992.

54. L. Polkowski and A. Skowron, editors. *Rough Sets in Knowledge Discovery*, volumes 1 and 2. Physica-Verlag, Heidelberg, 1998.

55. C. Cornelis, R. Jensen, G. H. Martin, and D. Slezak. Attribute Selection with Fuzzy Decision Reducts. *Information Sciences*, 180:209–224, 2010.

56. A. Skowron, R. W. Swiniarski, and P. Synak. Approximation Spaces and Information Granulation. *LNCS Transactions on Rough Sets*, 3:175–189, 2005.

57. A. Chouchoulas and Q. Shen. Rough Set-Aided Keyword Reduction for Text Categorisation. *Applied Artificial Intelligence*, 15(9):843–873, 2001.

58. R. Jensen and Q. Shen. Semantics-Preserving Dimensionality Reduction: Rough and Fuzzy-Rough-Based Approach. *IEEE Transactions on Knowledge and Data Engineering*, 16(12):1457–1471, 2004.

59. Y. Kudo, T. Murai, and S. Akama. A Granularity-Based Framework of Deduction, Induction, and Abduction. *International Journal of Approximate Reasoning*, 50(8):1215–1226, 2009.

60. Y. Qian, J. Liang, and C. Dang. Knowledge Structure, Knowledge Granulation and Knowledge Distance in a Knowledge Base. *International Journal of Approximate Reasoning*, 50(1):174–188, 2009.

61. J.-H. Chiang and S.-H. Ho. A Combination of Rough-Based Feature Selection and RBF Neural Network for Classification Using Gene Expression Data. *IEEE Transactions on NanoBioscience*, 7(1):91–99, 2008.

62. J. Fang and J. W. G. Busse. Mining of MicroRNA Expression Data: A Rough Set Approach. In *Proceedings of the 1st International Conference on Rough Sets and Knowledge Technology*, pages 758–765. Springer, Berlin, 2006.

63. P. Maji and S. Paul. Rough Set Based Maximum Relevance-Maximum Significance Criterion and Gene Selection from Microarray Data. *International Journal of Approximate Reasoning*, 52(3):408–426, 2011.

64. D. Slezak. Rough Sets and Few-Objects-Many-Attributes Problem: The Case Study of Analysis of Gene Expression Data Sets. In *Proceedings of the Frontiers in the Convergence of Bioscience and Information Technologies*, Jeju-Do, Korea, pages 233–240, 2007.

65. D. Slezak and J. Wroblewski. Roughfication of Numeric Decision Tables: The Case Study of Gene Expression Data. In *Proceedings of the 2nd International Conference on Rough Sets and Knowledge Technology*, pages 316–323. Springer, Berlin, 2007.

66. J. J. Valdes and A. J. Barton. Relevant Attribute Discovery in High Dimensional Data: Application to Breast Cancer Gene Expressions. In *Proceedings of the 1st International Conference on Rough Sets and Knowledge Technology*, pages 482–489. Springer, Berlin, 2006.

67. P. Maji and S. Paul. Rough Sets for Selection of Molecular Descriptors to Predict Biological Activity of Molecules. *IEEE Transactions on Systems Man and Cybernetics Part C-Applications and Reviews*, 40(6):639–648, 2010.

68. J. Jiang, D. Yang, and H. Wei. Image Segmentation Based on Rough Set Theory and Neural Networks. In *Proceedings of the 5th International Conference on Visual Information Engineering*, pages 361–365. IET, UK, 2008.

69. M. M. Mushrif and A. K. Ray. Color Image Segmentation: Rough-Set Theoretic Approach. *Pattern Recognition Letters*, 29(4):483–493, 2008.

70. S. K. Pal and P. Mitra. Multispectral Image Segmentation Using the Rough Set-Initialized-EM Algorithm. *IEEE Transactions on Geoscience and Remote Sensing*, 40(11):2495–2501, 2002.

71. S. Widz, K. Revett, and D. Slezak. A Hybrid Approach to MR Imaging Segmentation Using Unsupervised Clustering and Approximate Reducts. In *Proceedings of the 10th International Conference on Rough Sets, Fuzzy Sets, Data Mining, and Granular Computing*, pages 372–382, Regina, Canada, 2005.

72. S. Widz, K. Revett, and D. Slezak. A Rough Set-Based Magnetic Resonance Imaging Partial Volume Detection System. In *Proceedings of the 1st International Conference on Pattern Recognition and Machine Intelligence*, Kolkata, India, pages 756–761, 2005.

73. S. Widz and D. Slezak. Approximation Degrees in Decision Reduct-Based MRI Segmentation. In *Proceedings of the Frontiers in the Convergence of Bioscience and Information Technologies*, pages 431–436, Jeju-Do, Korea, 2007.

74. P. Lingras and C. West. Interval Set Clustering of Web Users with Rough K-Means. *Journal of Intelligent Information Systems*, 23(1):5–16, 2004.

75. Q. Shen and A. Chouchoulas. Combining Rough Sets and Data-Driven Fuzzy Learning for Generation of Classification Rules. *Pattern Recognition*, 32(12):2073–2076, 1999.

76. M. Inuiguchi, Y. Yoshioka, and Y. Kusunoki. Variable-Precision Dominance-Based Rough Set Approach and Attribute Reduction. *International Journal of Approximate Reasoning*, 50:1199–1214, 2009.

77. M. Kudo and T. Murai. Extended DNF Expression and Variable Granularity in Information Tables. *IEEE Transactions on Fuzzy Systems*, 16(2):285–298, 2008.

78. G. Xie, J. Zhang, K. K. Lai, and L. Yu. Variable Precision Rough Set for Group Decision-Making: An Application. *International Journal of Approximate Reasoning*, 49:331–343, 2008.

79. W. Ziarko. Variable Precision Rough Set Model. *Journal of Computer and System Sciences*, 46:39–59, 1993.

80. Y. Qian, J. Liang, and C. Dang. Incomplete Multigranulation Rough Set. *IEEE Transactions on Systems Man and Cybernetics Part A-Systems and Humans*, 40(2):420–431, 2010.

81. W. Zhu and F.-Y. Wang. On Three Types of Covering-Based Rough Sets. *IEEE Transactions on Knowledge and Data Engineering*, 19(8):1131–1144, 2007.

82. D. Kim. Data Classification Based on Tolerant Rough Set. *Pattern Recognition*, 34(8):1613–1624, 2001.

83. N. M. Parthalain and Q. Shen. Exploring the Boundary Region of Tolerance Rough Sets for Feature Selection. *Pattern Recognition*, 42(5):655–667, 2009.

84. Q. Hu, Z. Xie, and D. Yu. Hybrid Attribute Reduction Based on a Novel Fuzzy-Rough Model and Information Granulation. *Pattern Recognition*, 40:3577–3594, 2007.

85. Q. Hu, L. Zhang, D. Chen, W. Pedrycz, and D. Yu. Gaussian Kernel Based Fuzzy Rough Sets: Model, Uncertainty Measures and Applications. *International Journal of Approximate Reasoning*, 51:453–471, 2010.

86. R. Jensen and Q. Shen. Fuzzy-Rough Sets Assisted Attribute Selection. *IEEE Transactions on Fuzzy Systems*, 15:73–89, 2007.

87. W. Wu and W. Zhang. Constructive and Axiomatic Approaches of Fuzzy Approximation Operators. *Information Sciences*, 159:233–254, 2004.

88. S. Zhao, E. C. C. Tsang, and D. Chen. The Model of Fuzzy Variable Precision Rough Sets. *IEEE Transactions on Fuzzy Systems*, 17:451–467, 2009.

89. Q. Hu, D. Yu, J. Liu, and C. Wu. Neighborhood Rough Set Based Heterogeneous Feature Subset Selection. *Information Sciences*, 178:3577–3594, 2008.

90. J. Yao, Y. Yao, and W. Ziarko. Probabilistic Rough Sets: Approximations, Decision-Makings, and Applications. *International Journal of Approximate Reasoning*, 49(2):253–254, 2008.

91. Y. Yao. Probabilistic Rough Set Approximations. *International Journal of Approximate Reasoning*, 49(2):255–271, 2008.

92. W. Ziarko. Probabilistic Approach to Rough Sets. *International Journal of Approximate Reasoning*, 49(2):272–284, 2008.

93. Y. Tang, Y.-Q. Zhang, Z. Huang, X. Hu, and Y. Zhao. Recursive Fuzzy Granulation for Gene Subsets Extraction and Cancer Classification. *IEEE Transactions on Information Technology in Biomedicine*, 12(6):723–730, 2008.

94. S. K. Pal, B. U. Shankar, and P. Mitra. Granular Computing, Rough Entropy and Object Extraction. *Pattern Recognition Letters*, 26(16):2509–2517, 2005.

95. W. Pedrycz, A. Skowron, and V. Kreinovich, editors. *Handbook of Granular Computing*. John Wiley & Sons, Ltd., West Sussex, England, 2008.

96. A. Skowron and J. F. Peters. Rough-Granular Computing. In W. Pedrycz, A. Skowron, and V. Kreinovich, editors, *Handbook of Granular Computing*, pages 285–328. John Wiley & Sons, Ltd., West Sussex, England, 2008.

97. Y.-Q. Zhang. Constructive Granular Systems with Universal Approximation and Fast Knowledge Discovery. *IEEE Transactions on Fuzzy Systems*, 13(1):48–57, 2005.

98. Y. Q. Zhang, M. D. Fraser, R. A. Gagliano, and A. Kandel. Granular Neural Networks for Numerical-Linguistic Data Fusion and Knowledge Discovery. *IEEE Transactions on Neural Networks*, 11:658–667, 2000.

99. Y. Cao and G. Chen. A Fuzzy Petri-Nets Model for Computing with Words. *IEEE Transactions on Fuzzy Systems*, 18(3):486–499, 2010.

100. M. Inuiguchi, S. Tsumoto, and S. Hirano, editors. *Rough Set Theory and Granular Computing*. Springer-Verlag, Berlin, 2010.

101. T. Y. Lin. Granulation and Nearest Neighborhoods: Rough Set Approach. In W. Pedrycz, editor, *Granular Computing: An Emerging Paradigm*, pages 125–142. Physica-Verlag, Heidelberg, 2001.

102. L. A. Zadeh. A New Direction in AI: Toward a Computational Theory of Perceptions. *AI Magazine*, 22:73–84, 2001.

103. S. K. Pal. Computational Theory of Perception (CTP), Rough-Fuzzy Uncertainty Analysis and Mining in Bioinformatics and Web Intelligence: A Unified Framework. *LNCS Transactions on Rough Sets*, 5946:106–129, 2009.

104. S. K. Pal and P. Mitra. *Pattern Recognition Algorithms for Data Mining*. CRC Press, Boca Raton, FL, 2004.

105. M. Banerjee, S. Mitra, and S. K. Pal. Rough-Fuzzy MLP: Knowledge Encoding and Classification. *IEEE Transactions on Neural Networks*, 9(6):1203–1216, 1998.

106. D. Dubois and H. Prade. Rough Fuzzy Sets and Fuzzy Rough Sets. *International Journal of General Systems*, 17:191–209, 1990.

107. D. Dubois and H. Prade. Putting Fuzzy Sets and Rough Sets Together. In R. Slowiniski, editor, *Intelligent Decision Support: Handbook of Applications and Advances of Rough Sets Theory*, pages 203–232. Kluwer, Norwell, MA, 1992.

108. S. K. Pal, S. Mitra, and P. Mitra. Rough-Fuzzy MLP: Modular Evolution, Rule Generation, and Evaluation. *IEEE Transactions on Knowledge and Data Engineering*, 15(1):14–25, 2003.

109. M. Banerjee and S. K. Pal. Roughness of a Fuzzy Set. *Information Sciences*, 93(3/4):235–246, 1996.

110. K. Chakrabartya, R. Biswas, and S. Nanda. Fuzziness in Rough Sets. *Fuzzy Sets and Systems*, 110(2):247–251, 2000.

111. M. D. Cock, C. Cornelis, and E. E. Kerre. Fuzzy Rough Sets: The Forgotten Step. *IEEE Transactions on Fuzzy Systems*, 15(1):121–130, 2007.

112. R. Jensen and Q. Shen. *Computational Intelligence and Feature Selection: Rough and Fuzzy Approaches*. Wiley-IEEE Press, Hoboken, New Jersey, 2008.

113. X. Liu, W. Pedrycz, T. Chai, and M. Song. The Development of Fuzzy Rough Sets with the Use of Structures and Algebras of Axiomatic Fuzzy Sets. *IEEE Transactions on Knowledge and Data Engineering*, 21(3):443–462, 2009.

114. P. Maji and S. K. Pal. Rough-Fuzzy C-Medoids Algorithm and Selection of Bio-Basis for Amino Acid Sequence Analysis. *IEEE Transactions on Knowledge and Data Engineering*, 19(6):859–872, 2007.

115. P. Maji and S. K. Pal. Rough Set Based Generalized Fuzzy C-Means Algorithm and Quantitative Indices. *IEEE Transactions on Systems Man and Cybernetics Part B-Cybernetics*, 37(6):1529–1540, 2007.

116. P. Maji and S. K. Pal. Feature Selection Using f-Information Measures in Fuzzy Approximation Spaces. *IEEE Transactions on Knowledge and Data Engineering*, 22(6):854–867, 2010.

117. P. Maji and S. K. Pal. Fuzzy-Rough Sets for Information Measures and Selection of Relevant Genes from Microarray Data. *IEEE Transactions on Systems Man and Cybernetics Part B-Cybernetics*, 40(3):741–752, 2010.

118. S. K. Pal. Soft Data Mining, Computational Theory of Perceptions, and Rough-Fuzzy Approach. *Information Sciences*, 163(1–3):5–12, 2004.

119. S. K. Pal and A. Skowron, editors. *Rough-Fuzzy Hybridization: A New Trend in Decision Making*. Springer-Verlag, Singapore, 1999.

120. P. Srinivasan, M. E. Ruiz, D. H. Kraft, and J. Chen. Vocabulary Mining for Information Retrieval: Rough Sets and Fuzzy Sets. *Information Processing and Management*, 37(1):15–38, 1998.

121. E. C. C. Tsang, D. Chen, D. S. Yeung, X.-Z. Wang, and J. Lee. Attributes Reduction Using Fuzzy Rough Sets. *IEEE Transactions on Fuzzy Systems*, 16(5):1130–1141, 2008.

122. Y. F. Wang. Mining Stock Price Using Fuzzy Rough Set System. *Expert Systems with Applications*, 24(1):13–23, 2003.

123. H. Wu, Y. Wu, and J. Luo. An Interval Type-2 Fuzzy Rough Set Model for Attribute Reduction. *IEEE Transactions on Fuzzy Systems*, 17(2):301–315, 2009.

124. D. S. Yeung, D. Chen, E. C. C. Tsang, J. W. T. Lee, and W. Xizhao. On the Generalization of Fuzzy Rough Sets. *IEEE Transactions on Fuzzy Systems*, 13(3):343–361, 2005.

125. S. Zhao, E. C. C. Tsang, D. Chen, and X. Wang. Building a Rule-Based Classifier: A Fuzzy-Rough Set Approach. *IEEE Transactions on Knowledge and Data Engineering*, 22(5):624–638, 2010.

126. Q. Hu, D. Yu, Z. Xie, and J. Liu. Fuzzy Probabilistic Approximation Spaces and Their Information Measures. *IEEE Transactions on Fuzzy Systems*, 14(2):191–201, 2007.

127. R. Jensen and Q. Shen. Fuzzy-Rough Attribute Reduction with Application to Web Categorization. *Fuzzy Sets and Systems*, 141:469–485, 2004.

128. R. Jensen and Q. Shen. New Approaches to Fuzzy-Rough Feature Selection. *IEEE Transactions on Fuzzy Systems*, 17(4):824–838, 2009.

129. P. Maji. Fuzzy-Rough Supervised Attribute Clustering Algorithm and Classification of Microarray Data. *IEEE Transactions on Systems Man and Cybernetics Part B-Cybernetics*, 41(1):222–233, 2011.

130. P. Maji and S. K. Pal. RFCM: A Hybrid Clustering Algorithm Using Rough and Fuzzy Sets. *Fundamenta Informaticae*, 80(4):475–496, 2007.

131. R. Nowicki. On Combining Neuro-Fuzzy Architectures with the Rough Set Theory to Solve Classification Problems with Incomplete Data. *IEEE Transactions on Knowledge and Data Engineering*, 20(9):1239–1253, 2008.

132. R. Nowicki. Rough Neuro-Fuzzy Structures for Classification with Missing Data. *IEEE Transactions on Systems Man and Cybernetics Part B-Cybernetics*, 39(6):1334–1347, 2009.

133. A. E. Hassanien. Fuzzy Rough Sets Hybrid Scheme for Breast Cancer Detection. *Image and Vision Computing*, 25(2):172–183, 2007.

134. P. Maji and S. K. Pal. Maximum Class Separability for Rough-Fuzzy C-Means Based Brain MR Image Segmentation. *LNCS Transactions on Rough Sets*, 9:114–134, 2008.

135. S. K. Pal and J. F. Peter, editors. *Rough Fuzzy Image Analysis: Foundations and Methodologies*. Chapman & Hall/CRC Press, Boca Raton, FL, 2010.

136. D. Sen and S. K. Pal. Generalized Rough Sets, Entropy and Image Ambiguity Measures. *IEEE Transactions on Systems Man and Cybernetics Part B-Cybernetics*, 39(1):117–128, 2009.

137. F. Fernandez-Riverola, F. Diaz, and J. M. Corchado. Reducing the Memory Size of a Fuzzy Case-Based Reasoning System Applying Rough Set Techniques. *IEEE Transactions on Systems Man and Cybernetics Part C-Applications and Reviews*, 37(1):138–146, 2007.

138. Y. Li, S. C. K. Shiu, and S. K. Pal. Combining Feature Reduction and Case Selection in Building CBR Classifiers. *IEEE Transactions on Knowledge and Data Engineering*, 18(3):415–429, 2006.

139. S. K. Pal and P. Mitra. Case Generation Using Rough Sets with Fuzzy Representation. *IEEE Transactions on Knowledge and Data Engineering*, 16:293–300, 2004.

140. S. Asharaf and M. N. Murty. A Rough Fuzzy Approach to Web Usage Categorization. *Fuzzy Sets and Systems*, 148:119–129, 2004.

141. A. Skowron and J. Stepaniuk. Tolerance Approximation Spaces. *Fundamenta Informaticae*, 27(2–3):245–253, 1996.

142. T. Beaubouef, F. E. Petry, and G. Arora. Information-Theoretic Measures of Uncertainty for Rough Sets and Rough Relational Databases. *Information Sciences*, 109(1–4):185–195, 1998.

143. J. Liang, K. S. Chin, C. Dang, and R. C. M. Yam. A New Method for Measuring Uncertainty and Fuzziness in Rough Set Theory. *International Journal of General Systems*, 31(4):331–342, 2002.

144. J.-S. Mi, X.-M. Li, H.-Y. Zhao, and T. Feng. Information-Theoretic Measure of Uncertainty in Generalized Fuzzy Rough Sets. In *Rough Sets, Fuzzy Sets, Data Mining and Granular Computing, Lecture Notes in Computer Science*, pages 63–70. Springer, 2007.

145. M. J. Wierman. Measuring Uncertainty in Rough Set Theory. *International Journal of General Systems*, 28(4):283–297, 1999.

146. R. R. Yager. Entropy Measures under Similarity Relations. *International Journal of General Systems*, 20(4):341–358, 1992.

147. D. Sen and S. K. Pal. Histogram Thresholding Using Beam Theory and Ambiguity Measures. *Fundamenta Informaticae*, 75(1-4):483–504, 2007.

148. N. R. Pal and S. K. Pal. Entropy: A New Definition and Its Application. *IEEE Transactions on Systems Man and Cybernetics Part B-Cybernetics*, 21(5):1260–1270, 1991.

149. N. R. Pal and J. C. Bezdek. Measuring Fuzzy Uncertainty. *IEEE Transactions on Fuzzy Systems*, 2(2):107–118, 1994.

150. B. R. Ebanks. On Measures of Fuzziness and Their Representations. *Journal of Mathematical Analysis and Applications*, 94(1):24–37, 1983.

3

ROUGH-FUZZY CLUSTERING: GENERALIZED c-MEANS ALGORITHM

3.1 INTRODUCTION

Cluster analysis is a technique for finding natural groups present in the data. It divides a given data set into a set of clusters in such a way that two objects from the same cluster are as similar as possible and the objects from different clusters are as dissimilar as possible. In effect, it tries to mimic the human ability to group similar objects into classes and categories [1].

Clustering techniques have been effectively applied to a wide range of engineering and scientific disciplines such as pattern recognition, machine learning, psychology, biology, medicine, computer vision, web intelligence, communications, and remote sensing. A number of clustering algorithms have been proposed to suit different requirements [1, 2]. The hard clustering algorithms generate crisp clusters by assigning each object to exactly one cluster. When the clusters are not well defined, that is, when they are overlapping, one may desire fuzzy clusters. In this regard, the problem of pattern classification is formulated as the problem of interpolation of the membership function of a fuzzy set in Bellman et al. [3], and thereby a link with the basic problem of system identification is established. A seminal contribution to cluster analysis is Ruspini's concept of a fuzzy partition [4]. The application of fuzzy set theory to cluster analysis was initiated by Dunn and Bezdek by developing fuzzy ISODATA [5] and fuzzy c-means (FCM) algorithms [6].

Rough-Fuzzy Pattern Recognition: Applications in Bioinformatics and Medical Imaging,
First Edition. Pradipta Maji and Sankar K. Pal.
© 2012 John Wiley & Sons, Inc. Published 2012 by John Wiley & Sons, Inc.

One of the most widely used prototype-based partitional clustering algorithms is hard *c*-means (HCM) [7]. In HCM, each object is assigned to exactly one cluster. On the other hand, FCM relaxes this requirement by, allowing gradual memberships [6]. In effect, it offers the opportunity to deal with the data that belong to more than one cluster at the same time. It assigns memberships to an object that is inversely related to the relative distance of the object to cluster prototypes. Also, it can deal with the uncertainties arising from overlapping cluster boundaries.

Although FCM is a useful clustering method, the resulting membership values do not always correspond well to the degrees of belonging for the data, and it may be inaccurate in a noisy environment [8, 9]. However, in real data analysis, noise and outliers are unavoidable. To reduce this weakness of FCM and to produce memberships that have a good explanation of the degrees of belonging for the data, Krishnapuram and Keller [8, 9] proposed a possibilistic approach to clustering that used a possibilistic type of membership function to describe the degree of belonging. However, the possibilistic *c*-means (PCM) sometimes generates coincident clusters [10]. Recently, the use of both fuzzy (probabilistic) and possibilistic memberships in a clustering algorithm has been proposed [11–13].

Rough set theory is a new paradigm to deal with uncertainty, vagueness, and incompleteness. It is proposed for indiscernibility in classification or clustering according to some similarity [14]. Two of the early rough clustering algorithms are those due to Hirano and Tsumoto [15] and De [16]. Other notable algorithms include rough *c*-means (RCM) [17], rough self-organizing map [18], and rough support vector clustering [19]. In Pal and Mitra [20], the indiscernibility relation of rough sets has been used to initialize the expectation maximization algorithm. Lingras and West [17] have introduced a rough clustering method, called *rough c-means*, which describes a cluster by a prototype or center and a pair of lower and upper approximations. The lower and upper approximations are weighted different parameters to compute the new centers. Asharaf and Murty [21] extended this algorithm, which may not require specification of the number of clusters.

Combining fuzzy sets and rough sets provides an important direction in reasoning with uncertainty. Both fuzzy sets and rough sets provide a mathematical framework to capture uncertainties associated with the data. They are complementary in some aspects. Combining both rough sets and fuzzy sets, Mitra et al. [22] proposed rough-fuzzy *c*-means (RFCM), where each cluster consists of a fuzzy lower approximation and a fuzzy boundary. Each object in lower approximation takes a distinct weight, which is its fuzzy membership value. However, the objects in lower approximation of a cluster should have a similar influence on the corresponding centroid and cluster and their weights should be independent of other centroids and clusters. Hence, the concept of fuzzy lower approximation, introduced in RFCM of Mitra et al. [22], reduces the weights of objects of lower approximation. In effect, it drifts the cluster prototypes from their desired locations. Moreover, it is sensitive to noise and outliers.

Recently, a generalized hybrid algorithm, termed as *rough-fuzzy-possibilistic c-means* (RFPCM), was described by Maji and Pal [23, 24], based on rough sets and fuzzy sets. While the membership function of fuzzy sets enables efficient handling of overlapping partitions, the concept of lower and upper approximations of rough sets deals with uncertainty, vagueness, and incompleteness in class definition. The algorithm attempts to exploit the benefits of both probabilistic and possibilistic membership functions. Integration of probabilistic and possibilistic membership functions avoids the problems of noise sensitivity of FCM and the coincident clusters of PCM. Each partition is represented by a set of three parameters, namely, a cluster prototype or centroid, a crisp lower approximation, and a fuzzy boundary. The lower approximation influences the fuzziness of the final partition. The cluster prototype or centroid depends on the weighting average of the crisp lower approximation and fuzzy boundary. The algorithm is generalized in the sense that all the existing variants of c-means algorithms can be derived from the RFPCM algorithm as a special case.

The structure of the rest of this chapter is as follows: Section 3.2 briefly introduces the necessary notions of HCM, FCM, and RCM algorithms. In Section 3.3, the RFPCM algorithm is described in detail on the basis of the theory of rough sets and FCM. A mathematical analysis of the convergence property of the RFPCM algorithm is also presented. Section 3.4 establishes that the RFPCM algorithm is the generalization of existing c-means algorithms. Several quantitative performance measures are reported in Section 3.5 to evaluate the quality of different algorithms. A few case studies and an extensive comparison with other methods such as crisp, fuzzy, possibilistic, and RCM are presented in Section 3.6. Concluding remarks are given in Section 3.7.

3.2   EXISTING c-MEANS ALGORITHMS

This section presents the basic notions of FCM and RCM algorithms. The generalized RFPCM algorithm is developed on the basis of these algorithms.

3.2.1   Hard c-Means

The HCM [7] is one of the simplest unsupervised learning algorithms. The objective of the HCM algorithm is to assign n objects to c clusters. Each of the clusters β_i is represented by a centroid v_i, which is the cluster representative for that cluster. The process begins by randomly choosing c objects as the centroids or means. The objects are assigned to one of the c clusters based on the similarity or dissimilarity between the object x_j and the centroid v_i. After the assignment of all the objects to various clusters, the new centroids are calculated as follows:

$$v_i = \frac{1}{n_i} \sum_{x_j \in \beta_i} x_j, \tag{3.1}$$

where n_i represents the number of objects in cluster β_i. The main steps of the HCM algorithm are as follows:

1. Assign initial means or centroids v_i, $i = 1, 2, \ldots, c$.
2. For each object x_j, calculate distance d_{ij} between itself and the centroid v_i of cluster β_i.
3. If d_{ij} is minimum for $1 \leq i \leq c$, then $x_j \in \beta_i$.
4. Compute new centroid as per Equation (3.1).
5. Repeat steps 2–4 until no more new assignments can be made.

3.2.2 Fuzzy *c*-Means

Let $X = \{x_1, \ldots, x_j, \ldots, x_n\}$ and $V = \{v_1, \ldots, v_i, \ldots, v_c\}$ be the set of n objects and c centroids, respectively, having m dimensions where $x_j \in \Re^m$, $v_i \in \Re^m$, and $v_i \in X$. The FCM provides a fuzzification of the HCM [6]. It partitions X into c clusters by minimizing the following objective function

$$J = \sum_{j=1}^{n} \sum_{i=1}^{c} (\mu_{ij})^{\acute{m}_1} ||x_j - v_i||^2 \tag{3.2}$$

where $1 \leq \acute{m}_1 < \infty$ is the fuzzifier, v_i is the ith centroid corresponding to cluster β_i, $\mu_{ij} \in [0, 1]$ is the probabilistic membership of the pattern x_j to cluster β_i, and $||\cdot||$ is the distance norm such that

$$v_i = \frac{1}{n_i} \sum_{j=1}^{n} (\mu_{ij})^{\acute{m}_1} x_j, \text{ where } n_i = \sum_{j=1}^{n} (\mu_{ij})^{\acute{m}_1} \tag{3.3}$$

and

$$\mu_{ij} = \left(\sum_{k=1}^{c} \left(\frac{d_{ij}}{d_{kj}} \right)^{2/(\acute{m}_1 - 1)} \right)^{-1}, \text{ where } d_{ij}^2 = ||x_j - v_i||^2 \tag{3.4}$$

subject to

$$\sum_{i=1}^{c} \mu_{ij} = 1, \forall j, \text{ and } 0 < \sum_{j=1}^{n} \mu_{ij} < n, \forall i. \tag{3.5}$$

The process begins by randomly choosing c objects as the centroids or means of the c clusters. The memberships are calculated on the basis of the relative distance of the object x_j to the centroids $\{v_i\}$ by Equation (3.4). After computing memberships of all the objects, the new centroids of the clusters are calculated as per Equation (3.3). The process stops when the centroids stabilize. That is, the centroids from the previous iteration are identical to those generated in the current iteration. The basic steps are outlined as follows:

1. Assign initial means v_i, $i = 1, 2, \ldots, c$. Choose values for $\acute{m}_1$ and threshold ϵ. Set iteration counter $t = 1$.
2. Compute memberships μ_{ij} by Equation (3.4) for c clusters and n objects.
3. Update mean or centroid v_i by Equation (3.3).
4. Repeat steps 2–4, by incrementing t, until $|\mu_{ij}(t) - \mu_{ij}(t - 1)| > \epsilon$.

3.2.3 Possibilistic c-Means

In FCM, the memberships of an object are inversely related to the relative distance of the object to the cluster centroids. In effect, it is very sensitive to noise and outliers. Also, from the standpoint of compatibility with the centroid, the memberships of an object x_j in a cluster β_i should be determined solely by how close it is to the mean or centroid v_i of the class, and should not be coupled with its similarity with respect to other classes.

To alleviate this problem, Krishnapuram and Keller [8, 9] introduced PCM, where the objective function can be formulated as

$$J = \sum_{i=1}^{c} \sum_{j=1}^{n} (v_{ij})^{\acute{m}_2} ||x_j - v_i||^2 + \sum_{i=1}^{c} \eta_i \sum_{j=1}^{n} (1 - v_{ij})^{\acute{m}_2}, \qquad (3.6)$$

where $1 \le \acute{m}_2 \le \infty$ is the fuzzifier and η_i represents the scale parameter. The membership matrix v generated by PCM is not a partition matrix in the sense that it does not satisfy the constraint

$$\sum_{i=1}^{c} v_{ij} = 1, \forall j. \qquad (3.7)$$

The update equation of v_{ij} is given by

$$v_{ij} = \frac{1}{1 + D}, \quad \text{where } D = \left\{ \frac{||x_j - v_i||^2}{\eta_i} \right\}^{1/(\acute{m}_2 - 1)} \qquad (3.8)$$

subject to

$$v_{ij} \in [0, 1], \forall i, j; 0 < \sum_{j=1}^{n} v_{ij} \le n, \forall i; \text{ and } \max_{i} v_{ij} > 0, \forall j. \qquad (3.9)$$

The scale parameter η_i represents the zone of influence or size of the cluster β_i. The update equation for η_i is given by

$$\eta_i = K \cdot \frac{P}{Q}, \qquad (3.10)$$

where

$$P = \sum_{j=1}^{n} (v_{ij})^{\acute{m}_2} ||x_j - v_i||^2 \qquad (3.11)$$

and

$$Q = \sum_{j=1}^{n} (v_{ij})^{\acute{m}_2}. \qquad (3.12)$$

Typically, K is chosen to be 1. In each iteration, the updated value of v_{ij} depends only on the similarity between the object x_j and the centroid v_i. The resulting partition of the data can be interpreted as a possibilistic partition, and the membership values may be interpreted as degrees of possibility of the objects belonging to the classes, that is, the compatibilities of the objects with the means or centroids. The updating of the means proceeds exactly the same way as in the case of the FCM algorithm.

3.2.4 Rough c-Means

Let $\underline{A}(\beta_i)$ and $\overline{A}(\beta_i)$ be the lower and upper approximations of cluster β_i, and $B(\beta_i) = \{\overline{A}(\beta_i) \setminus \underline{A}(\beta_i)\}$ denotes the boundary region of cluster β_i. In the RCM algorithm, the concept of c-means algorithm is extended by viewing each cluster β_i as an interval or rough set [17]. However, it is possible to define a pair of lower and upper bounds $[\underline{A}(\beta_i), \overline{A}(\beta_i)]$ or a rough set for every set $\beta_i \subseteq U$, U being the set of objects of concern. The family of upper and lower bounds is required to follow some basic rough set properties such as the following:

1. An object x_j can be part of at most one lower bound.
2. $x_j \in \underline{A}(\beta_i) \Rightarrow x_j \in \overline{A}(\beta_i)$.
3. An object x_j is not part of any lower bound $\Rightarrow x_j$ belongs to two or more upper bounds.

Incorporating rough sets into the c-means algorithm, Lingras and West [17] introduced the RCM algorithm. It adds the concept of lower and upper bounds into the c-means algorithm. It classifies the object space into two parts, namely, lower approximation and boundary region. The mean or centroid is calculated on the basis of the weighting average of the lower bound and boundary region. All the objects in lower approximation take the same weight w, whereas all the objects in the boundary take another weighting index $\tilde{w}$ ($= 1 - w$) uniformly. Calculation of the centroid is modified to include the effects of lower as well as upper bounds. The modified centroid calculation for the RCM algorithm is given by

$$
v_i = \begin{cases} w \times \mathcal{A} + \tilde{w} \times \mathcal{B} & \text{if } \underline{A}(\beta_i) \neq \emptyset, B(\beta_i) \neq \emptyset \\ \mathcal{A} & \text{if } \underline{A}(\beta_i) \neq \emptyset, B(\beta_i) = \emptyset \\ \mathcal{B} & \text{if } \underline{A}(\beta_i) = \emptyset, B(\beta_i) \neq \emptyset, \end{cases} \tag{3.13}
$$

where

$$
\mathcal{A} = \frac{1}{|\underline{A}(\beta_i)|} \sum_{x_j \in \underline{A}(\beta_i)} x_j, \tag{3.14}
$$

and

$$
\mathcal{B} = \frac{1}{|B(\beta_i)|} \sum_{x_j \in B(\beta_i)} x_j. \tag{3.15}
$$

The main steps of the RCM algorithm are as follows:

1. Assign initial means v_i, $i = 1, 2, \ldots, c$. Choose a value for threshold δ.
2. For each object x_j, calculate the distance d_{ij} between itself and the centroid v_i of cluster β_i.
3. If d_{ij} is minimum for $1 \leq i \leq c$ and $(d_{kj} - d_{ij}) \leq \delta$, then $x_j \in \overline{A}(\beta_i)$ and $x_j \in \overline{A}(\beta_k)$. Furthermore, x_j is not part of any lower bound.
4. Otherwise, $x_j \in \underline{A}(\beta_i)$ such that d_{ij} is minimum for $1 \leq i \leq c$. In addition, by properties of rough sets, $x_j \in \overline{A}(\beta_i)$.
5. Compute a new centroid as per Equation (3.13).
6. Repeat steps 2–5 until no more new assignments can be made.

3.3 ROUGH-FUZZY-POSSIBILISTIC c-MEANS

Incorporating both fuzzy and rough sets, a generalized c-means algorithm, termed as *rough-fuzzy-possibilistic c-means*, is proposed by Maji and Pal [23, 24]. It adds the concept of fuzzy membership (both probabilistic and possibilistic) of fuzzy sets, and lower and upper approximations of rough sets into the c-means algorithm. While the membership of fuzzy sets enables efficient handling of overlapping partitions, the rough sets deal with uncertainty, vagueness, and incompleteness in class definition. Owing to the integration of both probabilistic and possibilistic memberships, the RFPCM avoids the problems of noise sensitivity of FCM and the coincident clusters of PCM.

3.3.1 Objective Function

The RFPCM algorithm partitions a set of n objects into c clusters by minimizing the following objective function:

$$
J_{\text{RFP}} = \begin{cases} w \times \mathcal{A}_1 + \tilde{w} \times \mathcal{B}_1 & \text{if } \underline{A}(\beta_i) \neq \emptyset,\, B(\beta_i) \neq \emptyset \\ \mathcal{A}_1 & \text{if } \underline{A}(\beta_i) \neq \emptyset,\, B(\beta_i) = \emptyset \\ \mathcal{B}_1 & \text{if } \underline{A}(\beta_i) = \emptyset,\, B(\beta_i) \neq \emptyset \end{cases}
\tag{3.16}
$$

$$
\mathcal{A}_1 = \sum_{i=1}^{c} \sum_{x_j \in \underline{A}(\beta_i)} \{a(\mu_{ij})^{\acute{m}_1} + b(\nu_{ij})^{\acute{m}_2}\}\|x_j - v_i\|^2 + \sum_{i=1}^{c} \eta_i \sum_{x_j \in \underline{A}(\beta_i)} (1 - \nu_{ij})^{\acute{m}_2}
$$

$$
\mathcal{B}_1 = \sum_{i=1}^{c} \sum_{x_j \in B(\beta_i)} \{a(\mu_{ij})^{\acute{m}_1} + b(\nu_{ij})^{\acute{m}_2}\}\|x_j - v_i\|^2 + \sum_{i=1}^{c} \eta_i \sum_{x_j \in B(\beta_i)} (1 - \nu_{ij})^{\acute{m}_2}.
$$

The parameters w and $\tilde{w}$ $(= 1 - w)$ correspond to the relative importance of the lower and boundary regions, respectively. The constants a and b $(= 1 - a)$ define the relative importance of probabilistic and possibilistic memberships, respectively. Note that μ_{ij} has the same meaning of membership as that in FCM. Similarly, ν_{ij} has the same interpretation of typicality as in PCM. Solving Equation (3.16) with respect to μ_{ij} and ν_{ij}, we get

$$
\mu_{ij} = \left(\sum_{k=1}^{c} \left(\frac{d_{ij}}{d_{kj}} \right)^{2/(\acute{m}_1 - 1)} \right)^{-1}, \quad \text{where } d_{ij}^2 = \|x_j - v_i\|^2
\tag{3.17}
$$

and

$$
\nu_{ij} = \frac{1}{1 + E}; \quad \text{where } E = \left\{ \frac{b\|x_j - v_i\|^2}{\eta_i} \right\}^{1/(\acute{m}_2 - 1)}.
\tag{3.18}
$$

Hence, the probabilistic membership μ_{ij} is independent of the constant a, while constant b has a direct influence on the possibilistic membership ν_{ij}. The scale parameter η_i has the same expression as that in Equation (3.10).

In RFPCM, each cluster is represented by a centroid, a crisp lower approximation, and a fuzzy boundary (Fig. 3.1). The lower approximation influences the fuzziness of final partition. According to the definitions of lower approximations and boundary of rough sets, if an object $x_j \in \underline{A}(\beta_i)$, then $x_j \notin \underline{A}(\beta_k)$, $\forall k \neq i$, and $x_j \notin B(\beta_i)$, $\forall i$. That is, the object x_j is contained in β_i definitely. Hence, the weights of the objects in lower approximation of a cluster should be independent of other centroids and clusters, and should not be coupled with their similarity with respect to other centroids. Also, the objects in the lower approximation of a cluster should have a similar influence on the corresponding centroid and cluster, whereas, if $x_j \in B(\beta_i)$, then the object x_j possibly belongs to cluster β_i and potentially belongs to another cluster. Hence, the objects in boundary regions

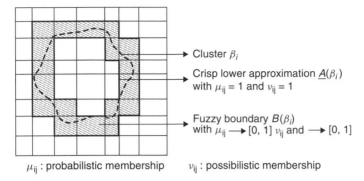

Cluster β_i

Crisp lower approximation $\underline{A}(\beta_i)$
with $\mu_{ij} = 1$ and $v_{ij} = 1$

Fuzzy boundary $B(\beta_i)$
with $\mu_{ij} \longrightarrow [0, 1]$ v_{ij} and $\longrightarrow [0, 1]$

μ_{ij} : probabilistic membership v_{ij} : possibilistic membership

Figure 3.1 Rough-fuzzy-possibilistic c-means: cluster β_i is represented by crisp lower bound and fuzzy boundary.

should have a different influence on the centroids and clusters. So, in RFPCM, the membership values of objects in the lower approximation are $\mu_{ij} = v_{ij} = 1$, while those in the boundary region are the same as FCM (Eq. (3.17)) and PCM (Eq. (3.18)). In other words, the RFPCM first partitions the data into two classes, namely, lower approximation and boundary. Only the objects in the boundary are fuzzified. Hence, $\mathcal{A}_1$ reduces to

$$\mathcal{A}_1 = \sum_{i=1}^{c} \sum_{x_j \in \underline{A}(\beta_i)} ||x_j - v_i||^2 \qquad (3.19)$$

and $\mathcal{B}_1$ has the same expression as that in Equation (3.16).

3.3.2 Cluster Prototypes

The new centroid is calculated on the basis of the weighting average of the crisp lower approximation and fuzzy boundary. Computation of the centroid is modified to include the effects of both fuzzy memberships (probabilistic and possibilistic) and lower and upper bounds. The modified centroid calculation for RFPCM is obtained by solving Equation (3.16) with respect to v_i:

$$v_i^{\text{RFP}} = \begin{cases} w \times \mathcal{C}_1 + \tilde{w} \times \mathcal{D}_1 & \text{if } \underline{A}(\beta_i) \neq \emptyset, B(\beta_i) \neq \emptyset \\ \mathcal{C}_1 & \text{if } \underline{A}(\beta_i) \neq \emptyset, B(\beta_i) = \emptyset \\ \mathcal{D}_1 & \text{if } \underline{A}(\beta_i) = \emptyset, B(\beta_i) \neq \emptyset \end{cases} \qquad (3.20)$$

$$\mathcal{C}_1 = \frac{1}{|\underline{A}(\beta_i)|} \sum_{x_j \in \underline{A}(\beta_i)} x_j, \qquad (3.21)$$

where $|\underline{A}(\beta_i)|$ represents the cardinality of $\underline{A}(\beta_i)$ and

$$\mathcal{D}_1 = \frac{1}{n_i} \sum_{x_j \in B(\beta_i)} \{a(\mu_{ij})^{\acute{m}_1} + b(v_{ij})^{\acute{m}_2}\} x_j, \tag{3.22}$$

where

$$n_i = \sum_{x_j \in B(\beta_i)} \{a(\mu_{ij})^{\acute{m}_1} + b(v_{ij})^{\acute{m}_2}\}.$$

Hence, the cluster prototypes or centroids depend on the parameters w and $\tilde{w}$, and constants a and b, and fuzzifiers $\acute{m}_1$ and $\acute{m}_2$ rule their relative influence. This shows that if b is higher than a, the centroids will be more influenced by the possibilistic memberships than the probabilistic memberships. Thus, to reduce the influence of noise and outliers, a bigger value for b than a should be used. The correlated influence of these parameters, constants, and fuzzifiers makes it somewhat difficult to determine their optimal values. Since the objects lying in the lower approximation definitely belong to a cluster, they are assigned a higher weight w compared to $\tilde{w}$ of the objects lying in the boundary region. Hence, for RFPCM, the values are given by

$$0 < \tilde{w} < w < 1, \quad 0 < a < 1, \text{ and } 0 < b < 1,$$

subject to

$$w + \tilde{w} = 1; \quad \text{and } a + b = 1.$$

3.3.3 Fundamental Properties

From the above discussions, the following properties of the RFPCM algorithm can be obtained:

1. Let $\bigcup \overline{A}(\beta_i) = U$, U be the set of objects of concern.
2. $\underline{A}(\beta_i) \cap \underline{A}(\beta_k) = \emptyset, \forall i \neq k$.
3. $\underline{A}(\beta_i) \cap B(\beta_i) = \emptyset, \forall i$.
4. $\exists i, k, B(\beta_i) \cap B(\beta_k) \neq \emptyset$.
5. $\mu_{ij} = v_{ij} = 1, \forall x_j \in \underline{A}(\beta_i)$.
6. $\mu_{ij} \in [0, 1], v_{ij} \in [0, 1], \forall x_j \in B(\beta_i)$.

Let us briefly comment on some properties of RFPCM. Property 2 says that if an object $x_j \in \underline{A}(\beta_i) \Rightarrow x_j \notin \underline{A}(\beta_k), \forall k \neq i$. That is, the object x_j is contained in cluster β_i definitely. Property 3 establishes the fact that if $x_j \in \underline{A}(\beta_i) \Rightarrow x_j \notin B(\beta_i)$, that is, an object may not be in both the lower and boundary regions of a cluster β_i. Property 4 says that if $x_j \in B(\beta_i) \Rightarrow \exists k, x_j \in B(\beta_k)$. It means an object $x_j \in B(\beta_i)$ possibly belongs to cluster β_i and potentially belongs to another cluster. Properties 5 and 6 are of great importance in computing the objective function J_{RFP} (Eq. (3.16)) and the cluster prototype v^{RFP} (Eq. (3.20)). They say that the membership values of objects in lower approximation are $\mu_{ij} = v_{ij} = 1$, while those in the boundary region are the same as FCM and PCM. That is, each cluster β_i consists of a crisp lower approximation $\underline{A}(\beta_i)$ and a fuzzy boundary $B(\beta_i)$.

Some limiting properties of the RFPCM algorithm related to probabilistic and possibilistic memberships are stated subsequently.

$$\lim_{\acute{m}_1 \to 1+} \{\mu_{ij}\} = \begin{cases} 1 & \text{if } d_{ij} < d_{kj}, \forall k \neq i \\ 0 & \text{otherwise} \end{cases} \tag{3.23}$$

$$\lim_{\acute{m}_2 \to 1+} \{v_{ij}\} = \begin{cases} 1 & \text{if } bd_{ij} < \eta_i \\ \frac{1}{2} & \text{if } bd_{ij} = \eta_i \\ 0 & \text{if } bd_{ij} > \eta_i \end{cases} \tag{3.24}$$

$$\lim_{\acute{m}_1 \to \infty} \{\mu_{ij}\} = \frac{1}{c}. \tag{3.25}$$

$$\lim_{\acute{m}_2 \to \infty} \{v_{ij}\} = \frac{1}{2}. \tag{3.26}$$

$$\lim_{\acute{m}_1,\acute{m}_2 \to \infty} \{v_i\} = \sum_{j=1}^{n} x_j = \bar{v}; \; 1 \leq i \leq c. \tag{3.27}$$

$$v_{ij} = \begin{cases} 1 & \text{if } b = 0 \\ \frac{1}{2} & \text{if } bd_{ij} = \eta_i \end{cases} \tag{3.28}$$

3.3.4 Convergence Condition

In this subsection, a mathematical analysis is presented on the convergence property of the RFPCM algorithm. According to Equation (3.20), the cluster prototype of the RFPCM algorithm is calculated on the basis of the weighting average of the crisp lower approximation and fuzzy boundary when both $\underline{A}(\beta_i) \neq \emptyset$ and $B(\beta_i) \neq \emptyset$, that is,

$$v_i^{\text{RFP}} = w \times \text{v}_i^{\text{RFP}} + \tilde{w} \times \tilde{\text{v}}_i^{\text{RFP}}, \tag{3.29}$$

where

$$\text{v}_i^{\text{RFP}} = \mathcal{C}_1 = \frac{1}{|\underline{A}(\beta_i)|} \sum_{x_j \in \underline{A}(\beta_i)} x_j \tag{3.30}$$

and

$$\tilde{v}_i^{RFP} = \mathcal{D}_1 = \frac{1}{n_i} \sum_{x_j \in B(\beta_i)} \{a(\mu_{ij})^{\acute{m}_1} + b(v_{ij})^{\acute{m}_2}\} x_j, \tag{3.31}$$

where

$$n_i = \sum_{x_j \in B(\beta_i)} \{a(\mu_{ij})^{\acute{m}_1} + b(v_{ij})^{\acute{m}_2}\}. $$

Now, as per property 3 of RFPCM, an object may not belong to both lower approximation and boundary regions of a cluster. Hence, the convergence of v_i^{RFP} depends on the convergence of v_i^{RFP} and $\tilde{v}_i^{RFP}$. Both Equations (3.30) and (3.31) can be rewritten as

$$(|\underline{A}(\beta_i)|) v_i^{RFP} = \sum_{x_j \in \underline{A}(\beta_i)} x_j, \tag{3.32}$$

and

$$(n_i) \tilde{v}_i^{RFP} = \sum_{x_j \in B(\beta_i)} \{a(\mu_{ij})^{\acute{m}_1} + b(v_{ij})^{\acute{m}_2}\} x_j. \tag{3.33}$$

Hence, Equations (3.32) and (3.33) represent a set of linear equations in terms of v_i^{RFP} and $\tilde{v}_i^{RFP}$ if both μ_{ij} and v_{ij} are kept constant. A simple way to analyze the convergence property of the algorithm is to view both Equations (3.30) and (3.31) as the Gauss–Seidel iterations for solving the set of linear equations. The Gauss–Seidel algorithm is guaranteed to converge if the matrix representing each equation is diagonally dominant [25, 26]. This is a sufficient condition, not a necessary one. The iteration may or may not converge if the matrix is not diagonally dominant [25, 26]. The matrix corresponding to Equation (3.30) is given by

$$A = \begin{bmatrix} |\underline{A}(\beta_1)| & 0 & \cdots & \cdots & 0 \\ 0 & |\underline{A}(\beta_2)| & 0 & \cdots & 0 \\ \cdots & \cdots & \cdots & \cdots & \cdots \\ 0 & 0 & \cdots & \cdots & |\underline{A}(\beta_c)| \end{bmatrix}, \tag{3.34}$$

where $|\underline{A}(\beta_i)|$ represents the cardinality of $\underline{A}(\beta_i)$. Similarly, the matrix corresponding to Equation (3.31) is given by

$$\tilde{A} = \begin{bmatrix} n_1 & 0 & \cdots & \cdots & 0 \\ 0 & n_2 & 0 & \cdots & 0 \\ \cdots & \cdots & \cdots & \cdots & \cdots \\ 0 & 0 & \cdots & \cdots & n_c \end{bmatrix} \tag{3.35}$$

where

$$n_i = \sum_{x_j \in B(\beta_i)} \{a(\mu_{ij})^{\acute{m}_1} + b(v_{ij})^{\acute{m}_2}\},$$

which represents the cardinality of $B(\beta_i)$. For both A and $\tilde{A}$ to be diagonally dominant, we must have

$$|\underline{A}(\beta_i)| > 0; \text{ and } n_i > 0. \tag{3.36}$$

This is the sufficient condition for matrices A and $\tilde{A}$ to be diagonally dominant. Under this condition, the iteration would converge if Equations (3.30) and (3.31) were applied repetitively with μ_{ij} and v_{ij} kept constant.

In practice, Equations (3.17), (3.18), and (3.20) are applied alternatively in the iterations. The condition in Equation (3.36) is still correct according to Bezdek's convergence theorem of the FCM algorithm [27, 28] and Yan's convergence analysis of the fuzzy curve-tracing algorithm [29]. Both the matrices A and $\tilde{A}$ are also the Hessian (second-order derivative) of $\mathcal{A}_1$ and $\mathcal{B}_1$ (Eq. (3.16)) with respect to v_i^{RFP} and $\tilde{v}_i^{\text{RFP}}$, respectively. As both A and $\tilde{A}$ are diagonally dominant, all their eigenvalues are positive. Also, the Hessian of $\mathcal{B}_1$ with respect to both μ_{ij} and v_{ij} can be easily shown to be diagonal matrix and are positive definite. Hence, according to the theorem derived by Bezdek [27, 28] and the analysis done by Yan [29], it can be concluded that the RFPCM algorithm converges, at least along a subsequence, to a local optimum solution as long as the condition in Equation (3.36) is satisfied. Intuitively, the objective function J_{RFP}, that is, Equation (3.16), reduces in all steps corresponding to Equations (3.17), (3.18), and (3.20), so the compound procedure makes the function J_{RFP} descent strictly.

3.3.5 Details of the Algorithm

Approximate optimization of the objective function J_{RFP}, that is, Equation (3.16), by RFPCM is based on Picard iteration through Equations (3.17), (3.18), and (3.20). This type of iteration is called *alternating optimization*. The process starts by randomly choosing c objects as the centroids of the c clusters. The probabilistic and possibilistic memberships of all the objects are calculated using Equations (3.17) and (3.18). The scale parameters η_i for c clusters are obtained using Equation (3.10).

Let $u_i = (u_{i1}, \ldots, u_{ij}, \ldots, u_{in})$ represent the fuzzy cluster β_i associated with the centroid v_i, and $u_{ij} = \{a\mu_{ij} + bv_{ij}\}$. After computing u_{ij} for c clusters and n objects, the values of u_{ij} for each object x_j are sorted and the difference of two highest memberships of x_j is compared with a threshold value δ. Let u_{ij} and u_{kj} be the highest and second highest memberships of x_j. If $(u_{ij} - u_{kj}) > \delta$, then $x_j \in \underline{A}(\beta_i)$ as well as $x_j \in \overline{A}(\beta_i)$, otherwise $x_j \in \overline{A}(\beta_i)$ and $x_j \in \overline{A}(\beta_k)$. After assigning each object in lower approximations or boundary regions of different clusters based on δ, both memberships μ_{ij} and v_{ij} of the objects are modified.

The values of μ_{ij} and v_{ij} are set to 1 for the objects in lower approximations, while those in the boundary regions remain unchanged. The new centroids of the clusters are calculated as per Equation (3.20). The main steps of the RFPCM algorithm proceed as follows:

1. Assign initial centroids v_i, $i = 1, 2, \ldots, c$. Choose values for fuzzifiers $\acute{m}_1$ and $\acute{m}_2$, and calculate threshold δ. Set iteration counter $t = 1$.
2. Compute μ_{ij} and v_{ij} from Equations (3.17) and (3.18), and finally u_{ij} for c clusters and n objects.
3. If u_{ij} and u_{kj} are the two highest memberships of x_j and $(u_{ij} - u_{kj}) \leq \delta$, then $x_j \in \overline{A}(\beta_i)$ and $x_j \in \overline{A}(\beta_k)$. Furthermore, x_j is not part of any lower bound.
4. Otherwise, $x_j \in \underline{A}(\beta_i)$. In addition, by properties of rough sets, $x_j \in \overline{A}(\beta_i)$.
5. Modify μ_{ij} and v_{ij} considering lower and boundary regions for c clusters and n objects.
6. Compute new centroid as per Equation (3.20).
7. Repeat steps 2 to 6, by incrementing t, until no more new assignments can be made.

3.3.6   Selection of Parameters

The parameter w has an influence on the performance of the RFPCM algorithm. Since the objects lying in lower approximation definitely belong to a cluster, they are assigned a higher weight w compared to $\tilde{w}$ of the objects lying in the boundary regions. On the other hand, the performance of RFPCM significantly reduces when $w \simeq 1.00$. In this case, since the clusters cannot see the objects of boundary regions, the mobility of the clusters and the centroids reduces. As a result, some centroids get stuck in local optimum. Hence, for the clusters and the centroids to have a greater degree of freedom to move, $0 < \tilde{w} < w < 1$.

The performance of RFPCM depends on the value of δ, which determines the class labels of all the objects. In other words, the RFPCM partitions the data set into two classes, namely, lower approximation and boundary, based on the value of δ. The δ represents the size of granules of rough-fuzzy clustering. In practice, the following definition works well:

$$\delta = \frac{1}{n} \sum_{j=1}^{n} (u_{ij} - u_{kj}), \tag{3.37}$$

where n is the total number of objects, u_{ij} and u_{kj} are the highest and second highest memberships of x_j, and $u_{ij} = \{a\mu_{ij} + bv_{ij}\}$. That is, the value of δ represents the average difference of two highest memberships of all the objects in the data set. A good clustering procedure should make the value of δ as high as possible.

3.4 GENERALIZATION OF EXISTING *c*-MEANS ALGORITHMS

Here, two derivatives of the RFPCM, namely, RFCM and rough-possibilistic *c*-means (RPCM), are described and it is proved that the RFPCM is the generalization of HCM, FCM, PCM, and RCM algorithms.

3.4.1 RFCM: Rough-Fuzzy *c*-Means

Let $0 < \delta < 1$ and $a = 1$, then Equations (3.16) and (3.20) become

$$J_{RF} = \begin{cases} w \times \mathcal{A}_2 + \tilde{w} \times \mathcal{B}_2 & \text{if } \underline{A}(\beta_i) \neq \emptyset, B(\beta_i) \neq \emptyset \\ \mathcal{A}_2 & \text{if } \underline{A}(\beta_i) \neq \emptyset, B(\beta_i) = \emptyset \\ \mathcal{B}_2 & \text{if } \underline{A}(\beta_i) = \emptyset, B(\beta_i) \neq \emptyset \end{cases} \tag{3.38}$$

where

$$\mathcal{A}_2 = \sum_{i=1}^{c} \sum_{x_j \in \underline{A}(\beta_i)} ||x_j - v_i||^2, \tag{3.39}$$

and

$$\mathcal{B}_2 = \sum_{i=1}^{c} \sum_{x_j \in B(\beta_i)} (\mu_{ij})^{\acute{m}_1} ||x_j - v_i||^2 \tag{3.40}$$

$$v_i^{RF} = \begin{cases} w \times \mathcal{C}_2 + \tilde{w} \times \mathcal{D}_2 & \text{if } \underline{A}(\beta_i) \neq \emptyset, B(\beta_i) \neq \emptyset \\ \mathcal{C}_2 & \text{if } \underline{A}(\beta_i) \neq \emptyset, B(\beta_i) = \emptyset \\ \mathcal{D}_2 & \text{if } \underline{A}(\beta_i) = \emptyset, B(\beta_i) \neq \emptyset \end{cases} \tag{3.41}$$

where

$$\mathcal{C}_2 = \frac{1}{|\underline{A}(\beta_i)|} \sum_{x_j \in \underline{A}(\beta_i)} x_j, \tag{3.42}$$

and

$$\mathcal{D}_2 = \frac{1}{n_i} \sum_{x_j \in B(\beta_i)} (\mu_{ij})^{\acute{m}_1} x_j; \quad n_i = \sum_{x_j \in B(\beta_i)} (\mu_{ij})^{\acute{m}_1}. \tag{3.43}$$

That is, if $0 < \delta < 1$ and $a = 1$, the RFPCM boils down to the RFCM, where each cluster consists of a crisp lower bound and a fuzzy boundary with probabilistic memberships. In Mitra et al. [22], a preliminary version of the RFCM has been proposed, where each cluster consists of a fuzzy lower approximation and a fuzzy boundary. If an object $x_j \in \underline{A}(\beta_i)$, then $\mu_{kj} = \mu_{ij}$ if $k = i$ and $\mu_{kj} = 0$ otherwise. That is, each object $x_j \in \underline{A}(\beta_i)$ takes a distinct weight, which is its fuzzy or probabilistic membership value. Hence, the weight of the object in lower approximation is inversely related to the relative distance of the object to all cluster prototypes. In fact, the objects in the lower approximation of a cluster should have a similar influence on the corresponding centroid and cluster. Also, their weights should be independent of other centroids and clusters and should not be coupled with their similarity with respect to other clusters. Thus, the concept of fuzzy lower approximation of the RFCM, introduced in Mitra et al. [22], reduces the weights of objects of lower approximation and effectively drifts the cluster centroids from their desired locations.

3.4.2   RPCM: Rough-Possibilistic c-Means

Let $0 < \delta < 1$ and $a = 0$, then Equation (3.16) becomes

$$
J_{\mathrm{RP}} = \begin{cases} w \times \mathcal{A}_3 + \tilde{w} \times \mathcal{B}_3 & \text{if } \underline{A}(\beta_i) \neq \emptyset,\, B(\beta_i) \neq \emptyset \\ \mathcal{A}_3 & \text{if } \underline{A}(\beta_i) \neq \emptyset,\, B(\beta_i) = \emptyset \\ \mathcal{B}_3 & \text{if } \underline{A}(\beta_i) = \emptyset,\, B(\beta_i) \neq \emptyset \end{cases} \tag{3.44}
$$

where

$$
\mathcal{A}_3 = \sum_{i=1}^{c} \sum_{x_j \in \underline{A}(\beta_i)} ||x_j - v_i||^2; \tag{3.45}
$$

$$
\mathcal{B}_3 = \sum_{i=1}^{c} \sum_{x_j \in B(\beta_i)} (v_{ij})^{\acute{m}_2} ||x_j - v_i||^2 + \sum_{i=1}^{c} \eta_i \sum_{x_j \in B(\beta_i)} (1 - v_{ij})^{\acute{m}_2}. \tag{3.46}
$$

Similarly, Equation (3.20) reduces to

$$
v_i^{\mathrm{RP}} = \begin{cases} w \times \mathcal{C}_3 + \tilde{w} \times \mathcal{D}_3 & \text{if } \underline{A}(\beta_i) \neq \emptyset,\, B(\beta_i) \neq \emptyset \\ \mathcal{C}_3 & \text{if } \underline{A}(\beta_i) \neq \emptyset,\, B(\beta_i) = \emptyset \\ \mathcal{D}_3 & \text{if } \underline{A}(\beta_i) = \emptyset,\, B(\beta_i) \neq \emptyset \end{cases} \tag{3.47}
$$

where

$$
\mathcal{C}_3 = \frac{1}{|\underline{A}(\beta_i)|} \sum_{x_j \in \underline{A}(\beta_i)} x_j; \tag{3.48}
$$

and

$$
\mathcal{D}_3 = \frac{1}{n_i} \sum_{x_j \in B(\beta_i)} (v_{ij})^{\acute{m}_2} x_j; \quad n_i = \sum_{x_j \in B(\beta_i)} (v_{ij})^{\acute{m}_2}. \tag{3.49}
$$

So, if $0 < \delta < 1$ and $a = 0$, the RFPCM reduces to the RPCM, where each cluster is represented by a crisp lower bound and a fuzzy boundary with possibilistic memberships.

3.4.3 RCM: Rough *c*-Means

If $0 < \delta < 1$, and for any nonzero μ_{ij} and ν_{ij}, if we set $\mu_{ij} = \nu_{ij} = 1, \forall i, j$, then Equations (3.16) and (3.20) become

$$
J_R = \begin{cases} w \times \mathcal{A}_4 + \tilde{w} \times \mathcal{B}_4 & \text{if } \underline{A}(\beta_i) \neq \emptyset, B(\beta_i) \neq \emptyset \\ \mathcal{A}_4 & \text{if } \underline{A}(\beta_i) \neq \emptyset, B(\beta_i) = \emptyset \\ \mathcal{B}_4 & \text{if } \underline{A}(\beta_i) = \emptyset, B(\beta_i) \neq \emptyset \end{cases} \tag{3.50}
$$

where

$$
\mathcal{A}_4 = \sum_{i=1}^{c} \sum_{x_j \in \underline{A}(\beta_i)} ||x_j - v_i||^2, \tag{3.51}
$$

and

$$
\mathcal{B}_4 = \sum_{i=1}^{c} \sum_{x_j \in B(\beta_i)} ||x_j - v_i||^2. \tag{3.52}
$$

$$
v_i^R = \begin{cases} w \times \mathcal{C}_4 + \tilde{w} \times \mathcal{D}_4 & \text{if } \underline{A}(\beta_i) \neq \emptyset, B(\beta_i) \neq \emptyset \\ \mathcal{C}_4 & \text{if } \underline{A}(\beta_i) \neq \emptyset, B(\beta_i) = \emptyset \\ \mathcal{D}_4 & \text{if } \underline{A}(\beta_i) = \emptyset, B(\beta_i) \neq \emptyset \end{cases} \tag{3.53}
$$

where

$$
\mathcal{C}_4 = \frac{1}{|\underline{A}(\beta_i)|} \sum_{x_j \in \underline{A}(\beta_i)} x_j, \tag{3.54}
$$

and

$$
\mathcal{D}_4 = \frac{1}{|B(\beta_i)|} \sum_{x_j \in B(\beta_i)} x_j. \tag{3.55}
$$

This is equivalent to the RCM of Lingras and West [17]. In case of the RCM, both the lower bound and boundary are crisp. Hence, the difference of the RCM with the RFPCM/RFCM/RPCM is that while in the RFPCM/RFCM/RPCM, each object in the boundary region takes a distinct weight, in the RCM the uniform weight is imposed on all the objects in the boundary region. In fact, the objects in boundary regions have different influences on the centroids or means and clusters.

3.4.4 FPCM: Fuzzy-Possibilistic c-Means

If we set $\delta = 1$, then $\underline{A}(\beta_i) = \emptyset$ and $\overline{A}(\beta_i) = B(\beta_i)$. Hence, for $0 < a < 1$, Equations (3.16) and (3.20) reduce to

$$J_{\mathrm{FP}} = \sum_{i=1}^{c} \sum_{j=1}^{n} \{a(\mu_{ij})^{\acute{m}_1} + b(\nu_{ij})^{\acute{m}_2}\} ||x_j - v_i||^2 + \sum_{i=1}^{c} \eta_i \sum_{j=1}^{n} (1 - \nu_{ij})^{\acute{m}_2}$$

$$(3.56)$$

and

$$v_i^{\mathrm{FP}} = \frac{1}{n_i} \sum_{j=1}^{n} \{a(\mu_{ij})^{\acute{m}_1} + b(\nu_{ij})^{\acute{m}_2}\} x_j \qquad (3.57)$$

where

$$n_i = \sum_{j=1}^{n} \{a(\mu_{ij})^{\acute{m}_1} + b(\nu_{ij})^{\acute{m}_2}\}.$$

So, for $\delta = 1$ and $0 < a < 1$, the RFPCM boils down to the fuzzy-possibilistic c-means (FPCM) algorithm of Pal et al. [12].

3.4.5 FCM: Fuzzy c-Means

If $\delta = 1$ and $a = 1$, then

$$J_{\mathrm{F}} = \sum_{i=1}^{c} \sum_{j=1}^{n} (\mu_{ij})^{\acute{m}_1} ||x_j - v_i||^2; \qquad (3.58)$$

and

$$v_i^{\mathrm{F}} = \frac{1}{n_i} \sum_{j=1}^{n} (\mu_{ij})^{\acute{m}_1} x_j; \ \ \text{where } n_i = \sum_{j=1}^{n} (\mu_{ij})^{\acute{m}_1} \qquad (3.59)$$

which is equivalent to the FCM algorithm proposed in Bezdek [6].

3.4.6 PCM: Possibilistic c-Means

If $\delta = 1$ and $a = 0$, then

$$J_{\mathrm{P}} = \sum_{i=1}^{c} \sum_{j=1}^{n} (\nu_{ij})^{\acute{m}_2} ||x_j - v_i||^2 + \sum_{i=1}^{c} \eta_i \sum_{j=1}^{n} (1 - \nu_{ij})^{\acute{m}_2}; \qquad (3.60)$$

and

$$v_i^{\mathrm{P}} = \frac{1}{n_i} \sum_{j=1}^{n} (v_{ij})^{\acute{m}_2} x_j; \quad \text{where } n_i = \sum_{j=1}^{n} (v_{ij})^{\acute{m}_2}. \tag{3.61}$$

Hence, for $\delta = 1$ and $a = 0$, the RFPCM reduces to the PCM algorithm introduced in Krishnapuram and Keller [8, 9].

3.4.7 HCM: Hard c-Means

If $\delta = 0$, then $B(\beta_i) = \emptyset$ and $\overline{A}(\beta_i) = \underline{A}(\beta_i) = \beta_i$. In effect, Equations (3.16) and (3.20) become

$$J_{\mathrm{H}} = \sum_{i=1}^{c} \sum_{x_j \in \beta_i} ||x_j - v_i||^2; \tag{3.62}$$

and

$$v_i^{\mathrm{H}} = \frac{1}{n_i} \sum_{x_j \in \beta_i} x_j. \tag{3.63}$$

Hence, for $\delta = 0$, the RFPCM boils down to the HCM.

Figure 3.2 shows the membership values of all the objects to two Gaussian-distributed clusters for the values $a = 0.0$, 0.5, and 1.0, and $\delta = 0.95$ and 1.0. The different shape of the memberships, especially for objects near to centers and objects far from the separating line, is apparent. Detailed analysis reported in this section confirms that the RFPCM algorithm is the generalization of existing c-means algorithms. It effectively integrates hard (crisp), fuzzy (probabilistic), and PCM using the concept of lower and upper approximations of rough sets.

3.5 QUANTITATIVE INDICES FOR ROUGH-FUZZY CLUSTERING

In this section, some quantitative indices are presented to evaluate the performance of different rough-fuzzy clustering algorithms incorporating the concepts of rough sets [14]. In this regard, it may be noted that some rough clustering quality indices are reported in Lingras et al. [30] based on the decision theory.

3.5.1 Average Accuracy, α Index

The α index is given by

$$\alpha = \frac{1}{c} \sum_{i=1}^{c} \frac{w A_i}{w A_i + \tilde{w} B_i} \tag{3.64}$$

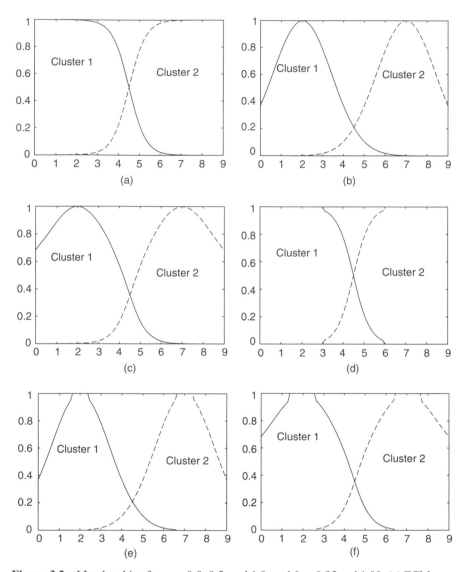

Figure 3.2   Memberships for $a = 0.0$, 0.5, and 1.0; and $\delta = 0.95$ and 1.00. (a) FCM: $a = 1.00$, $\delta = 1.00$; (b) PCM: $a = 0.00$, $\delta = 1.00$; (c) FPCM: $a = 0.50$, $\delta = 1.00$; (d) RFCM: $a = 1.00$, $\delta = 0.95$; (e) RPCM: $a = 0.00$, $\delta = 0.95$; and (f) RFPCM: $a = 0.50$; $\delta = 0.95$.

where

$$A_i = \sum_{x_j \in \underline{A}(\beta_i)} \{a(\mu_{ij})^{\acute{m}_1} + b(v_{ij})^{\acute{m}_2}\} = |\underline{A}(\beta_i)| \qquad (3.65)$$

and

$$B_i = \sum_{x_j \in B(\beta_i)} \{a(\mu_{ij})^{\acute{m}_1} + b(v_{ij})^{\acute{m}_2}\} \tag{3.66}$$

μ_{ij} and v_{ij} represent the probabilistic and possibilistic memberships of object x_j in cluster β_i, respectively. Constants a and b define the relative importance of probabilistic and possibilistic memberships, respectively, whereas the parameters w and $\tilde{w}$ correspond to the relative importance of lower and boundary regions, respectively.

The α index represents the average accuracy of c clusters. It is the average of the ratio of the number of objects in lower approximation to that in upper approximation of each cluster. In effect, it captures the average degree of completeness of knowledge about all clusters. A good clustering procedure should make all objects as similar to their centroids as possible. The α index increases with the increase in similarity within a cluster. Therefore, for a given data set and c value, the higher the similarity values within the clusters, the higher would be the α value. The value of α also increases with c. In an extreme case when the number of clusters is maximum, that is, $c = n$, the total number of objects in the data set, the value of $\alpha = 1$. When $\overline{A}(\beta_i) = \underline{A}(\beta_i)$, $\forall i$, that is, all the clusters $\{\beta_i\}$ are exact or definable, then we have $\alpha = 1$, whereas if $\overline{A}(\beta_i) = B(\beta_i)$, $\forall i$, the value of $\alpha = 0$. Hence, $0 \leq \alpha \leq 1$.

3.5.2 Average Roughness, ϱ Index

The ϱ index represents the average roughness of c clusters and is defined by subtracting the average accuracy α from 1:

$$\varrho = 1 - \alpha = 1 - \frac{1}{c} \sum_{i=1}^{c} \frac{w A_i}{w A_i + \tilde{w} B_i}, \tag{3.67}$$

where A_i and B_i are given by Equations (3.65) and (3.66), respectively. Note that the lower the value of ϱ, the better is the overall approximations of the clusters. Also, $0 \leq \varrho \leq 1$. Basically, ϱ index represents the average degree of incompleteness of knowledge about all clusters.

3.5.3 Accuracy of Approximation, $\alpha^\star$ Index

It can be defined as

$$\alpha^\star = \frac{\sum_{i=1}^{c} w A_i}{\sum_{i=1}^{c} \{w A_i + \tilde{w} B_i\}} \tag{3.68}$$

where A_i and B_i are given by Equations (3.65) and (3.66), respectively. The $\alpha^\star$ index represents the accuracy of approximation of all clusters. It captures the exactness of approximate clustering. A good clustering procedure should make the value of α^* as high as possible. The α^* index maximizes the exactness of approximate clustering.

3.5.4 Quality of Approximation, γ Index

It is the ratio of the total number of objects in lower approximations of all clusters to the cardinality of the universe of discourse U and is given by

$$\gamma = \frac{1}{|U|} \sum_{i=1}^{c} |\underline{A}(\beta_i)|. \tag{3.69}$$

The γ index basically represents the quality of approximation of a clustering algorithm. A good clustering procedure should make the value of γ index as high as possible.

3.6 PERFORMANCE ANALYSIS

The performance of three hybrid algorithms, namely, RFCM, RPCM, and RFPCM, is compared extensively with that of different c-means algorithms. These involve different combinations of the individual components of the hybrid scheme as well as some other kernel-based approaches. The algorithms compared are HCM [7], FCM [6], PCM [8, 9], FPCM [12], fuzzy-possibilistic c-means of Masulli and Rovetta (FPCM^{MR}) [11], kernel-based hard c-means (KHCM) [31–33], kernel-based fuzzy c-means (KFCM) [34–36], kernel-based possibilistic c-means (KPCM) [36, 37], kernel-based fuzzy-possibilistic c-means (KFPCM) [38], RCM [17], and RFCM of Mitra et al. (RFCM^{MBP}) [22]. All the algorithms are implemented in C language and run in LINUX environment having machine configuration Pentium IV, 3.2 GHz, 1 MB cache, and 1 GB RAM.

   To analyze the performance of different algorithms, the experimentation is done in two parts. In the first part, a synthetic data set is used, while the results on some benchmark data sets are presented in the second part. The values of δ for the RCM, RFCM, RPCM, and RFPCM algorithms are calculated using Equation (3.37), as described in Section 3.3.6. The final prototypes of the FCM are used to initialize the PCM, FPCM, KFCM, KPCM, and KFPCM, while the Gaussian function is used as the kernel function.

3.6.1 Quantitative Indices

The major metrics for evaluating the performance of different algorithms are the indices reported in Section 3.5 such as α, ϱ, α^*, and γ, as well as some existing measures such as Davies–Bouldin (DB) [39] and Dunn index [40], which are described subsequently.

3.6.1.1 *Davies–Bouldin Index* The Davies–Bouldin (DB) index [39] is a function of the ratio of sum of within-cluster distance to between-cluster separation and is given by

$$\text{DB} = \frac{1}{c} \sum_{i=1}^{c} \max_{i \neq k} \left\{ \frac{S(v_i) + S(v_k)}{d(v_i, v_k)} \right\} \quad \text{for } 1 \leq i, k \leq c. \tag{3.70}$$

The DB index minimizes the within-cluster distance $S(v_i)$ and maximizes the between-cluster separation $d(v_i, v_k)$. Therefore, for a given data set and c value, the higher the similarity values within the clusters and the between-cluster separation, the lower would be the DB index value. A good clustering procedure should make the value of the DB index as low as possible.

3.6.1.2 *Dunn Index* Dunn's index [40] is also designed to identify sets of clusters that are compact and well separated. Dunn's (D) index maximizes

$$D = \min_{i} \left\{ \min_{i \neq k} \left\{ \frac{d(v_i, v_k)}{\max_l S(v_l)} \right\} \right\} \quad \text{for } 1 \leq i, k, l \leq c. \tag{3.71}$$

3.6.2 Synthetic Data Set: $X32$

The synthetic data set $X32$ consists of $n = 32$ objects in $\Re^2$ with two clusters. Figure 3.3 depicts the scatter plot of the data set $X32$. The objects x_{30}, x_{31}, and x_{32} are outliers or noise, and the object x_7 is the so-called inlier or bridge. Also, the objects x_{30} and x_7 are equidistant from all corresponding pairs of points in

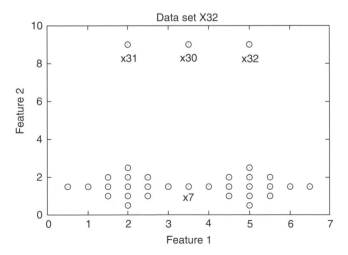

Figure 3.3 Scatter plot of example data set $X32$.

the two clusters. x_{31} is more closer to cluster 1, whereas x_{32} is more nearer to cluster 2.

After computing the membership values of each object x_j in two clusters, the differences of two memberships $(u_{ij} - u_{kj})$ are calculated. On the basis of the differences and the value of δ, all the objects are distributed in lower approximations and boundary regions of two clusters. For the RFPCM, the value of δ is 0.613409 considering $w = 0.9999$, $\acute{m}_1 = \acute{m}_2 = 2.0$, and $a = 0.5$. All the parameters are held constant across all runs. Two randomly generated initial centroids, along with two scale parameters and the final prototypes of different c-means algorithms, are reported in Table 3.1. Figure 3.4, 3.5, and 3.6 represent the scatter plots of the data set $X32$ along with the cluster prototypes obtained using different c-means algorithms. The objects of $X32$ are represented by $\odot$, while $\square$ depict the positions of cluster prototypes.

Table 3.2 reports the values of α, ϱ, $\alpha^\star$, γ, J_{RFP}, Dunn, and DB index of the RFPCM algorithm over different iterations. All the results reported in Table 3.2 show that as the number of iteration increases, the values of ϱ, J_{RFP}, and DB index decrease, while the values of α, $\alpha^\star$, γ, and Dunn index increase. Finally, all the indices are saturated when the RFPCM terminates after the fourth iteration. Hence, the indices such as α, ϱ, $\alpha^\star$, and γ can be used to act as the objective function of rough-fuzzy clustering as they reflect good quantitative measures such as the existing DB and Dunn indices.

Table 3.3 provides comparative results of different c-means algorithms. The rough-set-based clustering algorithms such as the RCM, RFCM^{MBP}, RFCM, RPCM, and RFPCM are found to improve the performance in terms of DB and Dunn indices over other algorithms. It is also observed that the RFCM, RPCM, and RFPCM algorithms perform better than the FCM, PCM, FPCM, FPCM^{MR}, KHCM, KFCM, KPCM, KFPCM, RCM, and RFCM^{MBP}, although it is expected that the PCM, FPCM, FPCM^{MR}, KHCM, KFCM, KPCM, and KFPCM perform well in a noisy environment.

Finally, Table 3.4 shows the comparative results of different rough-fuzzy clustering algorithms in terms of α, ϱ, $\alpha^\star$, and γ. The performance of the RFCM, RPCM, and RFPCM is better than that of the RFCM^{MBP}. Although the performance of the RFPCM is intermediate between the RFCM and RPCM with respect to α, ϱ, $\alpha^\star$, and γ, it is expected that the RFPCM will be more useful as it is naturally more immune to noise and outliers, and can avoid the problem of coincident clusters.

3.6.3 Benchmark Data Sets

This subsection demonstrates the performance of different c-means algorithms on some benchmark data sets. All the data sets are downloaded from http://www.ics.uci.edu/~mlearn. Details of these data sets are reported subsequently.

TABLE 3.1   Prototypes of Different c-Means Algorithms

Algorithms	Centroid 1		Centroid 2	
Initial	2.088235;	2.382353	5.100000;	2.000000
Scale	$\eta_1 = 4.553732$		$\eta_2 = 3.697741$	
HCM	2.088235;	2.382353	5.100000;	2.000000
FCM	2.025431;	1.943642	4.974481;	1.943406
PCM	2.087332;	1.500811	4.912668;	1.500811
FPCM	2.281087;	1.749121	4.782719;	1.765219
FPCM^{MR}	2.135505;	1.756445	4.864495;	1.756445
KHCM	2.060885;	2.111454	5.101695;	1.831358
KFCM	1.984085;	1.780511	5.015908;	1.780499
KPCM	2.088211;	1.500801	4.911789;	1.500801
KFPCM	2.103683;	1.662097	4.896317;	1.662097
RCM	1.807862;	1.500375	5.192139;	1.500375
RFCM^{MBP}	1.783023;	1.500408	5.216976;	1.500408
RFCM	1.727380;	1.636481	5.272620;	1.636481
RPCM	1.500126;	1.500002	5.499879;	1.500002
RFPCM	1.727400;	1.636394	5.499892;	1.500050

TABLE 3.2   Performance of RFPCM over Different Iterations

Iteration	ϱ Index	DB Index	J_{RFP}
1	0.000075	0.159549	15.540598
2	0.000042	0.126676	14.750732
3	0.000039	0.116704	10.130592
4	0.000036	0.116704	8.574513
5	0.000036	0.116704	8.574513

Iteration	α Index	$\alpha^\star$ Index	γ Index	Dunn Index
1	0.999925	0.999926	0.406250	11.483479
2	0.999958	0.999959	0.562500	14.770479
3	0.999961	0.999963	0.593750	16.578180
4	0.999964	0.999965	0.593750	16.578184
5	0.999964	0.999965	0.593750	16.578184

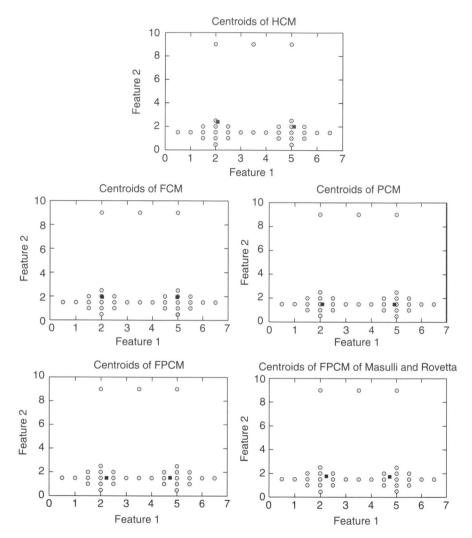

Figure 3.4   Cluster prototypes of different fuzzy c-means algorithms.

3.6.3.1   Iris Data Set   The iris data set is a four-dimensional data set containing 50 samples each of three types of iris flowers. One of the three clusters, that is, class 1, is well separated from the other two, whereas classes 2 and 3 have some overlap.

3.6.3.2   Glass Identification Database   The glass data set consists of six classes. Each sample has 10 continuous-valued attributes. The number of samples of this data set is 214.

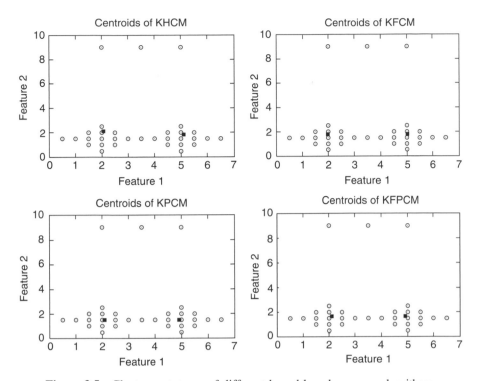

Figure 3.5   Cluster prototypes of different kernel-based c-means algorithms.

3.6.3.3 Ionosphere Database   It represents the autocorrelation functions of radar measurements. The task is to classify them into two classes denoting passage or obstruction in the ionosphere. There are 351 instances with 34 continuous-valued attributes.

3.6.3.4 Wine Recognition Database   It is a 13-dimensional data set with three classes. The number of instances per class are 59, 71, and 48, respectively. These data are the results of a chemical analysis of wines grown in the same region in Italy but derived from three different cultivars. The analysis determined the quantities of 13 constituents found in each of three wines.

3.6.3.5 Wisconsin Diagnostic Breast Cancer Database   It consists of 569 instances with 30 real-valued attributes. All the features are computed from a digitized image of a fine needle aspirate of a breast mass. They describe characteristics of the cell nuclei present in the image. Ten real-valued features are computed for each cell nucleus. The mean, standard error, and largest (mean of the three largest values) of these features are computed for each image, resulting in 30 features. All the feature values are recoded with four significant digits. This is a binary classification task with 357 instances benign and 212 instances malignant.

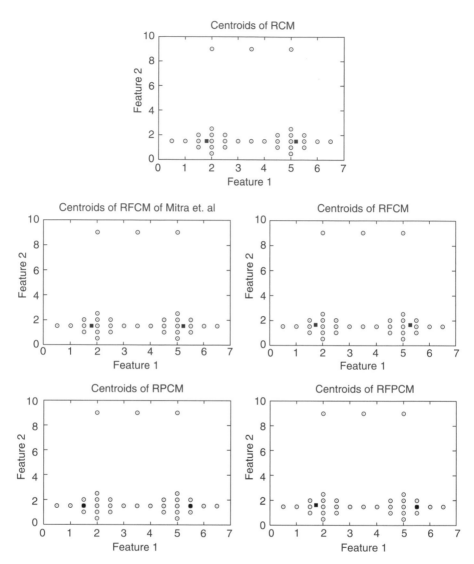

Figure 3.6   Cluster prototypes of rough-set-based different *c*-means.

The performance of different *c*-means algorithms on the iris data set is reported subsequently for $c = 2$ and 3. Several runs have been made with different initializations and different choices of parameters. The final prototypes of different *c*-means algorithms, along with initial centroids and scale parameters, are provided in Table 3.5 for $c = 2$ and 3. For the RFPCM, the values of δ are 0.51 and 0.48 for $c = 2$ and 3 considering $w = 0.99$, $\acute{m}_1 = \acute{m}_2 = 2.0$, and $a = b = 0.5$. The parameters are held constant across all runs.

TABLE 3.3 Performance of Different c-Means Algorithms

Algorithms	DB Index	Dunn Index
HCM	0.266063	5.934918
FCM	0.210139	9.022365
PCM	0.183069	10.309421
FPCM	0.276837	6.894526
FPCM^{MR}	0.218153	8.753471
KHCM	0.204109	8.149428
KFCM	0.172090	10.870337
KPCM	0.183359	10.294540
KFPCM	0.195370	9.703721
RCM	0.127478	14.252816
RFCM^{MBP}	0.125088	14.491059
RFCM	0.123709	14.621336
RPCM	0.113765	15.737709
RFPCM	0.116704	16.578184

TABLE 3.4 Quantitative Evaluation of Rough-Fuzzy Clustering

Methods	α Index	ϱ Index	$\alpha^\star$ Index	γ Index
RFCM^{MBP}	0.999953	0.000047	0.999942	0.625000
RFCM	0.999991	0.000009	0.999991	0.812500
RPCM	0.999947	0.000053	0.999947	0.500000
RFPCM	0.999964	0.000036	0.999965	0.593750

For $c = 3$, except the PCM and KPCM, all the c-means algorithms generate good prototypes. The final prototypes of the FCM are used to initialize the PCM, FPCM, KFCM, KPCM, and KFPCM. Even if three initial centroids belong to three different classes, both the PCM and KPCM generate coincident clusters. That is, two of three final prototypes are identical in case of both the PCM and KPCM. Tables 3.6 and 3.7 depict the best results obtained using different c-means algorithms for $c = 2$ and 3. In Table 3.6, the performance of different algorithms is reported with respect to the DB index, Dunn index, and CPU time (in milliseconds). The results reported in Table 3.6 establish the fact that although each c-means algorithm generates good prototypes with lower values of DB index and higher values of Dunn index for $c = 2$, the RFPCM provides the best result having lowest DB index and highest Dunn index. The results of other versions of rough clustering are quite similar to that of the RFPCM.

In iris data set, since classes 2 and 3 overlap, it may be thought of having two clusters. But, to design a classifier, at least three clusters have to be identified. Thus, for such applications, the RFCM, RPCM, and RFPCM will be more useful because they are not sensitive to noise, can avoid coincident clusters, and their DB and Dunn index values are far better than that of other algorithms, as reported in

TABLE 3.5 Initial Centroids, Scale Parameters, and Final Prototypes for Iris Data Set

Algorithms/ Methods	Prototypes $c = 2$		Prototypes $c = 3$		
	Centroid 1	Centroid 2	Centroid 1	Centroid 2	Centroid 3
Initial Scale	4.5 2.3 1.3 0.3 $\eta_1 = 0.651171$	6.3 2.7 4.9 1.8 $\eta_2 = 1.154369$	5.8 2.7 5.1 1.9 $\eta_1 = 0.689405$	5.0 2.0 3.5 1.0 $\eta_2 = 0.582364$	5.1 3.5 1.4 0.3 $\eta_3 = 0.344567$
FCM	5.02 3.37 1.57 0.29	6.34 2.91 5.01 1.73	6.78 3.05 5.65 2.05	5.89 2.76 4.36 1.40	5.00 3.40 1.49 0.25
PCM	5.04 3.41 1.47 0.24	6.17 2.88 4.76 1.60	6.17 2.88 4.76 1.61	6.17 2.88 4.76 1.61	5.04 3.41 1.47 0.24
FPCM	5.02 3.38 1.56 0.28	6.31 2.90 4.98 1.72	6.62 3.01 5.46 1.99	5.92 2.79 4.40 1.41	5.00 3.40 1.49 0.25
FPCM$^{\mathbf{MR}}$	5.01 3.38 1.53 0.27	6.35 2.91 5.04 1.74	6.85 3.08 5.71 2.09	5.86 2.72 4.37 1.40	5.00 3.42 1.46 0.24
KHCM	5.01 3.36 1.56 0.29	6.29 2.89 4.96 1.69	6.77 3.05 5.64 2.05	5.89 2.76 4.36 1.39	5.00 3.40 1.48 0.25
KFCM	5.02 3.37 1.57 0.29	6.33 2.90 5.01 1.73	6.85 3.08 5.71 2.05	5.88 2.74 4.39 1.43	5.01 3.42 1.46 0.24
KPCM	5.04 3.41 1.47 0.24	6.17 2.88 4.76 1.60	6.17 2.88 4.76 1.60	6.17 2.88 4.76 1.60	5.04 3.41 1.47 0.24
KFPCM	5.02 3.38 1.54 0.27	6.29 2.90 4.95 1.70	6.43 2.95 5.16 1.83	6.03 2.83 4.54 1.49	5.00 3.39 1.49 0.25
RCM	5.01 3.42 1.46 0.24	6.40 2.94 5.11 1.77	6.87 3.09 5.79 2.12	5.69 2.66 4.11 1.26	5.01 3.42 1.46 0.24
RFCM$^{\mathbf{MBP}}$	5.00 3.42 1.46 0.24	6.38 2.92 5.08 1.76	6.88 3.09 5.79 2.12	5.70 2.68 4.11 1.26	5.00 3.41 1.47 0.24
RFCM	5.01 3.42 1.46 0.24	6.40 2.94 5.11 1.77	6.87 3.09 5.79 2.12	5.69 2.66 4.11 1.26	5.01 3.42 1.46 0.24
RPCM	5.00 3.36 1.49 0.24	6.28 2.88 5.00 1.73	6.63 3.07 5.57 2.12	5.67 2.67 4.01 1.22	5.00 3.36 1.49 0.24
RFPCM	4.96 3.38 1.47 0.24	6.39 2.96 5.14 1.84	6.76 3.07 5.68 2.12	5.55 2.64 3.97 1.23	4.96 3.38 1.47 0.24

Table 3.6. Also, the execution time of different rough-fuzzy clustering algorithms is significantly lesser than that of different fuzzy and kernel-based fuzzy clustering algorithms.

Table 3.7 compares the performance of different rough-fuzzy clustering algorithms with respect to α, ϱ, $\alpha^\star$, and γ. The RFCM performs better than the RFCM^{MBP}. The performance of the RFPCM is intermediate between the RFCM and RPCM for $c = 2$ and better over other rough-fuzzy clustering algorithms having $c = 3$. However, it is expected that the RFPCM will be more useful as it is not sensitive to noise and outliers as well as does not produce coincident clusters.

Finally, Table 3.8 presents the comparative performance analysis of different c-means algorithms on some other benchmark data sets with respect to DB,

TABLE 3.6 Performance Analysis on Iris Data Set

Algorithms/ Methods	Prototypes $c = 2$			Prototypes $c = 3$		
	DB	Dunn	Time	DB	Dunn	Time
FCM	0.11	12.36	13	0.32	4.32	74
PCM	0.12	9.99	115	—	—	—
FPCM	0.11	11.87	63	0.48	2.63	68
FPCM^{MR}	0.16	9.07	4	0.35	4.36	20
KHCM	0.12	12.08	6	0.32	4.27	18
KFCM	0.11	12.36	22	0.29	4.89	85
KPCM	0.12	9.99	138	—	—	—
KFPCM	0.11	11.44	81	0.46	3.22	99
RCM	0.10	13.49	3	0.22	6.75	16
RFCM^{MBP}	0.11	13.42	17	0.22	6.91	33
RFCM	0.10	13.49	12	0.22	6.94	24
RPCM	0.10	11.67	37	0.22	7.13	59
RFPCM	0.10	13.61	25	0.21	7.75	24

TABLE 3.7 Quantitative Evaluation of Rough-Fuzzy Clustering

Prototypes	Methods	α Index	ϱ Index	$\alpha^\star$ Index	γ Index
$c = 2$	RFCM^{MBP}	0.999991	0.000009	0.999989	0.812500
	RFCM	0.999994	0.000006	0.999994	0.906667
	RPCM	0.999980	0.000020	0.999979	0.506667
	RFPCM	0.999986	0.000014	0.999985	0.686667
	RFCM^{MBP}	0.999971	0.000029	0.999963	0.625000
$c = 3$	RFCM	0.999986	0.000014	0.999988	0.800000
	RPCM	0.999983	0.000017	0.999985	0.553333
	RFPCM	0.999987	0.000013	0.999989	0.766667

TABLE 3.8 Performance of Different c-Means Algorithms on Glass, Ionosphere, Wine, and Wisconsin Breast Cancer Data Set

Algorithms/Methods	Glass Data Set			Ionosphere Data Set			Wine Data Set			Wisconsin Data Set		
	DB	Dunn	Time	DB	Dunn	Time	DB	Dunn	Time	DB	Dunn	Time
HCM	3.28	0.02	18	2.36	0.58	17	0.20	2.52	11	0.19	5.53	30
FCM	2.52	0.05	333	1.51	0.89	153	0.19	2.78	96	0.17	6.13	298
FPCM	—	—	—	—	—	—	0.45	1.40	204	—	—	—
FPCM^{MR}	2.48	0.11	29	1.52	0.89	24	0.88	1.78	17	0.13	8.29	13
KHCM	2.60	0.03	65	2.30	0.59	74	0.20	2.52	23	0.19	5.42	140
KFCM	2.51	0.05	132	1.51	0.89	360	0.19	2.76	196	0.18	5.84	723
KFPCM	—	—	—	—	—	—	0.48	1.54	459	—	—	—
RCM	1.96	0.37	53	1.01	1.57	27	0.20	3.08	15	0.16	7.85	17
RFCM^{MBP}	1.51	0.13	191	1.09	1.50	98	0.19	3.19	54	0.16	7.84	77
RFCM	1.45	0.49	124	0.99	1.60	96	0.19	3.13	44	0.12	8.82	62
RPCM	1.91	0.31	247	1.09	1.58	297	0.20	3.09	359	0.13	7.95	167
RFPCM	0.96	0.63	76	1.11	1.66	198	0.19	3.87	101	0.13	8.82	98
Methods	α	$\alpha^{\star}$	γ	α	$\alpha^{\star}$	γ	α	$\alpha^{\star}$	γ	α	$\alpha^{\star}$	γ
RFCM^{MBP}	0.9891	0.9719	0.5123	0.9918	0.9627	0.6217	0.8387	0.9251	0.5000	0.8977	0.9123	0.6349
RFCM	0.9942	0.9792	0.6250	0.9927	0.9712	0.6590	0.8918	0.9259	0.8275	0.8981	0.9386	0.8175
RPCM	0.9907	0.9775	0.5178	0.9913	0.9681	0.5125	0.8433	0.9306	0.6255	0.8990	0.9207	0.7724
RFPCM	0.9918	0.9804	0.6250	0.9987	0.9701	0.8271	0.9012	0.9258	0.7234	0.9192	0.9188	0.8125

Dunn, α, $\alpha^\star$, γ, and CPU time (in milliseconds). The following conclusions can be drawn from the results reported in Table 3.8:

- The performance of three algorithms, namely, RFCM, RPCM, and RFPCM, is significantly better than other fuzzy, kernel-based fuzzy, and rough algorithms with respect to DB, Dunn, α, $\alpha^\star$, and γ.
- The execution time required for different rough clustering algorithms is lesser compared to fuzzy and kernel-based fuzzy algorithms.
- Some of the existing algorithms such as the PCM, FPCM, KPCM, and KFPCM generate coincident clusters even when they have been initialized with the final prototypes of the FCM.
- The performance of kernel-based algorithms is better than their non-kernel-based counterparts, although they require more time to converge.

From the results reported in this chapter, the following conclusions can be drawn:

1. It is observed that the RFPCM is superior to other c-means algorithms. However, the RFPCM requires higher time compared to the HCM. But, the performance of the RFPCM is significantly higher than other c-means. The performance of both RFCM and RPCM is intermediate between the RFPCM and FCM/PCM. Also, the RFCM performs better than the RFCM^{MBP}.
2. Use of rough sets and fuzzy memberships (both probabilistic and possibilistic) adds a small computational load to the HCM algorithm; however, the corresponding integrated methods (RFCM, RPCM, and RFPCM) show a definite increase in the Dunn index and a decrease in DB index.
3. It is seen that the performance of the RFPCM is intermediate between both RFCM and RPCM with respect to α, ϱ, $\alpha^\star$, and γ. However, the RFPCM will be more useful than both RFCM and RPCM as its prototypes are not sensitive to outliers and can avoid coincident clusters.
4. The execution time required for different rough clustering algorithms is significantly lesser compared to different fuzzy and kernel-based fuzzy clustering algorithms.
5. The quantitative indices such as α, ϱ, $\alpha^\star$, and γ based on the theory of rough sets provide good quantitative measures for rough-fuzzy clustering. The values of these indices reflect the quality of clustering.

The best performance of the RFPCM algorithm in terms of α, ϱ, $\alpha^\star$, γ, DB, Dunn, and CPU time is achieved because of the following reasons:

1. The concept of crisp lower bound and fuzzy boundary of the RFPCM algorithm deals with uncertainty, vagueness, and incompleteness in class definition.

2. The membership function of the RFPCM handles efficiently overlapping partitions.

3. The probabilistic and possibilistic memberships of the RFPCM can avoid coincident clusters problem and make the algorithm insensitive to noise and outliers.

In effect, good cluster prototypes are obtained using the RFPCM algorithm with significantly less time.

3.7   CONCLUSION AND DISCUSSION

The problem of unsupervised clustering in the rough-fuzzy framework is addressed in this chapter. After explaining the basic characteristics of FCM and RCM, a generalized hybrid unsupervised learning algorithm, termed as *rough-fuzzy-possibilistic c-means* (RFPCM) is presented. It comprises a judicious integration of the principles of rough sets and fuzzy sets. While the concept of lower and upper approximations of rough sets deals with uncertainty, vagueness, and incompleteness in class definition, the membership function of fuzzy sets enables efficient handling of overlapping partitions. The RFPCM incorporates both probabilistic and possibilistic memberships simultaneously to avoid the problems of noise sensitivity of FCM and the coincident clusters of PCM. The concept of crisp lower bound and fuzzy boundary of a class, introduced herein, enables efficient selection of cluster prototypes. The algorithm is generalized in the sense that all the existing variants of c-means algorithms can be derived from the new algorithm as a special case.

Several quantitative indices are stated based on rough sets for evaluating the performance of different c-means algorithms. These indices may be used in a suitable combination to act as the objective function of an evolutionary algorithm, for rough-fuzzy clustering. The current formulation of the RFPCM is geared toward maximizing the utility of both rough sets and fuzzy sets with respect to knowledge discovery tasks. Although the methodology of integrating rough sets, fuzzy sets, and c-means algorithm has been efficiently demonstrated for synthetic and benchmark data sets, the concept can be applied to other unsupervised classification problems. Two important real-life applications of the rough-fuzzy clustering algorithm, namely, clustering of functionally similar genes from microarray gene expression data and segmentation of brain magnetic resonance images, are reported in Chapters 7 and 9, respectively.

So far, we have discussed the unsupervised clustering algorithm for object data. The term *object data* refers to the situation where the objects to be clustered are represented by vectors $x_i \in \Re^m$. In contrast, the relational data refers to the situation where we have only numerical values representing the degrees to which pairs of objects in the data set are related. The algorithms that generate partitions of relational data are usually referred to as *relational* or *pair-wise clustering algorithms*. As the relational data is less common than object data, relational pattern recognition methods are not as well developed as their object counterparts, particularly in the area of robust clustering. However, relational methods are becoming a necessity as relational data becomes more and more common in information retrieval, data mining, web mining, and bioinformatics. In this regard, Chapter 6 discusses a rough-fuzzy relational clustering algorithm, termed as *rough-fuzzy c-medoids algorithm*, and demonstrates its effectiveness in protein sequence analysis.

The next chapter deals with the problem of supervised classification in the rough-fuzzy framework. Here, the significance of integrating fuzzy granulation and neighborhood rough-set-based feature selection is addressed.

REFERENCES

1. A. K. Jain and R. C. Dubes. *Algorithms for Clustering Data*. Prentice Hall, Englewood Cliffs, NJ, 1988.

2. A. K. Jain, M. N. Murty, and P. J. Flynn. Data Clustering: A Review. *ACM Computing Surveys*, 31(3):264–323, 1999.

3. R. E. Bellman, R. E. Kalaba, and L. A. Zadeh. Abstraction and Pattern Classification. *Journal of Mathematical Analysis and Applications*, 13:1–7, 1966.

4. E. H. Ruspini. Numerical Methods for Fuzzy Clustering. *Information Sciences*, 2:319–350, 1970.

5. J. C. Dunn. A Fuzzy Relative of the ISODATA Process and Its Use in Detecting Compact, Well-Separated Clusters. *Journal of Cybernetics*, 3:32–57, 1974.

6. J. C. Bezdek. *Pattern Recognition with Fuzzy Objective Function Algorithm*. Plenum, New York, 1981.

7. J. McQueen. Some Methods for Classification and Analysis of Multivariate Observations. In *Proceedings of the 5th Berkeley Symposium on Mathematics, Statistics and Probability*, Berkeley, pages 281–297, 1967.

8. R. Krishnapuram and J. M. Keller. A Possibilistic Approach to Clustering. *IEEE Transactions on Fuzzy Systems*, 1(2):98–110, 1993.

9. R. Krishnapuram and J. M. Keller. The Possibilistic C-Means Algorithm: Insights and Recommendations. *IEEE Transactions on Fuzzy Systems*, 4(3):385–393, 1996.

10. M. Barni, V. Cappellini, and A. Mecocci. Comments on "A Possibilistic Approach to Clustering". *IEEE Transactions on Fuzzy Systems*, 4(3):393–396, 1996.

11. F. Masulli and S. Rovetta. Soft Transition from Probabilistic to Possibilistic Fuzzy Clustering. *IEEE Transactions on Fuzzy Systems*, 14(4):516–527, 2006.

12. N. R. Pal, K. Pal, J. M. Keller, and J. C. Bezdek. A Possibilistic Fuzzy C-Means Clustering Algorithm. *IEEE Transactions on Fuzzy Systems*, 13(4):517–530, 2005.

13. H. Timm, C. Borgelt, C. Doring, and R. Kruse. An Extension to Possibilistic Fuzzy Cluster Analysis. *Fuzzy Sets and Systems*, 147:3–16, 2004.

14. Z. Pawlak. *Rough Sets: Theoretical Aspects of Reasoning About Data*. Kluwer, Dordrecht, The Netherlands, 1991.

15. S. Hirano and S. Tsumoto. An Indiscernibility-Based Clustering Method with Iterative Refinement of Equivalence Relations: Rough Clustering. *Journal of Advanced Computational Intelligence and Intelligent Informatics*, 7(2):169–177, 2003.

16. S. K. De. A Rough Set Theoretic Approach to Clustering. *Fundamenta Informaticae*, 62(3–4):409–417, 2004.

17. P. Lingras and C. West. Interval Set Clustering of Web Users with Rough K-Means. *Journal of Intelligent Information Systems*, 23(1):5–16, 2004.

18. S. K. Pal, B. D. Gupta, and P. Mitra. Rough Self Organizing Map. *Applied Intelligence*, 21(3):289–299, 2004.

19. S. Asharaf, S. K. Shevade, and M. N. Murty. Rough Support Vector Clustering. *Pattern Recognition*, 38:1779–1783, 2005.

20. S. K. Pal and P. Mitra. Multispectral Image Segmentation Using the Rough Set-Initialized-EM Algorithm. *IEEE Transactions on Geoscience and Remote Sensing*, 40(11):2495–2501, 2002.

21. S. Asharaf and M. N. Murty. A Rough Fuzzy Approach to Web Usage Categorization. *Fuzzy Sets and Systems*, 148:119–129, 2004.

22. S. Mitra, H. Banka, and W. Pedrycz. Rough-Fuzzy Collaborative Clustering. *IEEE Transactions on Systems Man and Cybernetics Part B-Cybernetics*, 36:795–805, 2006.

23. P. Maji and S. K. Pal. RFCM: A Hybrid Clustering Algorithm Using Rough and Fuzzy Sets. *Fundamenta Informaticae*, 80(4):475–496, 2007.

24. P. Maji and S. K. Pal. Rough Set Based Generalized Fuzzy C-Means Algorithm and Quantitative Indices. *IEEE Transactions on Systems Man and Cybernetics Part B-Cybernetics*, 37(6):1529–1540, 2007.

25. G. James. *Modern Engineering Mathematics*. Addison-Wesley, Reading, MA, 1996.

26. D. W. Jordan and P. Smith. *Mathematical Techniques: An Introduction for Engineering, Physical, and Mathematical Sciences*. Oxford University Press, Oxford, UK, 1997.

27. J. Bezdek. A Convergence Theorem for the Fuzzy ISODATA Clustering Algorithm. *IEEE Transactions on Pattern Analysis and Machine Intelligence*, 2:1–8, 1980.

28. J. Bezdek, R. J. Hathaway, M. J. Sabin, and W. T. Tucker. Convergence Theory for Fuzzy C-Means: Counterexamples and Repairs. *IEEE Transactions on Systems, Man, and Cybernetics*, 17:873–877, 1987.

29. H. Yan. Convergence Condition and Efficient Implementation of the Fuzzy Curve-Tracing (FCT) Algorithm. *IEEE Transactions on Systems Man and Cybernetics Part B-Cybernetics*, 34(1):210–221, 2004.

30. P. Lingras, M. Chen, and D. Miao. Rough Cluster Quality Index Based on Decision Theory. *IEEE Transactions on Knowledge and Data Engineering*, 21(7):1014–1026, 2009.

31. I. Dhillon, Y. Guan, and B. Kulis. Kernel K-Means, Spectral Clustering and Normalized Cuts. In *Proceedings of the 10th ACM SIGKDD International Conference on Knowledge Discovery and Data Mining*, pages 551–556. ACM, New York, 2004.

32. M. Girolami. Mercer Kernel-Based Clustering in Feature Space. *IEEE Transactions on Neural Networks*, 13(3):780–784, 2002.

33. B. Scholkopf, A. Smola, and K.-R. Muller. Nonlinear Component Analysis as a Kernel Eigenvalue Problem. *Neural Computation*, 10:1299–1319, 1998.

34. S. Miyamoto and D. Suizu. Fuzzy C-Means Clustering Using Kernel Functions in Support Vector Machines. *Journal of Advanced Computational Intelligence and Intelligent Informatics*, 7(1):25–30, 2003.

35. Z. D. Wu and W. X. Xie. Fuzzy C-Means Clustering Algorithm Based on Kernel Method. In *Proceedings of the 5th International Conference on Computational Intelligence and Multimedia Applications*, Xi'an, China, pages 1–6, 2003.

36. D.-Q. Zhang and S.-C. Chen. Kernel Based Fuzzy and Possibilistic C-Means Clustering. In *Proceedings of the International Conference on Artificial Neural Network, Turkey*, pages 122–125, 2003.

37. K. Mizutani and S. Miyamoto. Possibilistic Approach to Kernel-Based Fuzzy C-Means Clustering with Entropy Regularization. In V. Torra, Y. Narukawa, and S. Miyamoto, editors, *Modeling Decisions for Artificial Intelligence, Lecture Notes in Computer Science*, volume 3558, pages 144–155. Springer, Berlin, 2005.

38. X.-H. Wu and J.-J. Zhou. Possibilistic Fuzzy C-Means Clustering Model Using Kernel Methods. In *Proceedings of the International Conference on Computational Intelligence for Modelling, Control and Automation*, volume 2, pages 465–470, IEEE Computer Society, Los Alamitos, CA, 2005.

39. D. L. Davies and D. W. Bouldin. A Cluster Separation Measure. *IEEE Transactions on Pattern Analysis and Machine Intelligence*, 1:224–227, 1979.

40. J. C. Bezdek and N. R. Pal. Some New Indexes for Cluster Validity. *IEEE Transactions on Systems Man and Cybernetics Part B-Cybernetics*, 28:301–315, 1988.

4

ROUGH-FUZZY GRANULATION AND PATTERN CLASSIFICATION

4.1 INTRODUCTION

Granular computing refers to computation and operations performed on information granules, that is, clumps of similar objects or points. Granular computing has been changed rapidly from a label to a conceptual and computational paradigm of study that deals with information and knowledge processing. Many researchers [1–3] have used the granular computing models to build efficient computational algorithms that can handle a huge amount of data, information, and knowledge. These models mainly deal with the efficiency, effectiveness, and robustness of using granules such as classes, clusters, subsets, groups, and intervals in problem solving [4].

Granular computing can be studied on the basis of its notions of representation and process. However, the main task to be focused on is to construct and describe information granules by a process called *information granulation* [5, 6] on which granular computing is oriented. Specifically, granulation is governed by the principles according to which the models should exploit the tolerance for imprecision and employ the coarsest level of granulation, which are consistent with the allowable level of imprecision. Modes of information granulation, in which the granules are crisp, play important roles in a wide variety of approaches and techniques. Although crisp information granulation has a wide range of applications, it has a major blind spot [7]. More particularly, it fails to reflect most of the processes of human reasoning and concept formation, where the granules are

Rough-Fuzzy Pattern Recognition: Applications in Bioinformatics and Medical Imaging,
First Edition. Pradipta Maji and Sankar K. Pal.
© 2012 John Wiley & Sons, Inc. Published 2012 by John Wiley & Sons, Inc.

more appropriately fuzzy rather than crisp. The fuzziness in granules and their values are characteristic of the ways in which human concepts are formed, organized, and manipulated [8]. In fact, fuzzy information granulation does not refer to a single fuzzy granule; rather, it is about a collection of fuzzy granules that results from granulating a crisp or fuzzy object. Depending on the problems, and whether the granules and computation performed are fuzzy or crisp, one may have operations such as granular fuzzy computing or fuzzy granular computing [9].

In the recent past, several research works have been carried out on the construction of fuzzy granules. The process of fuzzy granulation involves the basic idea of generating a family of fuzzy granules from numerical features and transforming them into fuzzy linguistic variables. These variables thus keep the semantics of the data and are easy to understand. Fuzzy information granulation has come up with an important concept in fuzzy set theories, rough set theories, and the combination of both in recent years [1, 2, 7–12]. In general, the process of fuzzy granulation can be broadly categorized as class dependent (CD) and class independent (CI). Fuzzy sets are used in both cases for linguistic representation of patterns and generation of fuzzy granulation of the feature space. With CI granulation, each feature of the pattern is described with three fuzzy membership functions representing three linguistic values over the whole space, as is done in Pal and Mitra [13]. Hence, an m-dimensional feature space is characterized by 3^m granules. However, this process of granulation does not take care of the class-belonging information of features to different classes. On the other hand, in CD granulation, each feature explores its class-belonging information to different classes. In this process, features are described by the fuzzy sets equal to the number of classes, and individual class information is preserved by the generated fuzzy granules. In other words, an m-dimensional feature space with c classes is characterized by c^m granules. Therefore, the CD granulation is expected to be more efficient in the classification of data sets with highly overlapping classes.

The rough set theory, as proposed by Pawlak, has been proved to be an effective tool for feature selection, knowledge discovery, and rule extraction from categorical data [14]. The theory enables the discovery of data dependency and performs the reduction or selection of features contained in a data set using the data alone, requiring no additional information. The rough sets can be used as an effective tool to deal with both vagueness and uncertainty in data sets and to perform granular computation. The rough-set-based feature selection not only retains the representational power of the data but also maintains its minimum redundancy [14]. A rough set learning algorithm is shown in Pawlak [14] to obtain a set of rules from a decision table in IF-THEN form. The rough set theoretic tree structure pattern classifier is reported in Ananthanarayana et al. [15] for efficient data mining. However, the rough set theory can be used for numerical data by discretizing them, which results in the loss of information and introduction of noise. The concept of fuzzy discretization of feature space for a rough set theoretic classifier is described in Roy and Pal [16]. Recently, rough-fuzzy computing has been successfully applied for the classification of incomplete data

[17, 18]. Also, the neighborhood rough set [19, 20] is found to be suitable for both numerical and categorical data sets. The advantage of the neighborhood rough set is that it facilitates gathering the possible local information through neighborhood granules that are useful for a better discrimination of patterns, particularly in an overlapping class environment.

In this chapter, a rough-fuzzy granular space (RFGS) is described using the CD fuzzy granulation and neighborhood-rough-set-based feature selection [21]. The model provides a synergistic integration of the merits of both CD fuzzy granulation and the theory of neighborhood rough sets. In the RFGS-based model, the fuzzy granulated features are first generated on the basis of the CD knowledge. The neighborhood-rough-set-based feature selection method is then applied to these fuzzy granular feature sets for computing the approximate reducts to select a subset of features. The resulting output of the RFGS-based model can be used as an input for any classifier. To demonstrate the effectiveness of the RFGS-based model, different classifiers such as K-nearest neighbor (K-NN) rule [22], maximum likelihood (ML) classifier [22], and multilayer perceptron (MLP) [23] are used. However, other classifiers may also be used.

The effectiveness of the RFGS-based model is demonstrated with seven completely labeled data sets including a synthetic remote sensing image and two partially labeled real multispectral remote sensing images. Various performance measures such as percentage of overall classification accuracy, precision, recall [22], kappa coefficient (KC) [24], and computation time are considered for completely labeled data sets. On the other hand, the β index [25] and Davies–Bouldin (DB) index [26] are used for partially labeled data sets. Another quantitative index, called *dispersion score* (DS), reflecting a different interpretation of the confusion matrix, is reported to measure the class-wise classifier performance. The dispersion measure quantifies the nature of distribution of the classified patterns among different classes. The statistical significance test is also performed using χ^2 analysis.

The structure of the rest of this chapter is as follows: In Section 4.2, the RFGS-based model for pattern classification is described, while several quantitative measures are presented in Section 4.3 to evaluate the performance of different pattern classification methods. A brief description of different data sets is reported in Section 4.4. Different case studies and a comparison among different methods are presented in Section 4.5. Concluding remarks are given in Section 4.6.

4.2 PATTERN CLASSIFICATION MODEL

The RFGS-based model has three main operational steps, as illustrated in Fig. 4.1. The first step of this model generates the CD fuzzy granulated feature space of input pattern vector. For fuzzy granulation, c number of fuzzy sets are used to characterize the feature values of each pattern vector, where c is the total number of classes. The π-type membership function can be considered for the

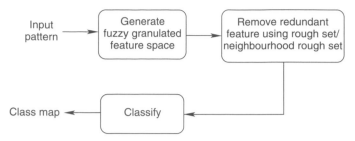

Figure 4.1 Schematic flow diagram of the RFGS-based model for classification.

fuzzification purpose. Each feature is thus represented by c number of $[0, 1]$-valued membership functions representing c fuzzy sets or characterizing c fuzzy granules along the axis. The π-type membership function explores the degree of belonging of a pattern into different classes based on individual features and the granules thus provide an improved class-wise representation of input patterns. The granules preserve the interrelated class information to build an informative granular space, which is potentially useful for improved classification for the data sets with overlapping classes.

In the granulation process, each feature value is represented with more than one membership value and thus the feature dimension increases. The increased dimension leads to great difficulty in solving many tasks of pattern recognition, machine learning, and data mining. In this regard, the neighborhood-rough-set-based feature selection method [19, 20] is used in the second step of the RFGS-based model (Fig. 4.1). The advantage in the use of neighborhood rough sets is that it can deal with both numerical and categorical data. The neighborhood rough set does not require any discretization of numerical data and is suitable for the fuzzy granulation of features. Further, the neighboring concept facilitates gathering of the possible local information through neighborhood granules that provide a better class discrimination information. Hence, the combination of these two steps of operations can be a better framework for the classification of patterns in an overlapping class environment. The RFGS-based model thus takes the advantages of both the CD fuzzy granulation and the neighborhood-rough-set-based feature selection method. After selecting the features, a classifier is used in the third step of Fig. 4.1 to classify the input pattern based on the selected features.

4.2.1 Class-Dependent Fuzzy Granulation

The CD fuzzy granulated feature space is generated using a fuzzy linguistic representation pattern. Only the case of numeric features is mentioned in this chapter. However, the features in descriptive and set forms can also be handled in this framework.

Let a pattern or object x be represented by m numeric features and expressed as $x = [x_1, \ldots, x_j, \ldots, x_m]$. Hence, the object x can be visualized as a point in the m-dimensional vector space. Each feature is described in terms of its fuzzy membership values corresponding to c linguistic fuzzy sets, where c is the total number of classes. Hence, an m-dimensional pattern vector is expressed as $(m \times c)$-dimensional vector and is given by

$$x = [\mu_1(x_1), \ldots, \mu_i(x_1), \ldots, \mu_c(x_1),$$

$$\ldots, \ldots, \ldots, \ldots, \ldots$$

$$\mu_1(x_j), \ldots, \mu_i(x_j), \ldots, \mu_c(x_j),$$

$$\ldots, \ldots, \ldots, \ldots, \ldots$$

$$\mu_1(x_m), \ldots, \mu_i(x_m), \ldots, \mu_c(x_m)], (i = 1, 2, \ldots, c), \qquad (4.1)$$

where $\mu_1(x_j), \ldots, \mu_i(x_j), \ldots, \mu_c(x_j)$ signify the membership values of x_j to c number of fuzzy sets along the jth feature axis and $\mu(x_j) \in [0, 1]$. It implies that each feature x_j is expressed separately by c number of membership functions expressing c fuzzy sets. In other words, each feature x_j characterizes c number of fuzzy granules along each axis and thus comprises c^m fuzzy granules in an m-dimensional feature space.

The π-type membership function [27] is used to characterize the fuzzy granules. The membership value $\mu_i(x_j)$ thus generated expresses the degree of belonging of the jth feature to the ith class of the pattern x. The π-type membership function is given by

$$\mu(x_j; a, r, b) = \begin{cases} 0 & x_j \le a \\ 2^{\eta-1}\left[\dfrac{x_j - a}{r - a}\right]^{\eta} & a < x_j \le p \\ 1 - 2^{\eta-1}\left[\dfrac{r - x_j}{r - a}\right]^{\eta} & p < x_j \le r \\ 1 - 2^{\eta-1}\left[\dfrac{x_j - r}{b - r}\right]^{\eta} & r < x_j \le q \\ 2^{\eta-1}\left[\dfrac{b - x_j}{b - r}\right]^{\eta} & q < x_j < b \\ 0 & x_j \ge b, \end{cases} \qquad (4.2)$$

where η is the fuzzifier of the membership function, as shown in Fig. 4.2. The membership function can be estimated with the center at r and $r = (p + q)/2$, where p and q are the two crossover points. The membership values at the crossover points are 0.5 and at the center r its value is maximum, that is, 1. Assignment of membership value is made in such a way that training data gets a value closer to 1 when it is nearer to the center of membership function and a value closer to 0.5 when it is away from the center. For the determination of

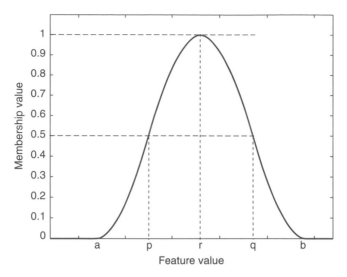

Figure 4.2   The π-type membership function.

the membership function, the position of the center is considered at mean point. The mean point is estimated as

$$r = \text{mean}(j), \tag{4.3}$$

that is, the mean point is the average value of the data set for the jth feature x_j. The two crossover points p and q are estimated as

$$p = \text{mean}(j) - \left(\frac{\max(j) - \min(j)}{2} \right), \tag{4.4}$$

$$q = \text{mean}(j) + \left(\frac{\max(j) - \min(j)}{2} \right), \tag{4.5}$$

where $\min(j)$ and $\max(j)$ are the minimum and maximum values of the data for the jth feature x_j, respectively. Hence, the feature-wise class membership of input pattern can be automatically determined from the training set.

4.2.2   Rough-Set-Based Feature Selection

This section presents some preliminaries relevant to feature selection methods using both rough sets and neighborhood rough sets [14, 19, 20].

4.2.2.1   Rough Sets   The theory of rough sets begins with the notion of an approximation space, which is a pair $<\mathbb{U}, \mathbb{A}>$, where $\mathbb{U}$ is a nonempty set, the universe of discourse, $\mathbb{U} = \{x_1, \ldots, x_i, \ldots, x_n\}$ and $\mathbb{A}$ is a family of attributes, also called knowledge in the universe. V is the value domain of $\mathbb{A}$ and f is an information function $f : \mathbb{U} \times \mathbb{A} \rightarrow V$. An approximation space is also called an *information system* [14]. Any subset $\mathbb{P}$ of knowledge $\mathbb{A}$ defines an equivalence, also called indiscernibility, relation $\text{IND}(\mathbb{P})$ on $\mathbb{U}$

$$\text{IND}(\mathbb{P}) = \{(x_i, x_j) \in \mathbb{U} \times \mathbb{U} | \forall a \in \mathbb{P}, f(x_i, a) = f(x_j, a)\}. \qquad (4.6)$$

If $(x_i, x_j) \in \text{IND}(\mathbb{P})$, then x_i and x_j are indiscernible by attributes from $\mathbb{P}$. The partition of $\mathbb{U}$ generated by $\text{IND}(\mathbb{P})$ is denoted as

$$\mathbb{U}/\text{IND}(\mathbb{P}) = \{[x_i]_\mathbb{P} : x_i \in \mathbb{U}\}, \qquad (4.7)$$

where $[x_i]_\mathbb{P}$ is the equivalence class containing x_i. The elements in $[x_i]_\mathbb{P}$ are indiscernible or equivalent with respect to knowledge $\mathbb{P}$. Equivalence classes, also termed as *information granules*, are used to characterize arbitrary subsets of $\mathbb{U}$. The equivalence classes of $\text{IND}(\mathbb{P})$ and the empty set $\emptyset$ are the elementary sets in the approximation space $<\mathbb{U}, \mathbb{A}>$.

Given an arbitrary set $X \subseteq \mathbb{U}$, in general, it may not be possible to describe X precisely in $<\mathbb{U}, \mathbb{A}>$. One may characterize X by a pair of lower and upper approximations, defined as follows [14]:

$$\underline{\mathbb{P}}(X) = \bigcup \{[x_i]_\mathbb{P} | [x_i]_\mathbb{P} \subseteq X\} \qquad (4.8)$$

and

$$\overline{\mathbb{P}}(X) = \bigcup \{[x_i]_\mathbb{P} | [x_i]_\mathbb{P} \cap X \neq \emptyset\}. \qquad (4.9)$$

Hence, the lower approximation $\underline{\mathbb{P}}(X)$ is the union of all the elementary sets that are subsets of X, and the upper approximation $\overline{\mathbb{P}}(X)$ is the union of all the elementary sets that have a nonempty intersection with X. The tuple $<\underline{\mathbb{P}}(X), \overline{\mathbb{P}}(X)>$ is the representation of an ordinary set X in the approximation space $<\mathbb{U}, \mathbb{A}>$ or simply called the rough set of X. The lower (respectively, upper) approximation $\underline{\mathbb{P}}(X)$ (respectively, $\overline{\mathbb{P}}(X)$) is interpreted as the collection of those elements of $\mathbb{U}$ that definitely (respectively, possibly) belong to X. The lower approximation is also called positive region, sometimes, denoted as $\text{POS}_\mathbb{P}(X)$. A set X is said to be definable or exact in $<\mathbb{U}, \mathbb{A}>$ iff $\underline{\mathbb{P}}(X) = \overline{\mathbb{P}}(X)$. Otherwise, X is indefinable and termed as a *rough set*. $BN_\mathbb{P}(X) = \overline{\mathbb{P}}(X) \setminus \underline{\mathbb{P}}(X)$ is called a *boundary set*.

Definition 1 *An information system $<\mathbb{U}, \mathbb{A}>$ is called a decision table if the attribute set $\mathbb{A} = \mathbb{C} \cup \mathbb{D}$, where $\mathbb{C}$ and $\mathbb{D}$ are the condition and decision attribute sets, respectively. The dependency between $\mathbb{C}$ and $\mathbb{D}$ can be defined as [14]*

$$\gamma_\mathbb{C}(\mathbb{D}) = \frac{|\text{POS}_\mathbb{C}(\mathbb{D})|}{|\mathbb{U}|}, \qquad (4.10)$$

where $POS_{\mathbb{C}}(\mathbb{D}) = \cup \underline{C}X_i$, X_i *is the ith equivalence class induced by* $\mathbb{D}$ *and* $|\cdot|$ *denotes the cardinality of a set.*

Let $\mathbb{I} = <\mathbb{U}, \mathbb{A}>$ be a decision table, where $\mathbb{U} = \{x_1, \ldots, x_7\}$ is a nonempty set of finite objects or the universe and $\mathbb{A} = \mathbb{C} \cup \mathbb{D}$ is a nonempty finite set of attributes, where $\mathbb{C} = \{Age, LEMS\}$ and $\mathbb{D} = \{Walk\}$ are the set of condition and decision attributes, respectively.

$x_i \in \mathbb{U}$	Age	LEMS	Walk
x_1	16–30	50	yes
x_2	16–30	0	no
x_3	31–45	1–25	no
x_4	31–45	1–25	yes
x_5	46–60	26–49	no
x_6	16–30	26–49	yes
x_7	46–60	26–49	no

IND({Age}) creates the following partition of $\mathbb{U}$:

$$\mathbb{U}/IND(\{Age\}) = \{\{x_1, x_2, x_6\}, \{x_3, x_4\}, \{x_5, x_7\}\}$$

as the objects x_1, x_2, and x_6 are indiscernible with respect to the condition attribute set {Age}. Similarly, the partition of $\mathbb{U}$ generated by the condition attribute set {LEMS} is given by

$$\mathbb{U}/IND(\{LEMS\}) = \{\{x_1\}, \{x_2\}, \{x_3, x_4\}, \{x_5, x_6, x_7\}\}$$

and the partition of $\mathbb{U}$ generated by the condition attribute set {Age, LEMS} is as follows:

$$\mathbb{U}/\{Age, LEMS\} = \{\{x_1\}, \{x_2\}, \{x_3, x_4\}, \{x_5, x_7\}, \{x_6\}\}.$$

Now, the partition of $\mathbb{U}$ generated by the decision attribute set {Walk} is given by

$$\mathbb{U}/\mathbb{D} = \mathbb{U}/\{Walk\} = \{\{x_1, x_4, x_6\}, \{x_2, x_3, x_5, x_7\}\}.$$

The positive region contains all objects of $\mathbb{U}$ that can be classified to classes of $\mathbb{U}/\mathbb{D}$ using the knowledge in attributes $\mathbb{C}$. Hence, for the above example, the positive region is as follows:

$$POS_{\mathbb{C}}(\mathbb{D}) = \bigcup\{\phi, \{x_1\}, \{x_2\}, \{x_5, x_7\}, \{x_6\}\} = \{x_1, x_2, x_5, x_6, x_7\}.$$

The dependency between $\mathbb{C}$ and $\mathbb{D}$ is, therefore, given by

$$\gamma_{\mathbb{C}}(\mathbb{D}) = \frac{5}{7}.$$

An important issue in data analysis is discovering the dependency between attributes. Intuitively, a set of attributes $\mathbb{D}$ depends totally on a set of attributes $\mathbb{C}$, denoted as $\mathbb{C} \Rightarrow \mathbb{D}$, if all attribute values from $\mathbb{D}$ are uniquely determined by values of attributes from $\mathbb{C}$. If there exists a functional dependency between values of $\mathbb{D}$ and $\mathbb{C}$, then $\mathbb{D}$ depends totally on $\mathbb{C}$. The dependency can be defined in the following way:

Definition 2 *Given* $\mathbb{C}, \mathbb{D} \subseteq \mathbb{A}$, *it is said that* $\mathbb{D}$ *depends on* $\mathbb{C}$ *in a degree* κ *($0 \leq \kappa \leq 1$), denoted as* $\mathbb{C} \Rightarrow_\kappa \mathbb{D}$, *if*

$$\kappa = \gamma_{\mathbb{C}}(\mathbb{D}) = \frac{|POS_{\mathbb{C}}(\mathbb{D})|}{|\mathbb{U}|}. \tag{4.11}$$

If $\kappa = 1$, $\mathbb{D}$ *depends totally on* $\mathbb{C}$, *if* $0 < \kappa < 1$, $\mathbb{D}$ *depends partially (in a degree κ) on* $\mathbb{C}$, *and if* $\kappa = 0$, *then* $\mathbb{D}$ *does not depend on* $\mathbb{C}$ *[14].*

To what extent an attribute contributes to the calculation of the dependency on decision attribute can be calculated by the significance of that attribute. The change in dependency when an attribute is removed from the set of condition attributes is a measure of the significance of the attribute. The higher the change in dependency, the more significant the attribute is. If the significance is 0, then the attribute is dispensable.

Definition 3 *Given* $\mathbb{C}$, $\mathbb{D}$ *and an attribute* $\mathcal{A} \in \mathbb{C}$, *the significance of the attribute* $\mathcal{A}$ *is defined as follows [14]:*

$$\sigma_{\mathbb{C}}(\mathbb{D}, \mathcal{A}) = \gamma_{\mathbb{C}}(\mathbb{D}) - \gamma_{\mathbb{C}-\{\mathcal{A}\}}(\mathbb{D}). \tag{4.12}$$

Considering the above table, let $\mathbb{C} = \{\mathcal{A}_1, \mathcal{A}_2\}$, where $\mathcal{A}_1 = \{Age\}$, $\mathcal{A}_2 = \{LEMS\}$, and $\mathbb{D} = \{Walk\}$. The significance of two attributes $\mathcal{A}_1$ and $\mathcal{A}_2$ is as follows:

$$\sigma_{\mathbb{C}}(\mathbb{D}, \mathcal{A}_1) = \gamma_{\mathbb{C}}(\mathbb{Q}) - \gamma_{\mathbb{C}-\{\mathcal{A}_1\}}(\mathbb{D}) = \frac{5}{7} - \frac{2}{7} = \frac{3}{7},$$

$$\sigma_{\mathbb{C}}(\mathbb{D}, \mathcal{A}_2) = \gamma_{\mathbb{C}}(\mathbb{Q}) - \gamma_{\mathbb{C}-\{\mathcal{A}_2\}}(\mathbb{D}) = \frac{5}{7} - \frac{2}{7} = \frac{3}{7}.$$

In this chapter, the quick reduct algorithm of Chouchoulas and Shen [28] is used to select features from a data set. It attempts to find an effective subset of features without exhaustively generating all possible feature subsets. It starts with an empty set and adds one feature at a time that results in the increase in dependency. The process goes on until it produces the maximum possible dependency value for a data set. In this work, the quick reduct algorithm is used to select features generated from the CD fuzzy granulation. The selected features are then used in a classifier for classifying the input pattern, as in the third step of Fig. 4.1.

4.2.2.2 Neighborhood Rough Sets Let $\mathbb{I} =< \mathbb{U}, \mathbb{A}>$ denote the information system, where $\mathbb{U} = \{x_1, \ldots, x_i, \ldots, x_n\}$ is the universal set, which is a nonempty and finite set of samples, and $\mathbb{A} = \{\mathbb{C} \cup \mathbb{D}\}$, where $\mathbb{C}$ and $\mathbb{D}$ are the sets of condition and decision features, respectively. Given an arbitrary $x_i \in \mathbb{U}$ and $\mathbb{B} \subseteq \mathbb{C}$, the neighborhood $\Phi_\mathbb{B}(x_i)$ of x_i with given Φ, in feature space $\mathbb{B}$, is defined as [19]

$$\Phi_\mathbb{B}(x_i) = \{x_j | x_j \in \mathbb{U}, \Delta^\mathbb{B}(x_i, x_j) \leq \Phi\} \qquad (4.13)$$

where Δ is a distance function. $\Phi_\mathbb{B}(x_i)$ in Equation (4.13) is the neighborhood information granule centered with sample x_i. In this study, three **p**-norm distances in Euclidean space are used. These are Manhattan distance ($\mathbf{p} = 1$), Euclidean distance ($\mathbf{p} = 2$), and Chebychev distance ($\mathbf{p} = \infty$). The neighborhood granule generation is effected by two key factors, namely, the used distance function Δ and parameter Φ. The first one determines the shape and the second one controls the size of the neighborhood granule. Both these factors play important roles in neighborhood rough sets and can be considered to control the granularity of data analysis. The significance of features varies with the granularity levels. Accordingly, the neighborhood-rough-set-based algorithm selects different feature subsets with the change in Δ function and Φ value. In this study, the effects of three **p**-norm distances are analyzed for a variation of Φ values, and the best one is selected on the basis of the performance with the present data sets. However, optimal parameter values can be obtained through an optimization technique such as genetic algorithm.

Hence, each sample generates granules with a neighborhood relation. For a metric space $\langle \mathbb{U}, \Delta \rangle$, the set of neighborhood granules $\{\Phi(x_i) | x_i \in \mathbb{U}\}$ forms an elemental granule system that covers the universal space rather than partitions it as in the case of rough sets. It is noted that the partition of the space generated by rough sets can be obtained from neighborhood rough sets with covering principle, while the other way round is not possible. Moreover, a neighborhood granule degrades to an equivalent class for $\Phi = 0$. In this case, the samples in the same neighborhood granule are equivalent to each other and the neighborhood rough set model degenerates to rough sets. Hence, the neighborhood rough sets can be treated as a generalized case of rough sets.

The dependency degree of decision feature $\mathbb{D}$ on condition feature set $\mathbb{B}$ in a neighborhood information system $<\mathbb{U}, \mathbb{C} \cup \mathbb{D}, N>$ with distance function Δ and neighborhood size Φ is defined as

$$\gamma_\mathbb{B}(\mathbb{D}) = \frac{|\text{POS}_\mathbb{B}(\mathbb{D})|}{|\mathbb{U}|}, \qquad (4.14)$$

where $| \bullet |$ denotes the cardinality of a set and $0 \leq \gamma_\mathbb{B}(\mathbb{D}) \leq 1$. The dependency function measures the approximation power of a condition feature set. Hence, it can be used to determine the significance of a subset of features, normally called *reduct*. The significance of a subset of features is calculated with the change in dependency, when a feature is removed from the set of considered condition features.

On the basis of the significance of a feature, the subset of features or reduct is evaluated. Many sets of reducts can be obtained on the basis of the significance and any of them will work for the feature reduction task. In this regard, Hu et al. [19] described a forward greedy search algorithm for feature selection using neighborhood rough sets. It begins with an empty reduct. In each step, one feature is added and the change in dependency or significance is determined, when a feature is removed from the set of considered condition features. The process is stopped when the significance of the reduct is less than a small threshold value. In this study, the forward greedy search algorithm of Hu et al. [19] is used for the selection of features in the RFGS-based model for classification. The selected features are then fed to a classifier for classifying the input pattern, as in the third step of Fig. 4.1.

4.3 QUANTITATIVE MEASURES

In this section, several quantitative indices such as DS measure [21], classification accuracy, precision, recall [22], and KC [24] for completely labeled data sets, and β index [25] and DB index [26] for partially labeled data sets, which are used to evaluate the performance of different classifiers, are reported. Among them, the DB index is briefly described in Chapter 3.

4.3.1 Dispersion Measure

The performance of a classifier can be analyzed with respect to its confusion matrix. The nature of distribution of the classified patterns among different classes reflects the overlapping characteristics of various regions. With this notion, the class-wise dispersion measure of a classifier, which quantifies the nature of dispersion of the misclassified patterns into different classes, can be defined. The dispersion score DS_i for the ith class is defined as [21]

$$\mathrm{DS}_i = \left[1 - \left(\lambda_1 \frac{V_i}{V_{\max}^i} + \lambda_2 \frac{Z_i}{c - 1} \right) \right]; \quad i, = 1, 2, ..., c, \qquad (4.15)$$

where c represents the total number of classes, V_i and $V_{\max}^i$ are the variance of the elements in ith row of the confusion matrix and maximum variance corresponding to ith row, respectively, Z_i is the number of elements in ith row of the confusion matrix with zero value and $Z_i \leq (c - 1)$, and λ_1 and $\lambda_2 (= 1 - \lambda_1)$ are the weight factors. The value of $V_{\max}^i$ is obtained when all the patterns are correctly classified to the ith class.

The first part of Equation (4.15) is the normalized variance for the class of interest and quantifies the nature of distribution of the classified patterns among the classes. The second part of Equation (4.15) is the normalized count for the number of classes where no patterns are classified. Both the aspects are combined to reflect the overlapping characteristics of one class with that of the others. Hence, the dispersion measure provides a measure quantifying the class-wise distribution of the classified patterns for a classifier. The measure can be viewed as an index in evaluating the class-wise performance of a classifier. In Equation (4.15), the weight factors λ_1 and λ_2 can be assigned according to the requirement at hand. That means, the more important an aspect, the higher the weight value.

According to the dispersion measure, the less the dispersion score value the better the agreement. The value of DS = 0 indicates perfect agreement between the two observers, namely, true and estimated, that is, all the test patterns of that class are correctly classified. If the patterns are misclassified to or confused into a minimum number of classes, the DS value would be less. Therefore, with a given number of overlapping classes for a particular class of interest in the feature space, the lower DS value is desirable. Lower value would also allow one to focus on the less number of confused classes in order to get some of the misclassified patterns corrected with available higher level information.

Apart from providing an index of class-wise performance, the DS measure provides a helpful clue for improving the class-wise performance with additional or higher level information. The smaller the DS value, that is, when the mis-classified patterns are mostly confined into a minimum number of classes, the larger the possibility that some of the misclassified patterns from the neighboring classes may get rectified with the higher level such as syntactic, semantic, and contextual information.

4.3.2 Classification Accuracy, Precision, and Recall

The percentage of overall classification accuracy is the percentage of samples that are correctly classified and can be evaluated from the confusion matrix. In this study, the significance of the confusion matrix is considered with respect to individual class. Sometimes a distinction is made between errors of omission and errors of commission, particularly when only a small number of class type is of interest. Hence, interpreting a confusion matrix from a particular class point of view, it is important to notice that different indications of class accuracies will result differently, according to whether the number of correct patterns for a class is divided by the total number of true or reference patterns for the class or the total number of patterns the classifier features to the class. The former is normally known as *recall* and the latter as *precision* [22]. Note that the overall classification accuracy does not provide the class-wise agreement between the true and estimated class labels, and the precision and recall measures give the results for individual classes only.

4.3.3 κ Coefficient

To get an overall class agreement based on the individual class accuracy, KC [24] estimation is used to validate the superiority of the classifiers effectively.

The KC measure was introduced by the psychologist Cohen [24] and adapted for accuracy assessment in the remote sensing field by Congalton and Mead [29]. The KC and classification accuracy are not proportional, that means a good percentage of accuracy may lead to a poor KC value because it provides the measurement of class-wise agreement between the true and estimated class labels. The higher the coefficient value, the better the agreement of the estimated data with the true one. The KC value is estimated from a confusion matrix as follows:

$$KC = \frac{M \sum_{i=1}^{r} Y_{ii} - \sum_{i=1}^{r}(Y_{i+} \times Y_{+i})}{M^2 - \sum_{i=1}^{r}(Y_{i+} \times Y_{+i})}, \qquad (4.16)$$

where r and M represent the number of rows in the error matrix and the total number of observations included in the matrix, respectively, while Y_{ii}, Y_{i+}, and Y_{+i} are the number of observations in row i and column i, total observation in row i, and total observation in column i, respectively. A KC value greater than 0 indicates the amount of agreement between the two observers, namely, true and estimated. A KC value of 1 indicates perfect agreement when all the values are falling on the diagonal [24].

4.3.4 β Index

The β index of Pal et al. [25] is defined as the ratio of the total variation and within-class or cluster variation, and is given by

$$\beta = \frac{N}{M}, \qquad (4.17)$$

where

$$N = \sum_{i=1}^{c} \sum_{j=1}^{n_i} ||x_{ij} - \bar{v}||^2, \qquad (4.18)$$

$$M = \sum_{i=1}^{c} \sum_{j=1}^{n_i} ||x_{ij} - v_i||^2; \text{ and } \sum_{i=1}^{c} n_i = n, \qquad (4.19)$$

where n_i is the number of objects in the ith class or cluster ($i = 1, 2, \ldots, c$), n the total number of objects, x_{ij} the jth object in cluster i, v_i the mean or centroid of the ith cluster, and $\bar{v}$ the mean of n objects. For a given image and c value, the higher the homogeneity within the segmented regions, the higher would be the β value. The value of β increases with c.

4.4 DESCRIPTION OF DATA SETS

In this study, seven completely labeled data sets, including a synthetic remote sensing image and two partially labeled real remote sensing images, collected from a satellite are chosen.

4.4.1 Completely Labeled Data Sets

This section presents a brief description of seven completely labeled data sets, which are used to evaluate the performance of the RFGS-based pattern classification model.

4.4.1.1 Vowel Data The vowel data is a set of Indian Telugu vowel sounds in a consonant–vowel–consonant context uttered by three male speakers in the age group of 30–35 years. As a whole, the data set consists of 871 patterns. It has three features and six classes /∂/, /a/, /i/, /u/, /e/, and /o/ with 72, 89, 172, 151, 207, and 180 samples, respectively. The data set, which is depicted in Fig. 4.3 with two dimensions for ease of understanding, has three features: F_1, F_2, and F_3 corresponding to the first, second, and third vowel formant frequencies obtained through spectrum analysis. The classes are highly overlapping and possess ill-defined boundaries.

4.4.1.2 Satimage Database The original Landsat data for this database is generated from data purchased from NASA by the Australian Centre for Remote Sensing, and used for research at the University of New South Wales. The sample database is generated taking a small section of 82 rows and 100 columns from the original data. The database is a tiny subarea of a scene, consisting of 82×100 pixels, each pixel covering an area on the ground of approximately 80×80 m. The information given for each pixel consists of the class value and the intensities in four spectral bands, from the green, red, and infrared regions of the spectrum. The data set contains 6435 examples: 4435 training and 2000 testing, with 36 real valued attributes and 6 classes.

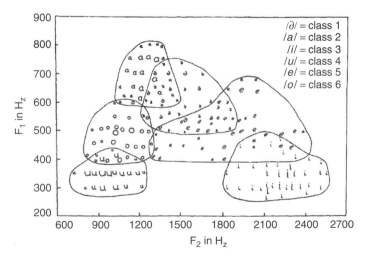

Figure 4.3   Scatter plot of vowel data in $F_1 - F_2$ plane.

4.4.1.3 Waveform Data   The Waveform data consists of three classes of waves with 21 features. Each class of the data set is generated from a combination of base waves. All the features are corrupted with noise (zero mean and unit variance). The data set contains 5000 patterns. The class distribution of the patterns present in the data set is made with 33% for each class.

4.4.1.4 Caldonazzo Data   The Caldonazzo data is obtained from a multispectral scanner satellite image. The data patterns used in this study is a subarea of a scene of 881×928 pixels. Each pixel value contains information from seven spectral bands. The data set contains 3884 patterns with the information from six different land cover classes.

4.4.1.5 Phoneme Data   The aim of this data is to distinguish between two classes, namely, nasal and oral vowels. It contains vowels coming from 1809 isolated syllables. The amplitudes of the five first harmonics are chosen as features to characterize each vowel. The data set has 5404 patterns.

4.4.1.6 Page-Block Data   The problem involved in this data is to classify the blocks of a page layout of a document that has been detected by a segmentation process. This is an essential step in document analysis in order to separate text from graphic areas. The five classes are text, horizontal line, picture, vertical line, and graphic. The page-block data set has 10 features and 5 classes with 5473 patterns.

4.4.1.7 Synthetic Image   A four-band synthetic image of size 512×512 has been generated with six major land cover classes similar to the Indian Remote Sensing Satellite (IRS-1A) image. Figure 4.4a shows the synthesized image in the near-infrared range (band-4). Different classification models are tested on various corrupted versions of the synthetic images. The synthetic image is corrupted with Gaussian noise (zero mean and standard deviation $\sigma = 1, 2, \ldots, 6$) in all four bands. Figure 4.4b shows the noisy version of the original image with $\sigma = 2$ as an example.

4.4.2 Partially Labeled Data Sets

This section provides a brief description of each of the two partially labeled data sets, which are used to evaluate the performance of the RFGS-based pattern classification model.

4.4.2.1 IRS-1A Image   The IRS-1A image of size 512×512 is obtained from IRS (Data Users Hand Book: Document No. IRS/NRSA/NDC/HB-02/89). The image taken from the Linear Imaging Self Scanner with spatial resolution of $36.25 \text{ m} \times 36.25 \text{ m}$ and wavelength range of $0.45-0.86 \text{ }\mu\text{m}$ is used. The whole spectrum range is decomposed into four spectral bands, namely, blue, green, red, and near infrared corresponding to band-1, band-2, band-3, and band-4,

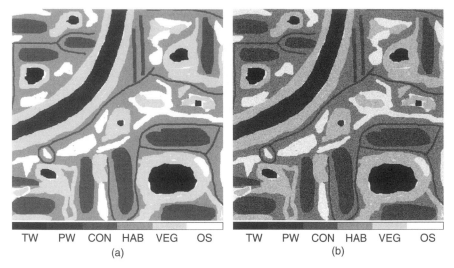

Figure 4.4   Synthetic remote sensing image (band-4). (a) Original and (b) noisy with $\sigma = 2$.

respectively. Since the image is of poor illumination, the enhanced image (band-xs4) is presented in Fig. 4.5a for the convenience of visualizing the content of the image. However, the algorithms are implemented on the original image. The image in Fig. 4.5a covers an area around the city of Kolkata, India, in the near-infrared band having six major land cover classes, namely, pure water (PW), turbid water (TW), concrete area (CA), habitation (HAB), vegetation (VEG), and open spaces (OS).

4.4.2.2   SPOT Image   The Systeme Pour d'Observation de la Terre (SPOT) image of size 512×512 shown in Fig. 4.5b (band-3) is obtained from the SPOT satellite. The image used here has been acquired in the wavelength range of 0.50–0.89 μm. The whole spectrum range is decomposed into three spectral bands, namely, green (band-1), red (band-2), and near infrared (band-3) of wavelengths 0.50–0.59 μm, 0.61–0.68 μm, and 0.79–0.89 μm, respectively. This image has a higher spatial resolution of 20 m $\times$ 20 m. The same six classes are considered as in the case of the IRS-1A image. Also, similar to the IRS-1A image, the classification models are applied on the original image.

4.5   EXPERIMENTAL RESULTS

In this chapter, the performance of the RFGS-based model is compared with that of different combinations of fuzzy granulation and rough feature selection methods. The CD granulation method is also compared with the CI granulation method. For the CI-based granulation, the whole feature space is used for granule

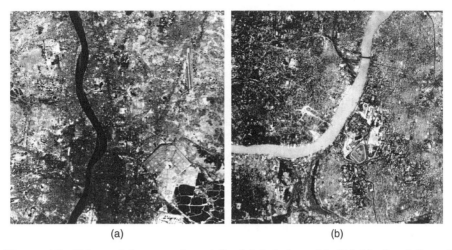

(a)   (b)

Figure 4.5 Enhanced images of partially labeled data. (a) IRS-1A (band-4) and (b) SPOT (band-3).

generation irrespective of the classes. Each feature of the pattern is represented by three fuzzy sets for characterizing three fuzzy granules along each axis, thereby providing 3^m fuzzy granules in an m-dimensional feature space.

Five different combinations of classification models using granular feature space and feature selection methods are considered for performance comparison as mentioned below. Patterns with their original feature representation are fed as input to these models.

1. *Model 1:* K-NN classifier;
2. *Model 2:* CI fuzzy granulation + rough-set-based feature selection + K-NN classifier;
3. *Model 3:* CI fuzzy granulation + neighborhood-rough-set-based feature selection + K-NN classifier;
4. *Model 4:* CD fuzzy granulation + rough-set-based feature selection + K-NN classifier; and
5. *Model 5:* CD fuzzy granulation + neighborhood-rough-set-based feature selection + K-NN classifier (RFGS-based model).

The value of K for the K-NN classifier is considered as 1. Apart from the performance comparison with different quantitative measures for both completely and partially labeled data sets, the efficacy of the RFGS-based model is also justified with the following types of analysis. However, the experimental results with these analyses are provided only for vowel data because a similar trend of comparative performance is observed for the remaining data sets.

- Variation in classification accuracy with different values of parameter Φ and distances used in neighborhood-rough-set-based feature selection for optimal value selection;
- Performance comparison in terms of total computation time (T_c);
- Precision- and recall-based analysis;
- Performance evaluation in terms of dispersion measure of different classes;
- Performance comparison of the RFGS-based model with other classifiers such as K-NN with K $= 3$ and 5, ML classifier, and MLP;
- Performance comparison of models in terms of β and DB indices; and
- Statistical significance test called χ^2-test for the RFGS-based model, which is described next.

4.5.1   Statistical Significance Test

The χ^2 analysis is normally used to test the goodness of fit of an observed distribution to an expected one. The χ^2 is defined as the sum of the squared difference between observed F^o and the expected F^e frequencies of samples, divided by the expected frequencies of samples in all possible categories, and is given by

$$\chi^2 = \sum_{i=1}^{c} \frac{(F_i^o - F_i^e)^2}{F_i^e} \tag{4.20}$$

where c is the number of categories. The step-wise procedure for interpreting the χ^2 value is as follows:

- Calculate the χ^2 value using Equation (4.20).
- Determine the degree of freedom $df = c - 1$.
- Determine the closest probability value p associated with χ^2 and df using χ^2 distribution table, where p denotes the probability that the deviation of the observed from that expected is due to chance alone.

The conclusion can be derived on the basis of the p-value, as follows:

- If the value of $p > 0.05$ for the calculated χ^2, then the significance of the model is accepted, that is, the deviation is small enough and it is because of chance alone. A p-value of 0.45, for example, means that there is a 45% probability that any deviation from expected is due to chance only.
- If the value of $p < 0.05$ for the calculated χ^2, then the significance of the model is not accepted, and it indicates that some factors other than chance are operating or are responsible for the concerned nonacceptable deviation. For example, a p-value of 0.01 means that there is only a 1% chance that this deviation is due to chance alone. Therefore, other factors must be involved.

4.5.2 Class Prediction Methods

The following three classifiers are used to evaluate the performance of different methods.

4.5.2.1 K-Nearest Neighbor Rule The K-NN rule [22] is used to evaluate the effectiveness of the reduced feature set for classification. It classifies samples based on closest training samples in the feature space. A sample is classified by a majority vote of its K-neighbors, with the sample being assigned to the class most common amongst its K-NN.

4.5.2.2 Maximum Likelihood Classifier The ML classifier [22] is one of the most powerful classifier in common use. It is a statistical decision rule that examines the probability function of an object for each of the classes. On the basis of the mean and variance, a probability function is calculated from the inputs for each class, and the object is assigned to the class with the highest probability.

4.5.2.3 Multilayer Perceptron The MLP [23] is a feed-forward neural network trained with the standard back-propagation algorithm. As a supervised network, it requires a desired response to be trained and learns how to transform input data into a desired response. Hence, it is widely used for pattern classification. With one or two hidden layers, it can approximate any input–output mapping. It has been shown to approximate the performance of optimal statistical classifiers in difficult problems. In this work, one hidden layer is used.

4.5.3 Performance on Completely Labeled Data

The selection of training and test samples for all classes, in case of completely labeled data sets including synthetic image, is made after splitting the whole data set into two parts as training and test sets. The 10%, 20%, and 50% data are taken as the training set and the rest 90%, 80%, and 50% are considered as test data. Training sets are selected randomly and an equal percentage of data is collected from each class. Each of these splitting sets is repeated 10 times, and the final results are then averaged over them.

4.5.3.1 Optimum Values of Different Parameters The classification results on a vowel data set with five different models using the K-NN classifier are depicted in Table 4.1 for three different percentages of the training sets. In this experiment, the classification performance of different models is compared with respect to three different aspects. These are performances based on granulated and non-granulated feature space, CD and CI fuzzy granulation, and rough-set-based and neighborhood-rough-set-based feature selection methods.

However, the performance of the neighborhood-rough-set-based feature selection method depends on the distance function Δ and parameter Φ of the neighborhood granules. In this study, the performance of Model 5 (RFGS-based model)

TABLE 4.1 Performance on Vowel Data for Different Training Sets

| Model | 10% Training Set | | | 20% Training Set | | | 50% Training Set | | |
	PA	KC	T_c	PA	KC	T_c	PA	KC	T_c
1	73.24	0.717	0.146	75.15	0.721	0.124	77.56	0.726	0.101
2	76.01	0.750	0.236	77.01	0.770	0.223	79.03	0.772	0.214
3	77.87	0.762	0.263	78.81	0.785	0.252	80.79	0.790	0.244
4	81.03	0.801	0.365	81.37	0.817	0.351	82.11	0.820	0.345
5	83.75	0.810	0.381	83.96	0.825	0.378	84.77	0.830	0.354

is analyzed for the variation of Δ and Φ with 20% training set of vowel data. The classification accuracy is plotted in Fig. 4.6 for three **p**-norm distances for a variation of Φ values ([0, 1]) in Euclidean space. These are Manhattan distance (**p** = 1), Euclidean distance (**p** = 2), and Chebychev distance (**p** = ∞). It is observed from Fig. 4.6 that the accuracy varies with Φ values for all types of distances. With the increase in Φ value, the accuracy increases at first, reaches a peak value, and then decreases. Roughly, for all the distances, the highest accuracy is obtained for $\Phi = [0.3, 0.5]$ with maximum for Euclidean distance. Beyond 0.5, the neighborhood-rough-set-based feature selection method cannot select the relevant features to distinguish patterns. Further, it can be seen that the numbers of selected features are different when Φ takes values in the interval [0.3, 0.5], although these features produce similar classification performance. Hence, the Φ value should be varied in [0.3, 0.5] to find the minimal subset of features with similar classification performance. Accordingly, for presenting the

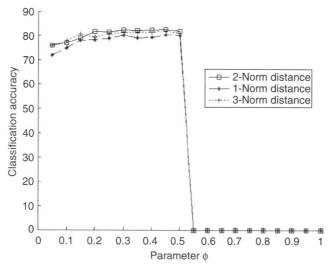

Figure 4.6 Variation of classification accuracy of Model 5 with parameter Φ.

results for the remaining data sets, the values of **p** and Φ are taken as 2 and 0.45, respectively.

4.5.3.2 Accuracy and κ Coefficient on Vowel Data   In a comparative analysis, it is observed from Table 4.1 that for all percentages of training sets the performance of the classifiers measured with percentage of accuracy (PA) value is more encouraging for models using granulated feature space. With 10% of training sets, Model 1 provides the PA value of 73.24, whereas with other models the values are nearly 4%–11% higher. In a comparison between the CD and CI fuzzy-granulation-based models (Table 4.1), the PA values for Models 4 and 5 (CD models) are superior compared to Models 2 and 3 (CI models) with improvement in accuracy by nearly 5%–6%, respectively. This clearly indicates that the CD granules efficiently explore the class-wise belongingness of features to classes and provide an improved class discrimination information responsible for enhanced classification accuracy.

In a performance comparison of models with both neighborhood rough sets and rough sets, it is observed from Table 4.1 that the PA values for Models 5 and 3 are higher compared to Models 4 and 2. This specifies that the neighborhood-rough-set-based feature selection method restores better local information from neighborhood granules that is helpful for improved performance. Hence, comparatively, among the five models of classification, Model 5, which explores and incorporates granular feature space, CD fuzzy granulation, and neighborhood-rough-set-based feature selection method, provides the best performance for all percentages of the training sets as observed in Table 4.1. Further, it is seen that the impact of the CD fuzzy granule generation is more effective compared to neighborhood-rough-set-based feature selection in the classification performance. That is, in an environment of rough set or neighborhood-rough-set-based feature selection methods, classification models with the CD fuzzy granulation provide the PA value increment of about 5%–6% over the CI (Model 4 over Model 2 and Model 5 over Model 3). In an environment of the CD or CI fuzzy-granulation-based methods, classification models with neighborhood-rough-set-based feature selection perform well than rough-set-based method with an increased PA value of about 2% (Model 3 over Model 2 and Model 5 over Model 4). Further, the performance with the combined effect of the CI fuzzy granulation and neighborhood rough sets (Model 3) is not comparable to the CD fuzzy granulation and rough sets (Model 4). These comparisons clearly justify the efficacy of the CD granulation. The superiority of Model 5 (RFGS-based model) to others is also validated with the KC measure, as shown in Table 4.1. The critically assessed improved performance obtained with the PA is justified and supports the superiority claim of the rough-fuzzy granulation and feature selection model with the KC measure.

Table 4.1 also reveals that the accuracy obtained with Model 5 for minimum percentage of training sets is higher compared to the model incorporating the CI fuzzy granulation and both rough-set-based and neighborhood-rough-set-based feature selection methods at 50% training set. This is particularly important when there is a scarcity of training sets such as land covers classification of real remote sensing images. This critically assessed improved performance claim is valid for both 20% and 50% training sets and is depicted in Table 4.1.

A comparative analysis in terms of total computational time T_c, that is, the sum of the training and testing times, as required by different models using K-NN classifier for all three percentages of training sets, is depicted in Table 4.1. All the simulations are done in a MATLAB (Matrix Laboratory) environment in a Pentium IV machine with 3.19 GHz processor speed. It is seen for all the cases that the T_c values (in seconds) for Model 5, as expected, are higher compared to others with the cost of improved performance.

4.5.3.3 Precision and Recall on Vowel Data   In addition to the above comparison of performance with the PA, measures such as precision and recall with the vowel data at 20% training set are calculated for different models. Models 1, 2, and 5 are selected for comparison in Table 4.2 as the first one is based on the nongranulated feature space, the second with the CI fuzzy granulation and rough sets, and third with the best combination of the CD fuzzy granulation and neighborhood rough sets. Hence, a feel of comparison with the models that provide different combinations of results can be obtained. Although the measurements are made for all percentages of training sets, the results shown in Table 4.2 only for 20% training set because the claim for improvement with Model 5 is similar for all training sets. It is observed from Table 4.2 that Model 5 performs better than others in a class-wise agreement comparison for all the classes and with both accuracy measurements.

4.5.3.4 β and DB Indices on Vowel Data   Further, the performance comparison of different models is made with vowel data set using β and DB indices. Here, the classifiers are trained with 20% of the data set and then the said trained classifiers are applied on the whole data to partition into six categories. Results in terms of β and DB values are depicted in Table 4.3, which reveal the superiority of Model 5 to others with respect to all aspects, that is, granulated over nongranulated feature space, CD over CI fuzzy granulation, and rough-set-based and neighborhood-rough-set-based feature selection.

TABLE 4.2   Precision and Recall Values on Vowel Data at 20% Training Set

Class	Model 1		Model 2		Model 5	
	Precision	Recall	Precision	Recall	Precision	Recall
1	66.05	75.28	69.43	78.77	75.68	92.71
2	74.67	77.80	76.62	82.13	83.31	89.56
3	83.08	89.29	86.03	92.79	90.39	96.75
4	80.02	87.86	84.65	93.44	93.39	95.75
5	68.44	78.21	70.89	80.77	76.64	91.01
6	78.93	86.21	82.12	92.23	86.76	93.79

TABLE 4.3 The β and DB Index Values on Vowel Data

Model	β Index	DB Index
Training Samples	6.5421	0.7213
1	4.8463	1.3231
2	4.9901	1.2847
3	5.1033	1.2603
4	5.5664	1.1565
5	5.7352	1.1264

4.5.3.5 Dispersion Score on Vowel Data   The efficiency of different classifiers is also studied in terms of the DS defined in Equation (4.15). Using the DS, the dispersion of classified patterns into various classes is estimated and its physical significance with vowel data set is highlighted. Also, the superiority of Model 5 is analyzed and justified with this measure. Comparative results of the DS of all the classes are depicted in Table 4.4 obtained using five classification models with the K-NN classifier for 20% training set.

Let us consider the results of confusion matrices and the corresponding DS values obtained with Models 3 and 5 for vowel data at 20% training set, as shown in Fig. 4.7. It is observed from the scatter plot of vowel data set (Fig. 4.3) that class 1 (/ə/) is highly overlapping with classes 5 (/e/) and 2 (/a/), and a small overlapping with class 6 (/o/). This is exactly reflected by the confusion matrix of Model 5 (Fig. 4.7b), whereas for Model 3 (Fig. 4.7a) this is not so. Accordingly, the DS value is lower for Model 5, signifying its superiority. The superiority of Models 5 to 3 and other models is similarly observed for all classes except class 4 (/u/), where the DS value of Model 5 is higher than those of Models 3 and 4 (Table 4.4). Therefore, the DS value may be viewed to provide an appropriate quantitative index in evaluating the overlapping characteristics of classes in terms of dispersion of misclassified patterns and to quantify the class-wise performance of classifiers accordingly.

4.5.3.6 Performance Analysis Using Different Classifiers   So far, the effectiveness of the RFGS-based model (Model 5) is described using the K-NN classifier with $K = 1$. In this section, the effectiveness of different models is described

TABLE 4.4 Dispersion Score on Vowel Data at 20% Training Set

Class	Model 1	Model 2	Model 3	Model 4	Model 5
1	0.9013	0.8298	0.8056	0.7846	0.7059
2	0.7134	0.5185	0.6295	0.4398	0.4248
3	0.4325	0.3467	0.2620	0.2601	0.2534
4	0.6132	0.5346	0.4386	0.4321	0.4422
5	0.7343	0.5338	0.5636	0.4979	0.4808
6	0.4262	0.3015	0.2021	0.2135	0.1950

	Class	C_1	C_2	C_3	C_4	C_5	C_6	Dispersion score
				Predicted class				
Actual class	C_1	29	7	0	1	20	7	0.8056
	C_2	10	56	1	0	3	10	0.6295
	C_3	0	0	133	0	21	0	0.2620
	C_4	2	0	0	102	0	31	0.4386
	C_5	13	0	29	1	140	3	0.5636
	C_6	0	0	0	14	2	146	0.2621

(a)

	Class	C_1	C_2	C_3	C_4	C_5	C_6	Dispersion score
				Predicted class				
Actual class	C_1	35	10	0	0	15	4	0.7059
	C_2	8	66	0	0	1	5	0.4248
	C_3	0	0	134	0	20	0	0.2534
	C_4	5	0	0	109	1	20	0.4422
	C_5	9	1	23	0	151	2	0.4908
	C_6	0	0	0	8	1	153	0.1950

(b)

Figure 4.7 Analysis of dispersion measure with confusion matrix for vowel data. (a) Model 3 and (b) Model 5.

using some other classifiers such as the K-NN with $K = 3$ and 5, MLP, and ML classifier.

The comparative results of all models with these classifiers are depicted in Table 4.5 for a training set of 20%, as an example. The superiority of Model 5 to others for different classifiers is evident. Also, a similar improvement in the performance of the models using different classifiers is observed with granulated over nongranulated, CD granulation over CI granulation, and neighborhood-rough-set-based feature selection over rough sets, as in the case of the K-NN classifier with $K = 1$.

4.5.3.7 Statistical Significance Analysis on Vowel Data To strengthen the claim of effectiveness of the RFGS-based model (Model 5) for classification,

TABLE 4.5 Classification Accuracy of Different Classifiers on Vowel Data

Model	K-NN ($K = 3$)	K-NN ($K = 5$)	ML	MLP
1	74.20	73.63	75.21	76.33
2	76.11	76.34	77.34	78.06
3	77.78	77.89	78.34	79.87
4	81.03	80.86	82.25	83.75
5	83.91	84.01	83.87	84.88

TABLE 4.6   Statistical Significance Test on Vowel Data

Model	p-Value
1	$0.10 > p > 0.05$
2	$0.20 > p > 0.10$
5	$0.50 > p > 0.30$

the statistical significance test is performed using χ^2. The comparative results for all models are depicted in Table 4.6 for test set of 80%, as an example. Models 1, 2, and 5 are selected for comparison, as is done for the evaluation in terms of precision and recall measures. It is observed from Table 4.6 that the p-value obtained for Model 5 is in the range of $0.30 < p < 0.50$, whereas it is $0.10 < p < 0.20$ and $0.05 < p < 0.10$, respectively for Models 2 and 1. This observation shows that the significance of Model 5 is much higher compared to Models 1 and 2 and therefore justifies its superiority.

4.5.3.8   Performance of Other Data Sets   Comparative analysis of five classification models is presented in Table 4.7 using the K-NN classifier on different data sets for different training sets and two measures, namely, PA and KC. It is seen that for all the training sets, the models with fuzzy granulated feature space provide greater PA values compared to Model 1, that is, with nongranulated feature space; justifying the use of granular-computing-based methods for improving the performance. The RFGS-based model (Model 5) is seen to be more effective with the CD fuzzy granulation and neighborhood-rough-set-based feature selection method. The improvement of Models 5 and 4 over Models 3 and 2 can be easily seen. Also, for all data sets, Model 5 (combination of the CD fuzzy-granulation-based and neighborhood-rough-set-based feature selection) performs the best. Other findings and issues as discussed in the case of vowel data, such as effect of changing **p** and Φ values, significance of the DS measure, performance with measures such as precision and recall, performance of models using other classifiers, are also found to be true for all these data sets.

4.5.3.9   Synthetic Image   Noisy synthetic remote sensing images with different σ values (Fig. 4.4b) are used to compare the performance of five models using the K-NN classifier, in terms of PA, and the corresponding results are shown in Table 4.8 for 20% training set. The table reveals the superiority of Model 5 to others for all the noise levels. Since a similar trend of observation, as discussed in the case of vowel data, is obtained with other measures for the synthetic remote sensing image, those results are not mentioned here. Figure 4.8 shows the resulting classified images obtained by Models 1 and 5 for the noisy input image with $\sigma = 2$. The superiority of Models 5 to 1, as indicated in Table 4.8, is further verified visually from Fig. 4.8. Here, the classified images obtained from these two models are shown, as an example, because one of them performs the worst and the other performs the best.

TABLE 4.7   Performance Analysis Using K-NN Classifier

Data		10% of Training		20% of Training		50% of Training	
Set	Model	PA	KC	PA	KC	PA	KC
	1	73.07	0.662	75.15	0.711	77.23	0.716
	2	77.05	0.699	78.06	0.735	79.27	0.738
Satimage	3	78.82	0.709	79.65	0.740	80.76	0.741
	4	81.67	0.782	81.98	0.787	82.07	0.789
	5	83.14	0.791	83.68	0.793	83.95	0.799
	1	72.89	0.616	74.46	0.630	76.95	0.668
	2	77.18	0.699	78.13	0.711	78.97	0.712
Waveform	3	79.30	0.701	79.84	0.711	80.02	0.714
	4	81.46	0.749	81.98	0.751	82.11	0.752
	5	83.87	0.753	84.21	0.755	84.95	0.758
	1	72.55	0.540	73.78	0.569	77.05	0.617
	2	74.01	0.601	76.16	0.611	78.18	0.668
Caldonazzo	3	77.71	0.612	78.65	0.623	79.91	0.671
	4	80.65	0.650	81.14	0.669	82.45	0.720
	5	82.18	0.665	82.71	0.682	83.91	0.732
	1	76.45	0.450	79.12	0.468	80.02	0.500
	2	78.37	0.498	79.76	0.521	80.11	0.562
Phoneme	3	80.01	0.542	80.54	0.563	80.98	0.579
	4	81.72	0.589	82.03	0.598	82.37	0.610
	5	83.03	0.597	83.89	0.614	84.11	0.657
	1	86.02	0.588	86.55	0.571	88.90	0.603
	2	88.39	0.612	88.88	0.624	90.03	0.647
Page-block	3	90.76	0.644	91.01	0.653	92.22	0.678
	4	91.85	0.653	92.73	0.668	93.15	0.701
	5	93.89	0.669	94.65	0.710	95.89	0.738

4.5.4   Performance on Partially Labeled Data

This section presents the performance of Model 5 for the classification of two partially labeled data, namely, IRS-1A and SPOT images. The classifiers are initially trained with labeled data of six land cover types and then the said trained classifiers are applied on the unlabeled image data to partition into six regions.

Both IRS-1A and SPOT images are classified with five different models using the K-NN classifier, and the performance comparison in terms of β and DB index values is shown in Table 4.9. As expected, the β value is the highest and DB value is the lowest for the training set (Table 4.9). It is also seen that Model 5 yields superior results in terms of both indices compared to other four models. As a whole, the gradation in the performance of the five models can be established with the following β relation:

$$\beta_{\text{Training}} > \beta_{\text{Model 5}} > \beta_{\text{Model 4}} > \beta_{\text{Model 3}} > \beta_{\text{Model 2}} > \beta_{\text{Model 1}}. \qquad (4.21)$$

TABLE 4.8 Classification Accuracy of Different Models on Synthetic Image

| Model | Classification Accuracy | | | |
	$\sigma = 1$	$\sigma = 2$	$\sigma = 3$	$\sigma = 4$
1	95.72	82.03	72.17	60.77
2	97.86	91.10	77.23	63.01
3	98.12	92.23	80.23	67.83
4	98.83	94.34	82.36	69.36
5	99.54	95.87	84.41	71.44

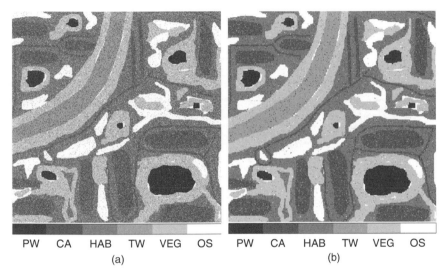

PW   CA   HAB   TW   VEG   OS   PW   CA   HAB   TW   VEG   OS

(a)   (b)

Figure 4.8 Classified synthetic image for $\sigma = 2$. (a) Model 1 and (b) Model 5.

A similar gradation in performance is also observed with the DB values, which further supports the superiority of Model 5, that is, the RFGS-based model.

In order to demonstrate the significance of granular computing visually, let us consider Fig. 4.9 depicting the output corresponding to Model 1 (without granulation) and Model 5 (with granulation). It is clear from the figure that Model 5 performs well in segregating different areas by properly classifying the land covers. A zoomed version of the Howrah bridge region is shown in Fig. 4.10 to have an improved visualization. Similarly, the regions such as the Salt Lake Stadium and water bodies are more distinct and well shaped with Model 5 as shown in Fig. 4.10. These observations further justify the significance of the β and DB indices in reflecting the performance of the models automatically without visual intervention.

The significance of Model 5 is further justified visually from Fig. 4.11 that illustrates the classified SPOT images corresponding to Models 1 and 5.

TABLE 4.9   Performance of Different Models on Partially Labeled Data

Model	β Index		DB Index	
	IRS-1A	SPOT	IRS-1A	SPOT
Training samples	9.4212	9.3343	0.5571	1.4893
1	6.8602	6.8745	0.9546	3.5146
2	7.1343	7.2301	0.9126	3.3413
3	7.3559	7.3407	0.8731	3.2078
4	8.1372	8.2166	0.7790	2.8897
5	8.4162	8.4715	0.7345	2.7338

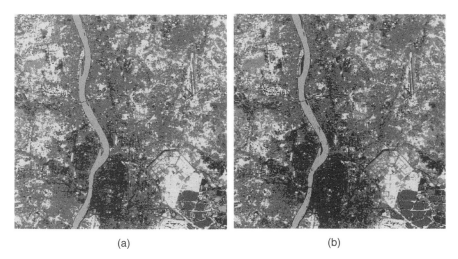

(a)   (b)

Figure 4.9   Classified IRS-1A images. (a) Model 1 and (b) Model 5.

Figure 4.11b is superior to Fig. 4.11a in the sense that the different structures such as roads and canals are more prominent.

4.6   CONCLUSION AND DISCUSSION

This chapter describes a rough-fuzzy model for pattern classification. The model formulates a CD fuzzy granulation of input feature space, where the membership functions explore the degree of belonging of features into different classes and make it more suitable for improved class label estimation. The advantage of neighborhood rough sets that deal with both numerical and categorical data without any discretization is also realized in the current model. The neighboring concept facilitates gathering the local or contextual information through neighborhood granules that provide improved class discrimination information. The dispersion measure for classifiers' performance reflects well the overlapping characteristics of a class with others and can be viewed as an appropriate index in

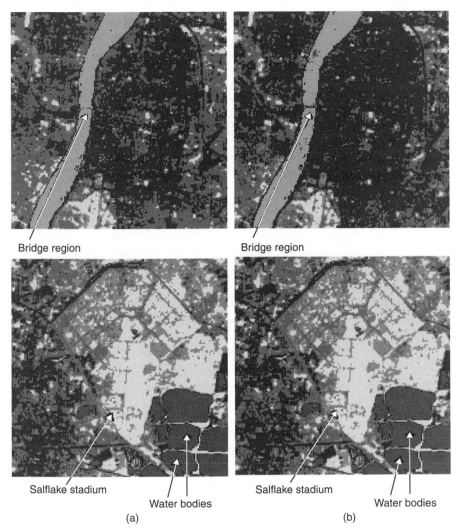

Bridge region   Bridge region

Salflake stadium   Salflake stadium

Water bodies   Water bodies

(a)   (b)

Figure 4.10 Zoomed version of two selected regions of classified IRS-1A image. (a) Model 1 and (b) Model 5.

evaluating the class-wise performance of a classifier. It may be mentioned here that fuzzy granulation of feature space described in Pal and Mitra [30] for case generation is similar to the method of CI granulation used here.

With extensive experimental results on various types of real life as well as synthetic data, both fully and partially labeled, it is found that the effect of CD fuzzy granulation is substantial compared to the rough feature selection methods in improving classification performance, and the combined effect is further encouraging for the data sets with highly overlapping classes. The statistical significance of the current model is also supported by the χ^2 test. The computational

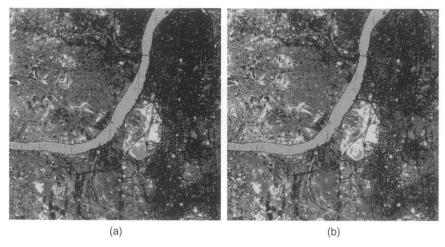

(a)   (b)

Figure 4.11 Classified SPOT images. (a) Model 1 and (b) Model 5.

complexity of this model is a little high. However, its learning ability with a small percentage of training samples makes it practicably applicable to problems with a large number of overlapping classes.

In this chapter and the previous one, we have discussed about the classification (both supervised and unsupervised) aspect of a pattern recognition system. In the next chapter, we address another task, namely, feature selection using fuzzy-rough sets.

REFERENCES

1. A. Bargiela and W. Pedrycz. *Granular Computing: An Introduction*. Kluwer Academic Publishers, Boston, MA, 2003.

2. A. Bargiela, W. Pedrycz, and K. Hirota. Granular Prototyping in Fuzzy Clustering. *IEEE Transactions on Fuzzy Systems*, 12(5):697–709, 2004.

3. A. Skowron and J. F. Peters. Rough-Granular Computing. In W. Pedrycz, A. Skowron, and V. Kreinovich, editors, *Handbook of Granular Computing*, pages 285–328. John Wiley & Sons, Ltd., West Sussex, 2008.

4. W. Pedrycz, A. Skowron, and V. Kreinovich, editors. *Handbook of Granular Computing*. John Wiley & Sons, Ltd., West Sussex, 2008.

5. J. F. Peters, Z. Pawlak, and A. Skowron. A Rough Set Approach to Measuring Information Granules. In *Proceedings of the Annual International Conference on Computer Software and Applications*, Oxford, England, pages 1355–1360, 2002.

6. L. A. Zadeh. Fuzzy Sets and Information Granularity. In M. Gupta, R. Ragade, and R. Yager, editors, *Advances in Fuzzy Set Theory and Applications*, pages 3–18. North-Holland Publishing Co., Amsterdam, The Netherlands, 1979.

7. L. A. Zadeh. Toward a Theory of Fuzzy Information Granulation and Its Centrality in Human Reasoning and Fuzzy Logic. *Fuzzy Sets and Systems*, 90:111–127, 1997.

8. A. Bargiela and W. Pedrycz. Toward a Theory of Granular Computing for Human-Centered Information Processing. *IEEE Transactions on Fuzzy Systems*, 16(2):320–330, 2008.

9. S. K. Pal and P. Mitra. *Pattern Recognition Algorithms for Data Mining*. CRC Press, Boca Raton, FL, 2004.

10. S. K. Pal and A. Skowron, editors. *Rough-Fuzzy Hybridization: A New Trend in Decision Making*. Springer-Verlag, Singapore, 1999.

11. Y. Tang, Y.-Q. Zhang, Z. Huang, X. Hu, and Y. Zhao. Recursive Fuzzy Granulation for Gene Subsets Extraction and Cancer Classification. *IEEE Transactions on Information Technology in Biomedicine*, 12(6):723–730, 2008.

12. Y.-Q. Zhang. Constructive Granular Systems with Universal Approximation and Fast Knowledge Discovery. *IEEE Transactions on Fuzzy Systems*, 13(1):48–57, 2005.

13. S. K. Pal and S. Mitra. Multilayer Perceptron, Fuzzy Sets, and Classification. *IEEE Transactions on Neural Networks*, 3:683–697, 1992.

14. Z. Pawlak. *Rough Sets: Theoretical Aspects of Resoning About Data*. Kluwer, Dordrecht, The Netherlands, 1991.

15. V. S. Ananthanarayana, M. N. Murty, and D. K. Subramanian. Tree Structure for Efficient Data Mining Using Rough Sets. *Pattern Recognition Letters*, 24(6):851–862, 2003.

16. A. Roy and S. K. Pal. Fuzzy Discretization of Feature Space for A Rough Set Classifier. *Pattern Recognition Letters*, 24(6):895–902, 2003.

17. R. Nowicki. On Combining Neuro-Fuzzy Architectures with the Rough Set Theory to Solve Classification Problems with Incomplete Data. *IEEE Transactions on Knowledge and Data Engineering*, 20(9):1239–1253, 2008.

18. S. Zhao, E. C. C. Tsang, D. Chen, and X. Wang. Building a Rule-Based Classifier: A Fuzzy-Rough Set Approach. *IEEE Transactions on Knowledge and Data Engineering*, 22(5):624–638, 2010.

19. Q. Hu, D. Yu, J. Liu, and C. Wu. Neighborhood Rough Set Based Heterogeneous Feature Subset Selection. *Information Sciences*, 178:3577–3594, 2008.

20. T. Y. Lin. Granulation and Nearest Neighborhoods: Rough Set Approach. In W. Pedrycz, editor, *Granular Computing: An Emerging Paradigm*, pages 125–142. Physica-Verlag, Heidelberg, Germany, 2001.

21. S. K. Pal, S. Meher, and S. Dutta. Class-Dependent Rough-Fuzzy Granular Space, Dispersion Index and Classification. *Pattern Recognition*, under revision.

22. R. O. Duda, P. E. Hart, and D. G. Stork. *Pattern Classification and Scene Analysis*. John Wiley & Sons, Inc., New York, 1999.

23. S. Haykin. *Neural Networks: A Comprehensive Foundation*. 2nd edition, Prentice Hall, Upper Saddle River, NJ, 1998.

24. J. Cohen. A Coefficient of Agreement for Nominal Scale. *Educational and Psychological Measurement*, 20:37–46, 1960.

25. S. K. Pal, A. Ghosh, and B. U. Shankar. Segmentation of Remotely Sensed Images with Fuzzy Thresholding and Quantitative Evaluation. *International Journal of Remote Sensing*, 21(11):2269–2300, 2000.

26. D. L. Davies and D. W. Bouldin. A Cluster Separation Measure. *IEEE Transactions on Pattern Analysis and Machine Intelligence*, 1:224–227, 1979.

27. S. K. Pal and D. D. Majumder. *Fuzzy Mathematical Approach to Pattern Recognition*. John Wiley, Halsted Press, New York, 1986.

28. A. Chouchoulas and Q. Shen. Rough Set-Aided Keyword Reduction for Text Categorisation. *Applied Artificial Intelligence*, 15(9):843–873, 2001.

29. R. G. Congalton and R. A. Mead. A Quantitative Method to Test for Consistency and Correctness in Photointerpretation. *Photogrammetric Engineering and Remote Sensing*, 49:69–74, 1983.

30. S. K. Pal and P. Mitra. Case Generation Using Rough Sets with Fuzzy Representation. *IEEE Transactions on Knowledge and Data Engineering*, 16:293–300, 2004.

5

FUZZY-ROUGH FEATURE SELECTION USING f-INFORMATION MEASURES

5.1 INTRODUCTION

Feature selection or dimensionality reduction of a data set is an essential pre-processing step used for pattern recognition, data mining, and machine learning. It is an important problem related to mining large data sets, both in dimension and size. Before the analysis of the data set, preprocessing the data to obtain a smaller set of representative features and retaining the optimal salient character-istics of the data not only decrease the processing time but also lead to more compactness of the models learned and better generalization. Hence, the general criterion for reducing the dimension is to preserve the most relevant information of the original data according to some optimality criteria.

Conventional methods of feature selection involve evaluating different fea-ture subsets using some indices and selecting the best among them. Depending on the way of computing the feature evaluation index, feature selection meth-ods are generally divided into two broad categories: filter approach [1, 2] and wrapper approach [1, 3]. In the filter approach, the algorithms do not perform classification of the data in the process of feature evaluation. Before application of the actual learning algorithm, the best subset of features is selected in one pass by evaluating some predefined criteria, which are independent of the actual generalization performance of the learning machine. Hence, the filter approach is computationally less expensive and more general [1, 2].

Rough-Fuzzy Pattern Recognition: Applications in Bioinformatics and Medical Imaging,
First Edition. Pradipta Maji and Sankar K. Pal.
© 2012 John Wiley & Sons, Inc. Published 2012 by John Wiley & Sons, Inc.

On the other hand, in its most general formulation, the wrapper approach consists of using the prediction performance of a given learning machine to assess the relative usefulness of different subsets of features. Since the wrapper approach uses the learning machine as a black box, it generally outperforms the filter approach in the aspect of final predictive accuracy of the learning machine. However, it is computationally more expensive than the filter approach [1, 3]. An efficient but less universal version of the wrapper approach is the embedded method, which performs feature selection in the process of training and is usually specific to the given learning machine. However, the embedded approach is more intricate and limited to a specific learning machine [4–7].

In the feature selection process, an optimal feature subset is always relative to a certain criterion. In general, different criteria may lead to different optimal feature subsets. However, every criterion tries to measure the discriminating ability of a feature or a subset of features to distinguish the different class labels. While the distance measure is a traditional discrimination or divergence measure, the dependence or correlation measure is mainly utilized to find the correlation between two features or a feature and a class [8]. As these two measures depend on the actual values of the training data, they are very much sensitive to the noise or outlier of the data set. On the other hand, information measures such as the entropy and mutual information [9, 10] compute the amount of information or the uncertainty of a feature for classification. As the information measure depends only on the probability distribution of a random variable rather than on its actual values, it has been widely used in feature selection [5–7, 9, 10].

Information measures are defined as the measures of the distance between a joint probability distribution and the product of the marginal distributions [11]. They constitute a subclass of the divergence measures, which are measures of the distance between two arbitrary distributions. A specific class of information (i.e., divergence) measures, of which mutual information is a member, is formed by the f-information (i.e., f-divergence) measures [11, 12]. Several f-information measures have been used in medical image registration [12], feature selection of real-valued data sets [13], and gene selection [14, 15] problems, and shown to yield significantly more accurate results than mutual information.

The rough set theory is a new paradigm to deal with uncertainty, vagueness, and incompleteness. It is proposed for indiscernibility in classification according to some similarity. The rough set theory has been applied successfully to feature selection of discrete valued data [16–18]. Given a data set with discretized attribute values, it is possible to find a subset of the original attributes using rough set theory that are the most informative; all other attributes can be removed from the data set with minimal information loss. From the dimensionality reduction perspective, informative features are those that are most useful in determining classifications from their values [19, 20].

One of the popular rough-set-based feature selection algorithms is the quick reduct (QR) algorithm [21, 22] in which the dependency or quality of approximation of a single attribute is first calculated with respect to the class labels or decision attribute. After selecting the best attribute, other attributes are added to it to produce better quality. Additions of attributes are stopped when the final subset of attributes has the same quality as that of maximum possible quality of the data set or the quality of the selected attributes remains the same. Other notable algorithms include the discernibility-matrix-based method [23, 24] and dynamic reducts [25]. Recently, a distance-measure-based approach was reported in Parthalain et al. [26] to explore the rough set boundary region for feature selection. However, all these approaches are computationally very costly. Different heuristic approaches based on the rough set theory are also developed for feature selection [27, 28]. Combining rough sets and genetic algorithms, different algorithms have been proposed [29–31] to discover an optimal or a close-to-optimal subset of features.

However, there are usually real-valued data and fuzzy information in real-world applications. Combining fuzzy sets and rough sets provides an important direction in reasoning with uncertainty for real-valued data sets. Both fuzzy sets and rough sets provide a mathematical framework to capture uncertainties associated with the data [32, 33]. They are complementary in some aspects. The generalized theories of rough-fuzzy sets and fuzzy-rough sets have been applied successfully to feature selection of real-valued data [34–38]. Both fuzzy-rough sets [39, 40] and neighborhood rough sets [41, 42] can handle continuous-valued attributes or features without any discretization.

Jensen and Shen [34, 35] introduced the fuzzy-rough QR algorithm for feature selection or dimensionality reduction of real-valued data sets. Hu et al. [43] have used the concept of fuzzy equivalence relation matrix to compute entropy and mutual information in fuzzy approximation spaces, which can be used for feature selection of real-valued data sets. However, many useful information measures such as several f-information measures cannot be computed from the fuzzy equivalence relation matrix introduced in Hu et al. [43] as it does not provide a way to compute marginal and joint distributions directly. Also, the fuzzy-rough-set-based feature selection methods proposed in Jensen and Shen [34, 35] and Hu et al. [43] select the relevant or predictive features of a data set without considering the redundancy among them. However, there exist a number of feature selection algorithms that group correlated features to reduce the redundancy among selected features [44–50].

Recently, Maji and Pal [13] described a feature selection method that employs fuzzy-rough sets to provide a means by which discrete or real-valued noisy data, or a mixture of both, can be effectively reduced without the need for user-specified information. Moreover, the method can be applied to data with continuous or nominal decision attributes, and can be applied to regression as well as classification of data sets. The method selects a subset of features from the whole feature set by maximizing the relevance and minimizing the redundancy of the selected features. The relevance and redundancy of the features are calculated using the f-information measures in fuzzy approximation spaces. Using the concept of fuzzy equivalence partition matrix (FEPM), the f-information measures are calculated for both condition and decision attributes. Hence, the only information required in the current feature selection method is in the form of fuzzy partitions for each attribute, which can be automatically derived from the given data set. This avoids the need for domain experts to provide information on the data involved and ties in with the advantage of rough sets in that it requires no information other than the data set itself.

The structure of the rest of this chapter is as follows: Section 5.2 briefly introduces the necessary notions of fuzzy-rough sets. In Section 5.3, the formulae of Shannon's entropy are reported for fuzzy approximation spaces with an FEPM. The f-information measures for fuzzy approximation spaces are presented in Section 5.4. The feature selection method [13] based on f-information measures for fuzzy approximation spaces is described in Section 5.5. Several quantitative measures are reported in Section 5.6 to evaluate the performance of different fuzzy-rough-set-based feature selection methods. A few case studies and a comparison with other methods are presented in Section 5.7. Concluding remarks are given in Section 5.8.

5.2   FUZZY-ROUGH SETS

As mentioned in Section 2.5 of Chapter 2, a crisp equivalence relation induces a crisp partition of the universe and generates a family of crisp equivalence classes. Correspondingly, a fuzzy equivalence relation generates a fuzzy partition of the universe and a series of fuzzy equivalence classes, which are also called *fuzzy knowledge granules* [32, 33, 35]. This means that the decision attributes and the condition attributes may all be fuzzy. The family of normal fuzzy sets produced by a fuzzy partitioning of the universe of discourse can play the role of fuzzy equivalence classes [33].

Let $<\mathbb{U}, \mathbb{A}>$ represent a fuzzy approximation space and X a fuzzy subset of $\mathbb{U}$. The fuzzy $\mathbb{P}$-lower and $\mathbb{P}$-upper approximations are then defined as follows [32, 33]:

$$\mu_{\underline{\mathbb{P}}X}(F_i) = \inf_x\{\max\{(1 - \mu_{F_i}(x)), \mu_X(x)\}\} \ \forall i \tag{5.1}$$

$$\mu_{\overline{\mathbb{P}}X}(F_i) = \sup_x\{\min\{\mu_{F_i}(x), \mu_X(x)\}\} \ \forall i, \tag{5.2}$$

where F_i represents a fuzzy equivalence class belonging to $\mathbb{U}/\mathbb{P}$ and $\mu_X(x)$ represents the membership of x in X. Note that although the universe of discourse in feature selection is finite, this is not the case in general, and hence the use of sup and inf. These definitions diverge a little from the crisp upper and lower approximations, as the memberships of individual objects to the approximations are not explicitly available. As a result of this, the fuzzy lower and upper approximations can be defined as [35]

$$\mu_{\underline{\mathbb{P}}X}(x) = \sup_{F_i \in \mathbb{U}/\mathbb{P}} \min\{\mu_{F_i}(x), \mu_{\underline{\mathbb{P}}X}(F_i)\}, \tag{5.3}$$

$$\mu_{\overline{\mathbb{P}}X}(x) = \sup_{F_i \in \mathbb{U}/\mathbb{P}} \min\{\mu_{F_i}(x), \mu_{\overline{\mathbb{P}}X}(F_i)\}. \tag{5.4}$$

The tuple $<\underline{\mathbb{P}}X, \overline{\mathbb{P}}X>$ is called a *fuzzy-rough set*. This definition degenerates to traditional rough sets when all equivalence classes are crisp. The membership of an object $x \in \mathbb{U}$, belonging to the fuzzy positive region is

$$\mu_{POS_{\mathbb{C}}(\mathbb{D})}(x) = \sup_{X \in \mathbb{U}/\mathbb{D}} \mu_{\underline{\mathbb{C}}X}(x), \tag{5.5}$$

where $\mathbb{A} = \mathbb{C} \cup \mathbb{D}$. Using the definition of the fuzzy positive region, the dependency function can be defined as follows [35]:

$$\gamma_{\mathbb{C}}(\mathbb{D}) = \frac{|\mu_{POS_{\mathbb{C}}(\mathbb{D})}(x)|}{|\mathbb{U}|} = \frac{1}{|\mathbb{U}|} \sum_{x \in \mathbb{U}} \mu_{POS_{\mathbb{C}}(\mathbb{D})}(x). \tag{5.6}$$

5.3 INFORMATION MEASURE ON FUZZY APPROXIMATION SPACES

Knowledge is thought to be the discernibility power of the attributes in the framework of rough set methodology. An attribute set forms an equivalence relation, and correspondingly generates a partition of the universe and a family of concepts. The quantity of knowledge measures the fineness degree of the partition. The finer the partition is, the more knowledge about the universe we have, and accordingly a finer approximation we will have. In this section, Shannon's information measure [51] is presented to compute the knowledge quantity of a fuzzy attribute set or a fuzzy partition of $\mathbb{U}$.

Shannon's information entropy [51] just works in the case where a crisp equivalence relation or a crisp partition is defined. That is, it is suitable for Pawlak's approximation space [16]. In this section, a formula to compute Shannon's entropy with an FEPM is presented, which can be used to compute the information measures on fuzzy approximation spaces [13].

5.3.1 Fuzzy Equivalence Partition Matrix and Entropy

Given a finite set $\mathbb{U}$, $\mathbb{A}$ is a fuzzy attribute set in $\mathbb{U}$, which generates a fuzzy equivalence partition on $\mathbb{U}$. If c denotes the number of fuzzy equivalence classes generated by the fuzzy equivalence relation and n is the number of objects in $\mathbb{U}$, then c-partitions of $\mathbb{U}$ are the sets of (cn) values $\{m_{ij}^{\mathbb{A}}\}$ that can be conveniently arrayed as a $(c \times n)$ matrix $\mathbb{M}_{\mathbb{A}} = [m_{ij}^{\mathbb{A}}]$. The matrix $\mathbb{M}_{\mathbb{A}}$ is termed as the *fuzzy equivalence partition matrix* and is denoted by

$$\mathbb{M}_{\mathbb{A}} = \begin{pmatrix} m_{11}^{\mathbb{A}} & m_{12}^{\mathbb{A}} & \cdots & m_{1n}^{\mathbb{A}} \\ m_{21}^{\mathbb{A}} & m_{22}^{\mathbb{A}} & \cdots & m_{2n}^{\mathbb{A}} \\ \cdots & \cdots & \cdots & \cdots \\ m_{c1}^{\mathbb{A}} & m_{c2}^{\mathbb{A}} & \cdots & m_{cn}^{\mathbb{A}} \end{pmatrix} \tag{5.7}$$

subject to $\displaystyle\sum_{i=1}^{c} m_{ij}^{\mathbb{A}} = 1, \forall j$, and for any value of i, if

$$k = \arg \max_{j}\{m_{ij}^{\mathbb{A}}\}, \text{ then } \max_{j}\{m_{ij}^{\mathbb{A}}\} = \max_{l}\{m_{lk}^{\mathbb{A}}\} > 0,$$

where $m_{ij}^{\mathbb{A}} \in [0, 1]$ represents the membership of object x_j in the ith fuzzy equivalence partition or class F_i. The above axioms should hold for every fuzzy equivalence partition, which correspond to the requirement that an equivalence class is nonempty. Obviously, this definition degenerates to the normal definition of equivalence classes when the equivalence relation is nonfuzzy.

Using the concept of FEPM, the dependency between a condition attribute set $\mathbb{C}$ and a decision attribute set $\mathbb{D}$ can be redefined as follows:

$$\gamma_{\mathbb{C}}(\mathbb{D}) = \frac{1}{n}\sum_{j=1}^{n} \kappa_j, \tag{5.8}$$

where $\mathbb{C} \cup \mathbb{D} = \mathbb{A}$ and

$$\kappa_j = \sup_{k}\{\sup_{i}\{\min\{m_{ij}^{\mathbb{C}}, \inf_{l}\{\max\{1 - m_{il}^{\mathbb{C}}, m_{kl}^{\mathbb{D}}\}\}\}\}\}. \tag{5.9}$$

A $c \times n$ FEPM $\mathbb{M}_{\mathbb{A}}$ represents the c-fuzzy equivalence partitions of the universe generated by a fuzzy equivalence relation. Each row of the matrix $\mathbb{M}_{\mathbb{A}}$ is a fuzzy equivalence partition or class. The ith fuzzy equivalence partition is, therefore, given by

$$F_i = \{m_{i1}^{\mathbb{A}}/x_1 + m_{i2}^{\mathbb{A}}/x_2 + \cdots + m_{in}^{\mathbb{A}}/x_n\}. \tag{5.10}$$

As to a fuzzy partition induced by a fuzzy equivalence relation, the equivalence class is a fuzzy set. "$+$" means the operator of union in this case. The cardinality of the fuzzy set F_i can be calculated with

$$|F_i| = \sum_{j=1}^{n} m_{ij}^{\mathbb{A}}, \tag{5.11}$$

which appears to be a natural generalization of the crisp set. The information quantity of a fuzzy attribute set $\mathbb{A}$ or fuzzy equivalence partition is then defined as

$$H(\mathbb{A}) = -\sum_{i=1}^{c} \lambda_{F_i} \log \lambda_{F_i}, \tag{5.12}$$

where $\lambda_{F_i} = |F_i|/n$, called a *fuzzy relative frequency* and c is the number of fuzzy equivalence partitions or classes. The measure $H(\mathbb{A})$ has the same form as

Shannon's entropy [51]. The information quantity or the entropy value increases monotonously with the discernibility power of the fuzzy attributes. Combining Equations (5.11) and (5.12), the form of Shannon's entropy, in terms of FEPM, on fuzzy approximation spaces is given by

$$H(\mathbb{A}) = -\sum_{i=1}^{c}\left[\frac{1}{n}\sum_{k=1}^{n}m_{ik}^{\mathbb{A}}\right]\log\left[\frac{1}{n}\sum_{k=1}^{n}m_{ik}^{\mathbb{A}}\right]. \tag{5.13}$$

5.3.2   Mutual Information

Given $< \mathbb{U}, \mathbb{A} >$, $\mathbb{P}$ and $\mathbb{Q}$ are two subsets of $\mathbb{A}$. The information quantity corresponding to $\mathbb{P}$ and $\mathbb{Q}$ are given by

$$H(\mathbb{P}) = -\sum_{i=1}^{p}\lambda_{P_i}\log\lambda_{P_i}, \tag{5.14}$$

$$H(\mathbb{Q}) = -\sum_{j=1}^{q}\lambda_{Q_j}\log\lambda_{Q_j}, \tag{5.15}$$

where p and q are the number of fuzzy equivalence partitions or classes generated by the fuzzy attribute sets $\mathbb{P}$ and $\mathbb{Q}$, respectively, and P_i and Q_j represent the corresponding ith and jth fuzzy equivalence partitions. The joint entropy of $\mathbb{P}$ and $\mathbb{Q}$ can be defined as follows:

$$H(\mathbb{PQ}) = -\sum_{k=1}^{r}\lambda_{R_k}\log\lambda_{R_k}, \tag{5.16}$$

where r is the number of resultant fuzzy equivalence partitions, R_k is the corresponding kth equivalence partition, and λ_{R_k} is the joint frequency of P_i and Q_j, which is given by

$$\lambda_{R_k} = \lambda_{P_iQ_j} = \frac{|P_i \cap Q_j|}{n}, \quad \text{where } k = (i-1)q + j. \tag{5.17}$$

Hence, the joint frequency λ_{R_k} can be calculated from the $r \times n$ FEPM $\mathbb{M}_{\mathbb{PQ}}$, where

$$\mathbb{M}_{\mathbb{PQ}} = \mathbb{M}_{\mathbb{P}} \cap \mathbb{M}_{\mathbb{Q}} \text{ and } m_{kl}^{\mathbb{PQ}} = m_{il}^{\mathbb{P}} \cap m_{jl}^{\mathbb{Q}}. \tag{5.18}$$

Similarly, the conditional entropy of $\mathbb{P}$ conditioned to $\mathbb{Q}$ is defined as

$$H(\mathbb{P}|\mathbb{Q}) = -\sum_{i=1}^{p}\sum_{j=1}^{q}\frac{|P_i \cap Q_j|}{n}\log\frac{|P_i \cap Q_j|}{|Q_j|} \tag{5.19}$$

$$= -\sum_{i=1}^{p}\sum_{j=1}^{q}\left\{\frac{|P_i \cap Q_j|}{n}\log\frac{|P_i \cap Q_j|}{n} - \frac{|P_i \cap Q_j|}{n}\log\frac{|Q_j|}{n}\right\}$$

$$= -\left\{\sum_{i=1}^{p}\sum_{j=1}^{q}\frac{|P_i \cap Q_j|}{n}\log\frac{|P_i \cap Q_j|}{n} - \sum_{j=1}^{q}\frac{|Q_j|}{n}\log\frac{|Q_j|}{n}\right\}.$$

That is, the conditional entropy of $\mathbb{P}$ conditioned to $\mathbb{Q}$ is

$$H(\mathbb{P}|\mathbb{Q}) = -\sum_{k=1}^{r}\lambda_{R_k}\log\lambda_{R_k} + \sum_{j=1}^{q}\lambda_{Q_j}\log\lambda_{Q_j}, \qquad (5.20)$$

where

$$\sum_{i=1}^{p}\sum_{j=1}^{q}\frac{|P_i \cap Q_j|}{n} = \sum_{j=1}^{q}\frac{|Q_j|}{n} \text{ and } \lambda_{Q_j} = \frac{|Q_j|}{n}.$$

Thus,

$$H(\mathbb{P}|\mathbb{Q}) = H(\mathbb{P}\mathbb{Q}) - H(\mathbb{Q}). \qquad (5.21)$$

Hence, the mutual information between two fuzzy attribute sets $\mathbb{P}$ and $\mathbb{Q}$ is given by

$$I(\mathbb{P}\mathbb{Q}) = H(\mathbb{P}) - H(\mathbb{P}|\mathbb{Q}) = H(\mathbb{P}) + H(\mathbb{Q}) - H(\mathbb{P}\mathbb{Q}). \qquad (5.22)$$

Combining Equations (5.11), (5.18), and (5.22), the mutual information between two fuzzy attribute sets $\mathbb{P}$ and $\mathbb{Q}$, in terms of FEPM, can be represented as

$$
\begin{aligned}
I(\mathbb{P}, \mathbb{Q}) = &-\sum_{i=1}^{p}\left[\frac{1}{n}\sum_{k=1}^{n}m_{ik}^{\mathbb{P}}\right]\log\left[\frac{1}{n}\sum_{k=1}^{n}m_{ik}^{\mathbb{P}}\right]\\
&-\sum_{j=1}^{q}\left[\frac{1}{n}\sum_{k=1}^{n}m_{jk}^{\mathbb{Q}}\right]\log\left[\frac{1}{n}\sum_{k=1}^{n}m_{jk}^{\mathbb{Q}}\right]\\
&+\sum_{i=1}^{p}\sum_{j=1}^{q}\left[\frac{1}{n}\sum_{k=1}^{n}(m_{ik}^{\mathbb{P}} \cap m_{jk}^{\mathbb{Q}})\right]\log\left[\frac{1}{n}\sum_{k=1}^{n}(m_{ik}^{\mathbb{P}} \cap m_{jk}^{\mathbb{Q}})\right]. \qquad (5.23)
\end{aligned}
$$

The mutual information $I(\mathbb{P}\mathbb{Q})$ between two fuzzy attribute sets $\mathbb{P}$ and $\mathbb{Q}$ quantifies the information shared by both of them. If $\mathbb{P}$ and $\mathbb{Q}$ do not share much information, the value of $I(\mathbb{P}\mathbb{Q})$ between them is small. While two highly nonlinearly correlated attribute sets will demonstrate a high mutual information value. The attribute sets can be both the condition attributes and the decision attributes in this study. The necessity for a fuzzy condition attribute to be an independent and informative feature can, therefore, be determined by the shared information between this attribute and the rest as well as the shared information between this attribute and the decision attribute or class labels.

5.4 f-INFORMATION AND FUZZY APPROXIMATION SPACES

The extent to which two probability distributions differ can be expressed by a so-called measure of divergence. Such a measure will reach a minimum value when the two probability distributions are identical and the value increases with the increasing disparity between the two distributions. A specific class of divergence measures is the set of f-divergence measures [11]. For two discrete probability distributions $P = \{p_i | i = 1, 2, \ldots, n\}$ and $Q = \{q_i | i = 1, 2, \ldots, n\}$, the f-divergence is defined as

$$f(P||Q) = \sum_i q_i f\left(\frac{p_i}{q_i}\right). \tag{5.24}$$

A special case of f-divergence measures are the f-information measures. These are defined similar to f-divergence measures, but apply only to specific probability distributions, namely, the joint probability of two variables and their marginal probabilities' product. Thus, the f-information is a measure of dependence: it measures the distance between a given joint probability and the joint probability when the variables are independent [11, 12].

The frequently used functions that can be used to form f-information measures include V-information, I_α-information, M_α-information, and χ^α-information. On the other hand, the Hellinger integral and Renyi's distance measures do not fall in the class of f-divergence measures as they do not satisfy the definition of f-divergence. However, they are divergence measures in the sense that they measure the distance between two distributions and they are directly related to f-divergence.

In this section, different f-information measures are reported for fuzzy approximation spaces based on the concept of fuzzy relative frequency. The f-information measures in fuzzy approximation spaces calculate the distance between a given joint frequency $\lambda_{R_k}(= \lambda_{P_i Q_j})$ and the joint frequency when the variables are independent ($\lambda_{P_i} \lambda_{Q_j}$). In the following analysis, it is assumed that all frequency distributions are complete, that is, $\sum \lambda_{P_i} = \sum \lambda_{Q_j} = \sum \lambda_{P_i Q_j} = 1$.

5.4.1 V-Information

On fuzzy approximation spaces, one of the simplest measures of dependence can be obtained using the function $V = |x - 1|$, which results in the V-information

$$V(R||P \times Q) = \sum_{i,j,k} |\lambda_{R_k} - \lambda_{P_i} \lambda_{Q_j}|, \tag{5.25}$$

where $P = \{\lambda_{P_i} | i = 1, 2, \ldots, p\}$, $Q = \{\lambda_{Q_j} | j = 1, 2, \ldots, q\}$, and $R = \{\lambda_{R_k} | k = 1, 2, \ldots, r\}$ represent two marginal frequency distributions and their joint frequency distribution, respectively. That is, the V-information calculates the absolute distance between the joint frequency of two fuzzy variables and the product of their marginal frequencies.

Combining Equations (5.11), (5.18), and (5.25), the V-information between two fuzzy attribute sets $\mathbb{P}$ and $\mathbb{Q}$, in terms of FEPM, can be represented by

$$V(\mathbb{P}, \mathbb{Q}) = \sum_{i=1}^{p} \sum_{j=1}^{q} \left| \frac{1}{n} \sum_{k=1}^{n} (m_{ik}^{\mathbb{P}} \cap m_{jk}^{\mathbb{Q}}) - \frac{1}{n^2} \sum_{k=1}^{n} m_{ik}^{\mathbb{P}} \sum_{k=1}^{n} m_{jk}^{\mathbb{Q}} \right|. \qquad (5.26)$$

5.4.2 I_α-Information

I_α-information can be defined as follows:

$$I_\alpha(R||P \times Q) = \frac{1}{\alpha(\alpha - 1)} \left(\sum_{i,j,k} \frac{(\lambda_{R_k})^\alpha}{(\lambda_{P_i} \lambda_{Q_j})^{\alpha-1}} - 1 \right) \qquad (5.27)$$

for $\alpha \neq 0, \alpha \neq 1$. Similar to the approach reported in Plum et al. [12], it can be proved that the class of I_α-information includes mutual information, which equals I_α for the limit $\alpha \to 1$, that is,

$$I_1(R||P \times Q) = \lim_{\alpha \to 1} I_\alpha(R||P \times Q)$$

$$= \lim_{\alpha \to 1} \frac{1}{\alpha(\alpha - 1)} \left(\sum_{i,j,k} (\lambda_{R_k})^\alpha (\lambda_{P_i} \lambda_{Q_j})^{1-\alpha} - 1 \right)$$

$$= \lim_{\alpha \to 1} \frac{1}{(2\alpha - 1)} \left\{ \sum_{i,j,k} (\lambda_{R_k})^\alpha (\lambda_{P_i} \lambda_{Q_j})^{1-\alpha} \log(\lambda_{R_k}) \right.$$

$$\left. - \sum_{i,j,k} (\lambda_{R_k})^\alpha (\lambda_{P_i} \lambda_{Q_j})^{1-\alpha} \log(\lambda_{P_i} \lambda_{Q_j}) \right\}$$

that is, $I_1(R||P \times Q) = \sum_{i,j,k} \lambda_{R_k} \log \left(\frac{\lambda_{R_k}}{\lambda_{P_i} \lambda_{Q_j}} \right)$ \qquad (5.28)

using l'Hospital's rule in the second step: if both $\lim_{\alpha \to 1} f(x)$ and $\lim_{\alpha \to 1} g(x)$ equal zero, $\lim_{\alpha \to 1} (f(x)/g(x)) = \lim_{\alpha \to 1} (f'(x)/g'(x))$.

Combining Equations (5.11), (5.18), and (5.27), the I_α-information between two fuzzy attribute sets $\mathbb{P}$ and $\mathbb{Q}$, in terms of FEPM, can be represented by

$$I_\alpha(\mathbb{P}, \mathbb{Q}) = \frac{1}{\alpha(\alpha - 1)} \left(\sum_{i=1}^{p} \sum_{j=1}^{q} \frac{\left[\frac{1}{n} \sum_{k=1}^{n} (m_{ik}^{\mathbb{P}} \cap m_{jk}^{\mathbb{Q}}) \right]^\alpha}{\left[\frac{1}{n^2} \sum_{k=1}^{n} m_{ik}^{\mathbb{P}} \sum_{k=1}^{n} m_{jk}^{\mathbb{Q}} \right]^{\alpha-1}} - 1 \right). \tag{5.29}$$

5.4.3 M_α-Information

The M_α-information, defined by Matusita [11, 12], is as follows:

$$M_\alpha(x) = |x^\alpha - 1|^{1/\alpha}, \quad 0 < \alpha \le 1. \tag{5.30}$$

When applying this function in the definition of an *f*-information measure on fuzzy approximation spaces, the resulting M_α-information measures are

$$M_\alpha(R||P \times Q) = \sum_{i,j,k} |(\lambda_{R_k})^\alpha - (\lambda_{P_i} \lambda_{Q_j})^\alpha|^{\frac{1}{\alpha}} \tag{5.31}$$

for $0 < \alpha \le 1$. These constitute a generalized version of V-information. That is, the M_α-information is identical to the V-information for $\alpha = 1$.

Combining Equations (5.11), (5.18), and (5.31), the M_α-information between two fuzzy attribute sets $\mathbb{P}$ and $\mathbb{Q}$, in terms of FEPM, can be represented by

$$M_\alpha(\mathbb{P}, \mathbb{Q}) = \sum_{i=1}^{p} \sum_{j=1}^{q} \left| \left[\frac{1}{n} \sum_{k=1}^{n} (m_{ik}^{\mathbb{P}} \cap m_{jk}^{\mathbb{Q}}) \right]^\alpha - \left[\frac{1}{n^2} \sum_{k=1}^{n} m_{ik}^{\mathbb{P}} \sum_{k=1}^{n} m_{jk}^{\mathbb{Q}} \right]^\alpha \right|^{1/\alpha}.$$

$$\tag{5.32}$$

5.4.4 χ^α-Information

The class of χ^α-information measures, proposed by Liese [11, 12], is as follows:

$$\chi^\alpha(x) = \begin{cases} |1 - x^\alpha|^{1/\alpha} & \text{for } 0 < \alpha \le 1 \\ |1 - x|^\alpha & \text{for } \alpha > 1. \end{cases} \tag{5.33}$$

For $0 < \alpha \le 1$, this function equals to the M_α-function. The χ^α- and M_α-information measures are, therefore, also identical for $0 < \alpha \le 1$. For $\alpha > 1$, χ^α-information can be written as

$$\chi^\alpha(R||P \times Q) = \sum_{i,j,k} \frac{|\lambda_{R_k} - \lambda_{P_i} \lambda_{Q_j}|^\alpha}{(\lambda_{P_i} \lambda_{Q_j})^{\alpha-1}}. \tag{5.34}$$

Combining Equations (5.11), (5.18), and (5.34), the χ^α-information between two fuzzy attribute sets $\mathbb{P}$ and $\mathbb{Q}$, in terms of FEPM, can be represented by

$$\chi^\alpha(\mathbb{P}, \mathbb{Q}) = \sum_{i=1}^{p} \sum_{j=1}^{q} \frac{\left| \frac{1}{n} \sum_{k=1}^{n} (m_{ik}^{\mathbb{P}} \cap m_{jk}^{\mathbb{Q}}) - \frac{1}{n^2} \sum_{k=1}^{n} m_{ik}^{\mathbb{P}} \sum_{k=1}^{n} m_{jk}^{\mathbb{Q}} \right|^\alpha}{\left(\frac{1}{n^2} \sum_{k=1}^{n} m_{ik}^{\mathbb{P}} \sum_{k=1}^{n} m_{jk}^{\mathbb{Q}} \right)^{\alpha-1}}. \quad (5.35)$$

5.4.5 Hellinger Integral

On fuzzy approximation spaces, the Hellinger integral of order α [11, 12] is defined as

$$H_\alpha(R||P \times Q) = 1 + \alpha(\alpha - 1)I_\alpha(R||P \times Q) \quad (5.36)$$

for $\alpha \neq 0, \alpha \neq 1$. That is, the Hellinger integral is directly related to the I_α-divergence.

5.4.6 Renyi Distance

The Renyi distance, a measure of information of order α [11, 12], can be defined as

$$\mathcal{R}_\alpha(R||P \times Q) = \frac{1}{\alpha - 1} \log \sum_{i,j,k} \frac{(\lambda_{R_k})^\alpha}{(\lambda_{P_i} \lambda_{Q_j})^{\alpha-1}} \quad (5.37)$$

for $\alpha \neq 0, \alpha \neq 1$. It reaches its minimum value when λ_{R_k} and $(\lambda_{P_i} \lambda_{Q_j})$ are identical, in which case the summation reduces to $\sum \lambda_{R_k}$. As we assume complete frequency distributions, the sum is 1 and the minimum value of the measure is, therefore, equal to zero. The limit of Renyi's measure for α approaching 1 equals $I_1(R||P \times Q)$, which is the mutual information.

Combining Equations (5.11), (5.18), and (5.37), the $\mathcal{R}_\alpha$-information between two fuzzy attribute sets $\mathbb{P}$ and $\mathbb{Q}$, in terms of FEPM, can be represented by

$$\mathcal{R}_\alpha(\mathbb{P}, \mathbb{Q}) = \frac{1}{\alpha - 1} \log \left(\sum_{i=1}^{p} \sum_{j=1}^{q} \frac{\left[\frac{1}{n} \sum_{k=1}^{n} (m_{ik}^{\mathbb{P}} \cap m_{jk}^{\mathbb{Q}}) \right]^\alpha}{\left[\frac{1}{n^2} \sum_{k=1}^{n} m_{ik}^{\mathbb{P}} \sum_{k=1}^{n} m_{jk}^{\mathbb{Q}} \right]^{\alpha-1}} \right). \quad (5.38)$$

The following relations hold between different information measures:

$$M_{0.5}(R||P \times Q) = \tfrac{1}{2} I_{0.5}(R||P \times Q)$$
$$= 2[1 - H_{0.5}(R||P \times Q)]$$
$$\chi^{2.0}(R||P \times Q) = 2I_{2.0}(R||P \times Q)$$

$$M_{1.0}(R||P \times Q) = V(R||P \times Q)$$

$$\mathcal{R}_{1.0}(R||P \times Q) = I_{1.0}(R||P \times Q)$$

$$\chi^{\alpha}(R||P \times Q) = M_{\alpha}(R||P \times Q) \ \ 0 < \alpha \leq 1.$$

5.5 *f*-INFORMATION FOR FEATURE SELECTION

In real data analysis, the data set may contain a number of redundant features with low relevance to the classes. The presence of such redundant and nonrelevant features leads to a reduction in the useful information. Ideally, the selected features should have high relevance with the classes, while the redundancy among them would be as low as possible. The features with high relevance are expected to be able to predict the classes of the samples. However, the prediction capability is reduced if many redundant features are selected. In contrast, a data set that contains features with high relevance not only with respect to the classes but also with low mutual redundancy is more effective in its prediction capability. Hence, to assess the effectiveness of the features, both relevance and redundancy need to be measured quantitatively. An information-measure-based criterion is chosen here to address this problem.

5.5.1 Feature Selection Using *f*-Information

Let $\mathbb{C} = \{\mathbb{C}_1, \ldots, \mathbb{C}_i, \ldots, \mathbb{C}_j, \ldots, \mathbb{C}_{\mathcal{D}}\}$ denote the set of $\mathcal{D}$ condition attributes or features of a given data set and $\mathbb{S}$ be the set of selected features. Define $\tilde{f}(\mathbb{C}_i, \mathbb{D})$ as the relevance of the fuzzy condition attribute $\mathbb{C}_i$ with respect to the fuzzy decision attribute $\mathbb{D}$, while $\tilde{f}(\mathbb{C}_i, \mathbb{C}_j)$ is the redundancy between two fuzzy condition attributes $\mathbb{C}_i$ and $\mathbb{C}_j$. The total relevance of all selected features is, therefore, given by

$$\mathcal{J}_{\text{relev}} = \sum_{\mathbb{C}_i \in \mathbb{S}} \tilde{f}(\mathbb{C}_i, \mathbb{D}), \tag{5.39}$$

while the total redundancy among the selected features is

$$\mathcal{J}_{\text{redun}} = \sum_{\mathbb{C}_i, \mathbb{C}_j \in \mathbb{S}} \tilde{f}(\mathbb{C}_i, \mathbb{C}_j). \tag{5.40}$$

Therefore, the problem of selecting a set $\mathbb{S}$ of nonredundant and relevant features from the whole set of condition features $\mathbb{C}$ is equivalent to maximizing $\mathcal{J}_{\text{relev}}$ and minimizing $\mathcal{J}_{\text{redun}}$, that is, to maximize the objective function $\mathcal{J}$, where

$$\mathcal{J} = \mathcal{J}_{\text{relev}} - \beta \mathcal{J}_{\text{redun}} = \sum_i \tilde{f}(\mathbb{C}_i, \mathbb{D}) - \beta \sum_{i,j} \tilde{f}(\mathbb{C}_i, \mathbb{C}_j), \tag{5.41}$$

where β is a weight parameter. To solve the above problem, a greedy algorithm proposed by Battiti [9] is followed, which is described as follows.

1. Initialize $\mathbb{C} \leftarrow \{\mathbb{C}_1, \ldots, \mathbb{C}_i, \ldots, \mathbb{C}_j, \ldots, \mathbb{C}_\mathcal{D}\}$ and $\mathbb{S} \leftarrow \emptyset$.
2. Calculate the relevance $\tilde{f}(\mathbb{C}_i, \mathbb{D})$ of each feature $\mathbb{C}_i \in \mathbb{C}$.
3. Select feature $\mathbb{C}_i$ as the first feature that has highest relevance $\tilde{f}(\mathbb{C}_i, \mathbb{D})$. In effect, $\mathbb{C}_i \in \mathbb{S}$ and $\mathbb{C} = \mathbb{C} \setminus \mathbb{C}_i$.
4. Calculate the redundancy between selected features of $\mathbb{S}$ and each of the remaining features of $\mathbb{C}$.
5. From the remaining features of $\mathbb{C}$, select feature $\mathbb{C}_j$ that maximizes

$$\tilde{f}(\mathbb{C}_j, \mathbb{D}) - \beta \frac{1}{|\mathbb{S}|} \sum_{\mathbb{C}_i \in \mathbb{S}} \tilde{f}(\mathbb{C}_i, \mathbb{C}_j).$$

As a result of that, $\mathbb{C}_j \in \mathbb{S}$ and $\mathbb{C} = \mathbb{C} \setminus \mathbb{C}_j$.
6. Repeat the above two steps until the desired number of features are selected.

The relevance of a fuzzy condition attribute with respect to the fuzzy decision attribute and the redundancy between two fuzzy condition attributes can be calculated using any one of the f-information measures on fuzzy approximation spaces.

5.5.2 Computational Complexity

The f-information-measure-based feature selection method has low computational complexity with respect to both the number of features and the number of samples or objects of the original data set. Before computing the relevance or redundancy of a fuzzy condition attribute, the FEPM for each condition and decision attribute is to be generated. The computational complexity to generate a $(c \times n)$ FEPM is $\mathcal{O}(cn)$, where c represents the number of fuzzy equivalence partitions and n is the total number of objects in the data set. However, two fuzzy equivalence partition matrices with size $(p \times n)$ and $(r \times n)$ have to be generated to compute the relevance of a fuzzy condition attribute with respect to the fuzzy decision attribute, where p and r represent the number of fuzzy equivalence partitions of fuzzy condition attribute and fuzzy decision attribute, respectively. Hence, the total time complexity to calculate the relevance of a fuzzy condition attribute using any one of the f-information measures is $(\mathcal{O}(pn) + \mathcal{O}(rn) + \mathcal{O}(prn)) = \mathcal{O}(prn)$. Similarly, the complexity in calculating the redundancy between two fuzzy condition attributes with p and q number of fuzzy equivalence partitions using any one of the f-information measures is $\mathcal{O}(pqn)$. Hence, the overall time complexity to calculate both relevance and redundancy of a fuzzy condition attribute is $(\mathcal{O}(prn) + \mathcal{O}(pqn)) = \mathcal{O}(n)$ as $p, q, r \ll n$. In effect, the selection of a set of d nonredundant and relevant features from the whole set of $\mathcal{D}$ features using the first-order incremental search method has an overall computational complexity of $\mathcal{O}(nd\mathcal{D})$.

5.5.3 Fuzzy Equivalence Classes

The family of normal fuzzy sets produced by a fuzzy partitioning of the universe of discourse can play the role of fuzzy equivalence classes [33]. In the current feature selection method, the π function in the one-dimensional form is used to assign membership values to different fuzzy equivalence classes for the input features. A fuzzy set with membership function $\pi(x; \bar{c}, \sigma)$ [52] represents a set of points clustered around $\bar{c}$, where

$$
\pi(x; \bar{c}, \sigma) = \begin{cases} 2\left(1 - \dfrac{||x - \bar{c}||}{\sigma}\right)^2 & \text{for } \dfrac{\sigma}{2} \leq ||x - \bar{c}|| \leq \sigma \\[2ex] 1 - 2\left(\dfrac{||x - \bar{c}||}{\sigma}\right)^2 & \text{for } 0 \leq ||x - \bar{c}|| \leq \dfrac{\sigma}{2} \\[2ex] 0 & \text{otherwise} \end{cases} \tag{5.42}
$$

where $\sigma > 0$ is the radius of the π function with $\bar{c}$ as the central point and $||\cdot||$ denotes the Euclidean norm. When the pattern x lies at the central point $\bar{c}$ of a class, then $||x - \bar{c}|| = 0$ and its membership value is maximum, that is, $\pi(\bar{c}; \bar{c}, \sigma) = 1$. The membership value of a point decreases as its distance from the central point $\bar{c}$, that is, $||x - \bar{c}||$ increases. When $||x - \bar{c}|| = (\frac{\sigma}{2})$, the membership value of x is 0.5 and this is called a *crossover point* [52].

Each input real-valued feature in quantitative and/or linguistic form can be assigned to different fuzzy equivalence classes in terms of membership values using the π fuzzy set with appropriate $\bar{c}$ and σ. The centers and radii of the π functions along each feature axis are determined automatically from the distribution of the training patterns or objects. In this context, it should be noted that other functions can also be used in generating fuzzy equivalence classes of the input features [52–54].

5.5.3.1 Choice of Parameters of π Function The parameters $\bar{c}$ and σ of each π fuzzy set are computed according to the following procedure [52]. Let $\overline{m}_i$ be the mean of the objects $x = \{x_1, \ldots, x_j, \ldots, x_n\}$ along the ith feature $\mathbb{C}_i$. Then $\overline{m}_{i_l}$ and $\overline{m}_{i_h}$ are defined as the means along the ith feature of the objects having coordinate values in the range $[\mathbb{C}_{i_{\min}}, \overline{m}_i)$ and $(\overline{m}_i, \mathbb{C}_{i_{\max}}]$, respectively, where $\mathbb{C}_{i_{\max}}$ and $\mathbb{C}_{i_{\min}}$ denote the upper and lower bounds of the dynamic range of feature $\mathbb{C}_i$ for the training set. For the three linguistic property sets low, medium, and high, the centers and the corresponding radii are defined as follows:

$$
\bar{c}_{\text{low}}(\mathbb{C}_i) = \overline{m}_{i_l}
$$
$$
\bar{c}_{\text{medium}}(\mathbb{C}_i) = \overline{m}_i \tag{5.43}
$$
$$
\bar{c}_{\text{high}}(\mathbb{C}_i) = \overline{m}_{i_h}
$$

$$\sigma_{\text{low}}(\mathbb{C}_i) = 2(\overline{c}_{\text{medium}}(\mathbb{C}_i) - \overline{c}_{\text{low}}(\mathbb{C}_i))$$

$$\sigma_{\text{high}}(\mathbb{C}_i) = 2(\overline{c}_{\text{high}}(\mathbb{C}_i) - \overline{c}_{\text{medium}}(\mathbb{C}_i)) \qquad (5.44)$$

$$\sigma_{\text{medium}}(\mathbb{C}_i) = \eta \times \frac{A}{B}$$

where

$$A = \big\{ \sigma_{\text{low}}(\mathbb{C}_i)(\mathbb{C}_{i_{\text{max}}} - c_{\text{medium}}(\mathbb{C}_i))$$

$$+ \sigma_{\text{high}}(\mathbb{C}_i)(c_{\text{medium}}(\mathbb{C}_i) - \mathbb{C}_{i_{\text{min}}}) \big\} ; \qquad (5.45)$$

and

$$B = \{ \mathbb{C}_{i_{\text{max}}} - \mathbb{C}_{i_{\text{min}}} \}, \qquad (5.46)$$

where η is a multiplicative parameter controlling the extent of the overlapping. The distribution of the patterns or objects along each feature axis is taken into account while computing the corresponding centers and radii of the linguistic properties. Also, the amount of overlap between the three linguistic properties can be different along the different axis, depending on the distribution of the objects or patterns.

5.5.3.2 Fuzzy Equivalence Partition Matrix The $c \times n$ FEPM $\mathbb{M}_{\mathbb{C}_i}$, corresponding to the ith feature $\mathbb{C}_i$, can be calculated from the c-fuzzy equivalence classes of the objects $x = \{x_1, \ldots, x_j, \ldots, x_n\}$, where

$$m_{kj}^{\mathbb{C}_i} = \frac{\pi(x_j; \overline{c}_k, \sigma_k)}{\sum_{l=1}^{c} \pi(x_j; \overline{c}_l, \sigma_l)}. \qquad (5.47)$$

Corresponding to three fuzzy sets low, medium, and high ($c = 3$), the following relations hold:

$$\overline{c}_1 = \overline{c}_{\text{low}}(\mathbb{C}_i); \overline{c}_2 = \overline{c}_{\text{medium}} \overline{c}_2 = \overline{c}_{\text{medium}}(\mathbb{C}_i); \overline{c}_3 = \overline{c}_{\text{high}}(\mathbb{C}_i)$$

$$\sigma_1 = \sigma_{\text{low}}(\mathbb{C}_i); \sigma_2 = \sigma_{\text{medium}}(\mathbb{C}_i); \sigma_3 = \sigma_{\text{high}}(\mathbb{C}_i).$$

In effect, each position $m_{kj}^{\mathbb{C}_i}$ of the FEPM $\mathbb{M}_{\mathbb{C}_i}$ must satisfy the following conditions:

$$m_{kj}^{\mathbb{C}_i} \in [0, 1]; \sum_{k=1}^{c} m_{kj}^{\mathbb{C}_i} = 1, \forall j \text{ and for any value of } k, \text{ if}$$

$$s = \arg \max_{j} \{ m_{kj}^{\mathbb{C}_i} \}, \text{ then } \max_{j} \{ m_{kj}^{\mathbb{C}_i} \} = \max_{l} \{ m_{ls}^{\mathbb{C}_i} \} > 0.$$

5.6 QUANTITATIVE MEASURES

In this section, two new quantitative indices are presented, along with some existing indices, to evaluate the performance of different feature selection methods. Two new indices are based on the concept of fuzzy-rough sets.

5.6.1 Fuzzy-Rough-Set-Based Quantitative Indices

Using the definition of fuzzy positive region, two indices, namely, $\mathbb{RELEV}$ and $\mathbb{REDUN}$ index, are reported next for evaluating quantitatively the quality of selected features.

5.6.1.1 $\mathbb{RELEV}$ Index The $\mathbb{RELEV}$ index is defined as

$$\mathbb{RELEV} = \frac{1}{|\mathbb{S}|} \sum_{\mathbb{C}_i \in \mathbb{S}} \gamma_{\mathbb{C}_i}(\mathbb{D}), \tag{5.48}$$

where $\gamma_{\mathbb{C}_i}(\mathbb{D})$ represents the degree of dependency of the decision attribute $\mathbb{D}$ on the condition attribute $\mathbb{C}_i$, which can be calculated using Equation (5.8). That is, $\mathbb{RELEV}$ index is the average relevance of all selected features. A good feature selection algorithm should make all selected features as relevant as possible. The $\mathbb{RELEV}$ index increases with the increase in relevance of each selected feature. Therefore, for a given data set and number of selected features, the higher the relevance of each selected feature, the higher would be the $\mathbb{RELEV}$ index value.

5.6.1.2 $\mathbb{REDUN}$ Index It can be defined as

$$\mathbb{REDUN} = \frac{1}{2|\mathbb{S}||\mathbb{S}-1|} \sum_{\mathbb{C}_i,\mathbb{C}_j} \{\gamma_{\mathbb{C}_i}(\mathbb{C}_j) + \gamma_{\mathbb{C}_j}(\mathbb{C}_i)\}, \tag{5.49}$$

where $\gamma_{\mathbb{C}_i}(\mathbb{C}_j)$ represents the degree of dependency of the condition attribute $\mathbb{C}_j$ on another condition attribute $\mathbb{C}_i$. The $\mathbb{REDUN}$ index calculates the amount of redundancy among the selected features. A good feature selection algorithm should make the redundancy between all selected features as low as possible. The $\mathbb{REDUN}$ index minimizes the redundancy between the selected features.

5.6.2 Existing Feature Evaluation Indices

Some existing indices that are considered for evaluating the effectiveness of the selected feature subsets are described next. While the first three indices, namely, class separability [1], K-NN classification error [55], and C4.5 classification error [6], do need class information of the samples, the remaining two, namely, entropy [45] and representation entropy [1], do not. The class separability, K-NN, and C4.5 classification error measure the effectiveness of the feature subsets for classification. The entropy evaluates the clustering performance of the feature subsets, while the representation entropy measures the amount of redundancy present in the feature subset. However, the K-NN classifier is briefly described in Chapter 4. The value of K, chosen for the K-NN, is the square root of the number of samples in the training set.

5.6.2.1 Class Separability The class separability index S of a data set is defined as [1]

$$S = trace(S_b^{-1} S_w), \tag{5.50}$$

where S_w and S_b represent the within-class and between-class scatter matrix, respectively, and are defined as follows:

$$S_w = \sum_{j=1}^{C} p_j E\{(X - \mu_j)(X - \mu_j)^T | w_j\} = \sum_{j=1}^{C} p_j \Sigma_j \tag{5.51}$$

$$S_b = \sum_{j=1}^{C} (\mu_j - \mu_0)(\mu_j - \mu_0)^T, \text{ where } \mu_0 = \sum_{j=1}^{C} p_j \mu_j, \tag{5.52}$$

where C is the number of classes, p_j is a priori probability that a pattern belongs to class w_j, X is a feature vector, μ_0 is the sample mean vector for the entire data points, μ_j is the sample mean vector of class w_j, Σ_j is the sample covariance matrix of class w_j, and $E\{\cdot\}$ is the expectation operator. A lower value of the separability criteria ensures that the classes are well separated by their scatter means.

5.6.2.2 C4.5 Classification Error The C4.5 [6] is a popular decision-tree-based classification algorithm. It is used for evaluating the effectiveness of a reduced feature set for classification. The selected feature set is fed to the C4.5 for building classification models. The C4.5 is used here because it performs feature selection in the process of training and the classification models it builds are represented in the form of decision trees, which can be further examined.

5.6.2.3 Entropy Let the distance between two data points x_i and x_j be

$$D_{ij} = \left[\sum_{k=1}^{d} \left(\frac{x_{ik} - x_{jk}}{max_k - min_k} \right)^2 \right]^{1/2}, \tag{5.53}$$

where x_{ik} denotes feature value for x_i along the kth direction, and max_k, min_k are the maximum and minimum values computed over all the samples along the kth axis, and d is the number of selected features. The similarity between x_i and x_j is given by $sim(i, j) = e^{-\alpha D_{ij}}$, where α is a positive constant. A possible value of α is $-\ln 0.5 / \bar{D}$. $\bar{D}$ is the average distance between data points computed over the entire data set. Entropy is then defined as [45]

$$E = - \sum_{i=1}^{n} \sum_{j=1}^{n} (\text{sim}(i, j) \times \log(\text{sim}(i, j))$$

$$+ (1 - \text{sim}(i, j)) \times \log(1 - \text{sim}(i, j)). \quad (5.54)$$

If the data is uniformly distributed in the feature space, entropy is maximum. When the data has well-formed clusters, uncertainty is low and so is entropy.

5.6.2.4 Representation Entropy Let the eigenvalues of the $d \times d$ covariance matrix of a feature set of size d be $\lambda_j, j = 1, \ldots, d$. Let

$$\tilde{\lambda}_j = \frac{\lambda_j}{\sum_{j=1}^{d} \lambda_j}, \quad (5.55)$$

$\tilde{\lambda}_j$ has similar properties such as probability, namely, $0 \leq \tilde{\lambda}_j \leq 1$ and $\sum_{j=1}^{d} \tilde{\lambda}_j = 1$. Hence, an entropy function can be defined as [1]

$$H_R = - \sum_{j=1}^{d} \tilde{\lambda}_j \log \tilde{\lambda}_j. \quad (5.56)$$

The function H_R attains a minimum value zero when all the eigenvalues except one are zero or, in other words, when all the information is present along a single coordinate direction. If all the eigenvalues are equal, that is, information is equally distributed among all the features, H_R is maximum and so is the uncertainty involved in feature reduction. The above measure is known as *representation entropy*. It is a property of the data set as represented by a particular set of features, and is a measure of the amount of information compression possible by dimensionality reduction. This is equivalent to the amount of redundancy present in that particular representation of the data set. Since the current feature selection method takes into account the redundancy among the selected features, it is expected that the reduced feature set attains a high value of representation entropy.

5.7   EXPERIMENTAL RESULTS

The performance of the FEPM -based feature selection method is extensively studied. On the basis of the argumentation given in Section 5.4, the following information measures are chosen for inclusion in the study:

MI:	mutual information;	VI:	V-information;
I_α:	for $\alpha \neq 0, \alpha \neq 1$;	M_α:	for $0 < \alpha \leq 1$;
χ^α:	for $\alpha > 1$;	$\mathcal{R}_\alpha$:	for $\alpha \neq 0, \alpha \neq 1$;
C:	fuzzy approximation;	D:	crisp approximation.

These measures are applied to calculate both relevance and redundancy of the features. The values of α investigated are 0.2, 0.5, 0.8, 1.5, 2.0, 3.0, 4.0, and 5.0. The values close to 1.0 are excluded, either because the measures resemble mutual information for such values (I_α, $\mathcal{R}_\alpha$) or because they resemble another measure (M_1 and χ^1 equal VI). The performance of the FEPM-based feature selection method is also compared with that of the popular QR algorithm, both in fuzzy [35] (fuzzy-rough QR) and crisp [21] (rough QR) approximation spaces. All the algorithms are implemented in C language and run in LINUX environment having machine configuration Pentium IV, 3.2 GHz, 1 MB cache, and 1 GB RAM.

To analyze the performance of different feature selection methods, the experimentation is done on some benchmark real-life data sets. The major metrics for evaluating the performance of different algorithms are the indices reported in Section 5.6 such as RELEV index, REDUN index, class separability [1], K-NN [55] and C4.5 classification error [6], entropy [45], and representation entropy [1].

5.7.1 Description of Data Sets

This section reports the brief description of some benchmark data sets. All the data sets are downloaded from http://www.ics.uci.edu/~mlearn.

5.7.1.1 Iris Data Set It is a four-dimensional data set containing 50 samples each of three types of iris flowers. One of the three clusters (class 1) is well separated from the other two, while classes 2 and 3 have some overlap.

5.7.1.2 E. coli Database It is a seven-dimensional data set with eight classes related to the localization site of protein. The number of samples is 336.

5.7.1.3 Wine Recognition Database It is a 13-dimensional data set with three classes. The number of instances per class are 59, 71, and 48, respectively. These data are the result of a chemical analysis of wines grown in the same region in Italy but derived from three different cultivars. The analysis determined the quantities of 13 constituents found in each of three wines.

5.7.1.4 Letter Database It is a 16-dimensional data set with 26 classes related to the 26 capital letters of the English alphabet. The data set contains 20,000 examples: 15,000 training and 5000 testing, which are produced from the images of 20 different fonts. For each image, 16 numerical attributes are calculated using edge counts and measures of statistical moments.

5.7.1.5 Ionosphere Database It represents autocorrelation functions of radar measurements. The task is to classify them into two classes denoting passage or obstruction in the ionosphere. There are 351 instances with 34 continuous-valued attributes.

5.7.1.6 Satimage Database   The original Landsat data for this database is generated from data purchased from NASA by the Australian Centre for Remote Sensing, and used for research at the University of New South Wales. The sample database is generated taking a small section of 82 rows and 100 columns from the original data. The database is a tiny subarea of a scene, consisting of 82×100 pixels, each pixel covering an area on the ground of approximately 80×80 m. The information given for each pixel consists of the class value and the intensities in four spectral bands (from the green, red, and infrared regions of the spectrum). The data set contains 6435 examples: 4435 training and 2000 testing, with 36 real-valued attributes and 6 classes.

5.7.1.7 Isolet   It consists of several spectral coefficients of utterances of English alphabets by 150 subjects. There are 617 real-valued features with 7797 instances and 26 classes.

To compute the classification error of both K-NN rule and C4.5-based decision tree algorithm, both leave-one-out cross-validation (LOOCV) and training-testing are performed. The LOOCV is performed on *E. coli*, Wine, Isolet, and Ionosphere data, whereas the training-testing is done on Letter and Satimage data.

5.7.2 Illustrative Example

Consider the four-dimensional data set iris with 150 samples. The parameters generated in the FEPM-based feature selection method and the relevance of each feature are shown next, as an example. The values of input parameters used are also presented here. The mutual information is chosen to calculate the relevance and redundancy of the features.

$$\text{Number of samples or objects, } n = 150$$
$$\text{Number of dimensions or features, } \mathcal{D} = 4$$
$$\text{Value of weight parameter, } \beta = 0.5$$
$$\text{Value of multiplicative parameter, } \eta = 1.5$$

Feature 1:
$\bar{c}_{low} = 0.2496; \bar{c}_{medium} = 0.4287; \bar{c}_{high} = 0.6333$
$\sigma_{low} = 0.3581; \sigma_{medium} = 0.5701; \sigma_{high} = 0.4093$
Feature 2:
$\bar{c}_{low} = 0.3138; \bar{c}_{medium} = 0.4392; \bar{c}_{high} = 0.5945$
$\sigma_{low} = 0.2508; \sigma_{medium} = 0.4157; \sigma_{high} = 0.3107$
Feature 3:
$\bar{c}_{low} = 0.1192; \bar{c}_{medium} = 0.4676; \bar{c}_{high} = 0.6811$
$\sigma_{low} = 0.6967; \sigma_{medium} = 0.8559; \sigma_{high} = 0.4269$
Feature 4:
$\bar{c}_{low} = 0.1146; \bar{c}_{medium} = 0.4578; \bar{c}_{high} = 0.6866$
$\sigma_{low} = 0.6864; \sigma_{medium} = 0.8725; \sigma_{high} = 0.4576$

Relevance of each feature:

Feature 1 : 0.2669;   Feature 2 : 0.1488

Feature 3 : 0.3793;   Feature 4 : 0.3739

In the FEPM-based feature selection method, Feature 3 will be selected first as it has the highest relevance value. After selecting Feature 3, the redundancy and the objective function of each feature are calculated and are given next.

Redundancy of each feature:

Feature 1: 0.1295; Feature 2: 0.0572; Feature 4: 0.1522

Value of objective function:

Feature 1: 0.2021; Feature 2: 0.1202; Feature 4: 0.2978

On the basis of the value of the objective function, Feature 4 will be selected next as the second feature. The values of different quantitative indices for these two features (Features 3 and 4) are reported next, along with that for whole feature sets.

Measures/Features	3 and 4	1 to 4
Classification error C4.5	2.0%	2.0%
Class separability S	0.0909	0.2343
Entropy E	0.6904	0.7535
Representation Entropy H_R	0.9973	0.8785
RELEV Index	0.5126	0.4407
REDUN Index	0.4149	0.4440

The results reported above establish the fact that the FEPM-based feature selection method selects the most significant features from the whole feature sets by maximizing the relevance and minimizing the redundancy of the selected features.

5.7.3 Effectiveness of the FEPM-Based Method

To better understand the effectiveness of the FEPM-based feature selection method using f-information measures in fuzzy approximation spaces, extensive experimental results are reported in Tables 5.1 and 5.2 for different data sets. Subsequent discussions analyze the results with respect to the classification error obtained using the C4.5.

Tables 5.1 and 5.2 report the classification error of the C4.5 for mutual-information-based feature selection method both in fuzzy and crisp approximation spaces. Results are presented for different values of the number of selected features d, weight parameter β, and multiplicative parameter η. All the results reported here confirm that the mutual-information-based feature selection method is more effective in fuzzy approximation spaces than in crisp approximation spaces with smaller number of features. The FEPM-based method in fuzzy

TABLE 5.1 Classification Error of C4.5 for Mutual-Information-Based Feature Selection on *E. coli* and Satimage

Data Sets	Selected Features	Measures/ Methods	$\eta = 0.7/\beta$			$\eta = 1.1/\beta$			$\eta = 1.5/\beta$			$\eta = 1.9/\beta$		
			0.0	0.5	1.0	0.0	0.5	1.0	0.0	0.5	1.0	0.0	0.5	1.0
E. coli ($D=7$, $n=336$)	2	MI-C	23.7	17.1	17.1	23.2	15.8	15.8	23.2	15.8	20.2	23.2	15.8	15.8
		MI-D	23.7	18.4	18.9	23.7	18.4	18.4	23.7	17.1	18.3	23.7	17.6	18.9
	4	MI-C	10.1	8.5	9.5	9.5	8.5	9.3	9.5	8.1	9.5	9.5	9.1	9.7
		MI-D	11.7	11.7	11.7	10.1	10.1	10.1	10.1	10.1	10.1	10.1	10.1	10.1
	6	MI-C	6.5	6.5	6.5	6.5	6.5	6.5	6.5	6.5	6.5	6.5	6.5	6.5
		MI-D	6.5	6.5	6.5	6.5	6.5	6.5	6.5	6.5	6.5	6.5	6.5	6.5
Satimage ($D=36$, $n=4435$)	5	MI-C	19.9	19.1	19.1	18.6	17.9	18.6	19.1	18.9	18.9	17.2	17.2	17.2
		MI-D	23.8	22.9	22.9	22.3	21.4	21.6	21.8	20.2	21.1	21.8	20.6	21.0
	10	MI-C	20.6	18.8	18.8	20.6	18.8	18.1	19.8	17.3	17.3	18.7	18.1	18.6
		MI-D	22.9	22.7	22.9	21.9	19.8	20.7	20.5	18.1	19.9	22.9	20.1	20.7
	15	MI-C	18.1	17.9	18.0	18.0	17.6	17.9	22.1	17.4	21.6	19.3	16.9	16.9
		MI-D	21.9	21.6	21.6	21.6	19.1	20.2	22.6	21.5	21.8	22.7	21.5	21.8
	20	MI-C	18.1	17.9	17.9	17.6	17.6	17.9	19.8	18.1	18.9	19.1	16.6	18.1
		MI-D	20.3	20.3	20.3	20.3	20.2	20.0	20.1	19.6	19.7	20.4	19.8	20.2
	25	MI-C	17.9	17.9	17.9	17.6	16.6	16.6	16.2	15.9	16.1	16.2	16.0	16.1
		MI-D	18.8	18.8	18.0	18.1	17.4	17.8	17.9	17.2	17.3	18.0	17.2	17.6
	30	MI-C	19.6	18.9	19.2	18.7	18.6	18.6	18.6	18.6	18.6	18.8	18.8	18.8
		MI-D	19.6	19.6	19.6	18.9	18.9	18.9	18.9	18.9	18.9	18.9	18.9	18.9
	35	MI-C	17.6	17.6	17.6	17.6	17.6	17.6	17.6	17.6	17.6	17.2	17.2	17.2
		MI-D	17.6	17.6	17.6	17.6	17.6	17.6	17.6	17.6	17.6	17.2	17.2	17.2

TABLE 5.2 Classification Error of C4.5 for Mutual-Information-Based Feature Selection on Letter and Isolet

Data Sets	Selected Features	Measures/Methods	$\eta = 0.7/\beta$			$\eta = 1.1/\beta$			$\eta = 1.5/\beta$			$\eta = 1.9/\beta$		
			0.0	0.5	1.0	0.0	0.5	1.0	0.0	0.5	1.0	0.0	0.5	1.0
Letter	5	MI-C	27.5	26.1	26.1	29.7	24.6	24.6	24.6	24.6	24.6	27.5	26.1	26.1
($\mathcal{D}=16$,		MI-D	34.5	31.6	33.9	33.9	30.1	30.1	29.9	29.9	25.0	34.5	31.6	32.5
$n=15000$)	10	MI-C	14.7	13.9	14.1	13.9	13.6	13.7	13.7	12.9	13.1	13.9	13.4	13.4
		MI-D	15.2	14.1	14.1	13.9	13.9	13.9	13.9	13.2	13.6	13.9	13.6	13.6
	15	MI-C	12.4	12.4	12.4	12.4	12.4	12.4	12.4	12.1	12.4	12.4	12.4	12.4
		MI-D	12.4	12.4	12.4	12.4	12.4	12.4	12.4	12.4	12.4	12.4	12.4	12.4
	15	MI-C	16.3	11.9	12.1	16.3	11.5	12.1	15.7	11.3	12.0	15.1	11.4	12.1
		MI-D	23.7	23.7	22.3	23.7	19.1	22.3	21.4	18.5	18.9	22.0	18.8	18.8
	20	MI-C	18.7	14.0	14.7	18.9	13.1	15.7	18.3	12.7	13.3	18.4	12.8	13.9
		MI-D	24.3	22.8	22.8	24.3	19.6	21.0	22.3	19.1	19.6	22.5	19.3	20.5
Isolet	25	MI-C	15.7	14.5	14.5	12.9	9.6	11.7	12.1	8.8	11.6	12.5	9.1	11.9
($\mathcal{D}=617$,		MI-D	21.1	19.2	21.1	21.1	17.1	18.3	20.0	12.3	19.5	19.6	14.1	19.1
$n=7797$)	30	MI-C	11.7	8.2	11.2	11.7	8.2	10.6	10.9	8.0	10.4	11.4	8.3	10.5
		MI-D	17.3	14.5	14.5	17.3	14.5	16.9	14.3	11.7	12.8	15.0	11.8	12.9
	35	MI-C	11.2	7.6	11.5	11.5	6.9	10.6	10.9	6.8	10.3	11.2	6.8	10.7
		MI-D	14.7	9.2	11.2	14.7	9.2	11.3	13.2	8.3	10.7	13.3	9.0	10.9

approximation spaces improves the classification accuracy of the C4.5 significantly over its crisp counterpart, especially with small number of features. As the number of selected features d increases, the difference between the fuzzy and crisp approximation spaces decreases. For a given data set with n samples and $\mathcal{D}$ features, the classification error of the C4.5 remains unchanged for any combination of β and η when the number of selected features d approaches $\mathcal{D}$. In case of *E. coli* and Letter data sets, the error becomes almost the same for $d = 6$ and 15 as the values of corresponding $\mathcal{D} = 7$ and 16, respectively. Similarly, for the Satimage data set, the classification error remains almost the same at $d = 35$ as the corresponding $\mathcal{D} = 36$. However, for feature selection, a small feature set is of practical importance. Also, for a given data set and fixed d and η values, the classification error would be lower for nonzero β values. In other words, if the redundancy between the selected feature sets is taken into consideration, the performance of the FEPM-based method would be better both in fuzzy and crisp approximation spaces.

5.7.4 Optimum Value of Weight Parameter β

The parameter β regulates the relative importance of the redundancy between the candidate feature and the already selected features with respect to the relevance with the output class. If β is zero, only the relevance with the output class is considered for each feature selection. If β increases, this measure is discounted by a quantity proportional to the total redundancy with respect to the already selected features. The presence of a β value larger than zero is crucial in order to obtain good results. If the redundancy between features is not taken into account, selecting the features with the highest relevance with respect to the output class tends to produce a set of redundant features that may leave out useful complementary information.

Tables 5.3, 5.4, and 5.5 present the performance of the feature section method based on minimum redundancy and maximum relevance using both mutual information and V-information for different values of β. The results and subsequent discussions are presented in these tables with respect to various quantitative indices for both fuzzy and crisp approximation spaces. In Tables 5.3, 5.4, and 5.5, it is seen that as the value of β increases, the values of RELEV index and representative entropy H_R increase, whereas the classification error of both C4.5 and K-NN, and the values of REDUN index, class separability S, and entropy E decrease. The mutual information and V-information achieve their best performance for $0.5 \leq \beta < 1$ with respect to all these quantitative indices. In other words, the best performance of mutual information and V-information is achieved when the relevance of each feature is discounted by at least 50% of the total redundancy with respect to the already selected features.

5.7.5 Optimum Value of Multiplicative Parameter η

The η is a multiplicative parameter controlling the extent of the overlapping between the linguistic property sets low and medium or medium and high.

TABLE 5.3 Performance on Satimage Database for Different Values of Weight Parameter β Considering $d = 10$ and $\eta = 1.5$

Evaluation Criteria	Algorithms/ Methods	Value of Weight Parameter β										
		0.0	0.1	0.2	0.3	0.4	0.5	0.6	0.7	0.8	0.9	1.0
Classification error (C4.5)	MI-C	19.8	19.8	19.8	19.1	18.5	17.3	17.3	17.3	17.3	17.3	17.3
	MI-D	20.5	20.5	20.5	20.5	19.7	18.1	18.1	18.8	19.9	19.9	19.9
	VI-C	26.6	26.6	26.6	22.9	17.4	17.4	17.4	17.4	16.8	16.8	16.8
	VI-D	29.6	29.6	29.6	25.1	18.1	18.1	18.1	18.1	23.7	23.7	25.1
Classification error (K-NN)	MI-C	18.8	18.8	18.8	18.8	17.3	17.3	17.3	17.3	17.3	17.3	17.3
	MI-D	20.5	20.5	20.5	20.7	19.8	18.8	18.8	18.8	19.1	20.5	20.5
	VI-C	26.6	26.6	20.5	19.1	16.6	16.6	16.6	16.6	16.8	17.4	17.4
	VI-D	29.6	29.6	29.6	19.9	18.1	18.1	18.1	18.1	18.1	19.8	19.8
Class separability	MI-C	0.393	0.393	0.393	0.393	0.384	0.366	0.366	0.366	0.366	0.366	0.366
	MI-D	0.597	0.597	0.597	0.597	0.593	0.467	0.467	0.467	0.467	0.464	0.464
	VI-C	0.417	0.417	0.397	0.397	0.366	0.366	0.366	0.366	0.366	0.366	0.366
	VI-D	0.643	0.643	0.643	0.617	0.495	0.467	0.495	0.495	0.560	0.560	0.560
Entropy (E)	MI-C	0.828	0.828	0.828	0.828	0.828	0.824	0.824	0.824	0.824	0.827	0.827
	MI-D	0.833	0.833	0.833	0.833	0.830	0.832	0.832	0.830	0.830	0.830	0.830
	VI-C	0.817	0.817	0.811	0.802	0.802	0.802	0.802	0.809	0.809	0.809	0.809
	VI-D	0.832	0.832	0.832	0.832	0.830	0.830	0.829	0.829	0.830	0.827	0.830
Representation entropy (H_R)	MI-C	3.366	3.366	3.366	3.366	3.366	3.366	3.399	3.399	3.399	3.399	3.399
	MI-D	3.260	3.260	3.263	3.263	3.263	3.263	3.263	3.298	3.298	3.298	3.298
	VI-C	3.344	3.344	3.344	3.420	3.420	3.420	3.420	3.406	3.366	3.366	3.366
	VI-D	3.198	3.198	3.198	3.217	3.217	3.217	3.226	3.226	3.226	3.226	3.226

TABLE 5.4 Performance on Isolet Database for Different Values of Weight Parameter β Considering $d = 25$ and $\eta = 1.5$

Evaluation Criteria	Algorithms/ Methods	Value of Weight Parameter β										
		0.0	0.1	0.2	0.3	0.4	0.5	0.6	0.7	0.8	0.9	1.0
Classification error (C4.5)	MI-C	12.1	12.1	10.4	9.7	8.8	8.8	8.7	8.7	10.1	10.3	11.6
	MI-D	20.0	20.0	18.6	12.3	12.3	12.3	11.9	13.7	16.2	19.5	19.5
	VI-C	11.5	10.2	9.7	9.7	9.7	8.4	8.3	8.5	8.4	8.4	8.4
	VI-D	18.6	18.6	14.9	13.1	11.4	11.4	11.4	15.0	16.4	17.3	17.3
Classification error (K-NN)	MI-C	12.1	9.7	9.7	9.7	6.1	6.1	6.1	6.1	6.1	6.8	6.8
	MI-D	18.5	18.5	16.2	13.7	9.3	9.3	9.3	9.3	11.5	12.3	12.3
	VI-C	10.4	9.7	9.7	8.8	5.8	5.8	5.8	5.8	5.8	6.1	6.1
	VI-D	18.6	18.6	13.7	12.1	10.7	10.7	10.7	10.7	12.3	12.3	12.3
Class separability	MI-C	0.158	0.158	0.147	0.140	0.126	0.113	0.113	0.113	0.113	0.113	0.113
	MI-D	0.371	0.371	0.371	0.371	0.344	0.344	0.344	0.344	0.344	0.344	0.344
	VI-C	0.138	0.138	0.138	0.127	0.127	0.097	0.097	0.097	0.114	0.114	0.123
	VI-D	0.362	0.362	0.358	0.358	0.358	0.358	0.359	0.357	0.357	0.357	0.357
Entropy (E)	MI-C	0.276	0.276	0.276	0.276	0.276	0.276	0.275	0.275	0.275	0.275	0.275
	MI-D	0.313	0.313	0.313	0.313	0.313	0.313	0.313	0.313	0.311	0.309	0.309
	VI-C	0.276	0.276	0.276	0.276	0.276	0.276	0.273	0.273	0.273	0.273	0.273
	VI-D	0.304	0.304	0.304	0.304	0.303	0.303	0.301	0.301	0.301	0.301	0.301
Representation entropy (H_R)	MI-C	4.619	4.619	4.619	4.619	4.629	4.629	4.629	4.629	4.629	4.629	4.629
	MI-D	4.402	4.402	4.402	4.407	4.407	4.407	4.410	4.410	4.410	4.410	4.410
	VI-C	4.637	4.641	4.641	4.644	4.647	4.647	4.647	4.630	4.630	4.630	4.612
	VI-D	4.441	4.441	4.441	4.441	4.445	4.445	4.445	4.445	4.439	4.439	4.439

TABLE 5.5 RELEV and REDUN Index Values on Satimage and Isolet Database for $\eta = 1.5$

Evaluation Criteria	Algorithms/ Methods	Value of Weight Parameter β										
		0.0	0.1	0.2	0.3	0.4	0.5	0.6	0.7	0.8	0.9	1.0
		Satimage Database Considering $d = 10$										
RELEV index	MI-C	0.427	0.427	0.427	0.427	0.427	0.427	0.425	0.425	0.425	0.425	0.425
	MI-D	0.338	0.338	0.336	0.334	0.334	0.334	0.334	0.330	0.330	0.330	0.330
	VI-C	0.439	0.439	0.439	0.438	0.434	0.434	0.434	0.434	0.431	0.431	0.430
	VI-D	0.343	0.343	0.343	0.343	0.340	0.340	0.340	0.339	0.339	0.339	0.339
REDUN index	MI-C	0.350	0.350	0.350	0.350	0.350	0.345	0.345	0.345	0.345	0.345	0.345
	MI-D	0.419	0.419	0.419	0.419	0.402	0.350	0.350	0.350	0.350	0.350	0.348
	VI-C	0.423	0.423	0.423	0.419	0.419	0.419	0.406	0.406	0.399	0.399	0.399
	VI-D	0.428	0.428	0.428	0.428	0.411	0.411	0.411	0.418	0.408	0.408	0.408
		Isolet Database Considering $d = 25$										
RELEV index	MI-C	0.417	0.417	0.417	0.417	0.417	0.417	0.419	0.419	0.419	0.419	0.419
	MI-D	0.311	0.311	0.311	0.311	0.314	0.314	0.314	0.314	0.314	0.314	0.314
	VI-C	0.416	0.416	0.417	0.417	0.417	0.418	0.418	0.415	0.415	0.411	0.409
	VI-D	0.313	0.313	0.313	0.313	0.316	0.316	0.316	0.316	0.316	0.316	0.316
REDUN index	MI-C	0.427	0.427	0.427	0.427	0.425	0.425	0.425	0.425	0.425	0.425	0.425
	MI-D	0.496	0.496	0.496	0.490	0.490	0.490	0.490	0.490	0.490	0.490	0.490
	VI-C	0.427	0.427	0.427	0.424	0.424	0.424	0.424	0.424	0.424	0.425	0.425
	VI-D	0.462	0.462	0.462	0.462	0.461	0.461	0.461	0.458	0.458	0.458	0.458

Keeping the values of σ_{low} and σ_{high} fixed, the amount of overlapping among the three π functions can be altered varying σ_{medium}. As η is decreased, the radius σ_{medium} decreases around $\bar{c}_{\text{medium}}$ such that ultimately there is insignificant overlapping between the π functions low and medium or medium and high. This implies that certain regions along the ith feature axis $\mathbb{C}_i$ go underrepresented such that the three membership values corresponding to three fuzzy sets low, medium, and high attain small values. Note that the particular choice of the values of the σs and $\bar{c}$s ensure that for any pattern x_j along the ith feature axis $\mathbb{C}_i$, at least one of the membership values should be greater than 0.5. On the other hand, as η is increased, the radius σ_{medium} increases around $\bar{c}_{\text{medium}}$ such that the amount of overlapping between the π functions increases.

Tables 5.6, 5.7, and 5.8 represent the performance of the FEPM-based feature selection method in terms of various quantitative indices for different values of η. Results are presented for different data sets considering the information measure as both mutual information and V-information. It is seen from the results reported in Tables 5.6, 5.7, and 5.8 that, in case of both mutual information and V-information, the FEPM-based method achieves consistently better performance for $1.1 < \eta < 1.7$. In fact, very large or very small amounts of overlapping among the three linguistic properties of the input feature are found to be undesirable.

5.7.6 Performance of Different f-Information Measures

Furthermore, extensive experiments are done to evaluate the performance of different f-information measures, both in fuzzy and crisp approximation spaces. Tables 5.9, 5.10, and 5.11 report the values of the classification error of both C4.5 and K-NN, class separability, entropy, representation entropy, RELEV index, and REDUN index of different f-information measures for the data sets reported in Section 5.7.1. Results are presented for different values of α considering $\beta = 0.5$ and $\eta = 1.5$. For each data set, the value of d (number of selected features) is chosen through extensive experimentation in such a way that the classification error of both C4.5 and K-NN becomes almost equal to that of the original feature set.

From the results reported in Tables 5.9, 5.10, and 5.11, it is seen that most of the f-information measures achieve consistently better performance than the mutual information or $I_{1.0}$-information or $\mathcal{R}_{1.0}$-information for different values of α, both in fuzzy and crisp approximation spaces. Some of the f-information measures are shown to perform poorly on all aspects for certain values of α. The majority of measures produces results similar to those of mutual information. An important finding, however, is that several measures, although slightly more difficult to optimize, can potentially yield significantly better results than the mutual information. For Satimage and Isolet data sets, V-information or $M_{1.0}$-information, I_{α}-information and $\mathcal{R}_{\alpha}$-information for $0.8 \leq \alpha \leq 4.0$, and χ^{α}-information for $\alpha = 2.0$ and 3.0 perform better than the mutual information in fuzzy approximation spaces, while for Letter data set, $I_{0.5}$-information, $M_{0.5}$-information, and V-information measures yield best results with respect to most of the quantitative

TABLE 5.6 Performance on Satimage Database for Different Values of Multiplicative Parameter η Considering $d = 10$ and $\beta = 0.5$

Evaluation Criteria	Algorithms/ Methods	Value of Multiplicative Parameter η											
		0.8	0.9	1.0	1.1	1.2	1.3	1.4	1.5	1.6	1.7	1.8	1.9
Classification error (C4.5)	MI-C	18.8	18.8	18.8	18.8	19.3	18.6	17.2	17.3	17.1	18.5	18.6	18.1
	MI-D	22.7	22.7	20.9	19.8	19.8	19.8	18.4	18.1	18.2	19.7	20.1	20.1
	VI-C	17.8	17.8	17.8	17.8	17.8	17.6	17.4	17.4	18.6	17.2	17.2	17.4
	VI-D	21.3	21.3	21.3	20.8	20.8	18.6	18.2	18.1	19.7	19.2	19.2	19.0
Classification error (K-NN)	MI-C	19.3	19.3	19.3	18.6	18.6	18.6	17.3	17.3	17.2	18.1	18.3	18.8
	MI-D	22.7	22.7	21.3	19.8	19.8	18.8	18.8	18.8	18.6	19.7	20.1	20.1
	VI-C	22.7	22.7	19.7	19.7	17.8	17.8	16.6	16.6	17.3	18.1	19.2	21.5
	VI-D	21.3	21.3	20.8	20.8	20.8	20.8	18.6	18.1	18.2	18.6	22.7	22.7
Class separability	MI-C	0.449	0.467	0.458	0.465	0.367	0.367	0.367	0.366	0.310	0.408	0.421	0.461
	MI-D	0.617	0.611	0.611	0.608	0.563	0.563	0.467	0.467	0.460	0.499	0.586	0.586
	VI-C	0.447	0.463	0.417	0.417	0.398	0.398	0.361	0.366	0.355	0.328	0.391	0.437
	VI-D	0.614	0.610	0.632	0.617	0.559	0.510	0.483	0.467	0.491	0.491	0.512	0.580
Entropy (E)	MI-C	0.836	0.832	0.832	0.832	0.830	0.829	0.828	0.824	0.832	0.839	0.840	0.838
	MI-D	0.861	0.863	0.844	0.844	0.844	0.834	0.830	0.832	0.837	0.841	0.841	0.846
	VI-C	0.835	0.835	0.834	0.829	0.829	0.816	0.816	0.802	0.802	0.811	0.834	0.833
	VI-D	0.858	0.858	0.851	0.850	0.841	0.841	0.837	0.830	0.827	0.832	0.832	0.839
Representation entropy (H_R)	MI-C	3.284	3.284	3.284	3.284	3.284	3.282	3.313	3.366	3.295	3.295	3.299	3.299
	MI-D	3.217	3.217	3.246	3.248	3.248	3.263	3.263	3.263	3.263	3.254	3.259	3.259
	VI-C	3.321	3.321	3.299	3.325	3.327	3.327	3.420	3.420	3.420	3.421	3.417	3.417
	VI-D	3.208	3.208	3.208	3.212	3.219	3.214	3.214	3.217	3.217	3.217	3.211	3.211

TABLE 5.7 Performance on Isolet Database for Different Values of Multiplicative Parameter η Considering $d = 25$ and $\beta = 0.5$

Evaluation Criteria	Algorithms/ Methods	Value of Multiplicative Parameter η											
		0.8	0.9	1.0	1.1	1.2	1.3	1.4	1.5	1.6	1.7	1.8	1.9
Classification error (C4.5)	MI-C	14.5	14.5	12.7	9.6	12.4	11.5	9.3	8.8	9.2	8.9	9.1	9.1
	MI-D	19.2	19.2	18.4	17.1	17.0	13.8	12.6	12.3	12.3	13.9	13.9	14.1
	VI-C	12.4	12.3	10.9	9.4	9.4	8.8	8.7	8.4	8.3	8.5	9.1	9.0
	VI-D	18.7	18.7	17.4	16.3	16.1	13.9	12.0	11.4	11.5	11.4	12.8	13.7
Classification error (K-NN)	MI-C	12.4	12.4	11.5	11.5	11.5	8.8	8.8	6.1	7.2	8.5	9.1	9.1
	MI-D	18.7	18.7	18.7	18.7	16.2	13.8	9.3	9.3	9.3	10.9	13.9	13.9
	VI-C	13.1	12.3	12.3	10.9	9.4	8.8	6.1	5.8	6.1	7.8	9.6	11.7
	VI-D	17.6	17.6	16.2	14.5	14.5	11.4	11.4	10.7	14.5	17.9	18.0	19.2
Class separability	MI-C	0.259	0.241	0.268	0.255	0.218	0.189	0.147	0.113	0.126	0.169	0.179	0.208
	MI-D	0.411	0.403	0.395	0.374	0.368	0.351	0.350	0.344	0.341	0.363	0.398	0.418
	VI-C	0.153	0.158	0.131	0.118	0.113	0.105	0.099	0.097	0.097	0.102	0.134	0.162
	VI-D	0.411	0.417	0.404	0.396	0.388	0.364	0.361	0.358	0.357	0.391	0.399	0.412
Entropy (E)	MI-C	0.286	0.286	0.287	0.284	0.284	0.281	0.279	0.276	0.276	0.278	0.283	0.289
	MI-D	0.329	0.328	0.329	0.322	0.322	0.318	0.314	0.313	0.316	0.317	0.320	0.324
	VI-C	0.286	0.286	0.286	0.283	0.279	0.279	0.278	0.276	0.276	0.277	0.281	0.285
	VI-D	0.317	0.316	0.316	0.314	0.315	0.308	0.308	0.303	0.304	0.311	0.312	0.317
Representation entropy (H_R)	MI-C	4.403	4.417	4.329	4.461	4.466	4.607	4.633	4.629	4.617	4.502	4.495	4.327
	MI-D	4.017	4.004	4.097	4.176	4.205	4.289	4.331	4.407	4.416	4.228	4.109	4.082
	VI-C	4.414	4.427	4.326	4.475	4.483	4.591	4.603	4.647	4.644	4.428	4.417	4.445
	VI-D	4.016	4.017	4.073	4.128	4.362	4.360	4.447	4.445	4.446	4.437	4.308	4.184

TABLE 5.8 RELEV and REDUN Index Values on Satimage and Isolet Database for $\beta = 0.5$

Evaluation Criteria	Algorithms/ Methods	Value of Multiplicative Parameter η											
		0.8	0.9	1.0	1.1	1.2	1.3	1.4	1.5	1.6	1.7	1.8	1.9
		Satimage Database Considering $d = 10$											
RELEV index	MI-C	0.257	0.283	0.307	0.331	0.358	0.383	0.407	0.427	0.407	0.387	0.393	0.403
	MI-D	0.254	0.254	0.278	0.305	0.314	0.327	0.334	0.334	0.334	0.303	0.318	0.318
	VI-C	0.421	0.419	0.421	0.431	0.431	0.434	0.434	0.434	0.437	0.437	0.429	0.429
	VI-D	0.261	0.260	0.285	0.308	0.313	0.330	0.337	0.340	0.344	0.326	0.319	0.318
REDUN index	MI-C	0.369	0.393	0.419	0.443	0.369	0.392	0.412	0.345	0.408	0.409	0.424	0.437
	MI-D	0.417	0.423	0.428	0.441	0.397	0.378	0.365	0.350	0.389	0.414	0.451	0.455
	VI-C	0.451	0.437	0.434	0.426	0.425	0.422	0.417	0.419	0.424	0.437	0.440	0.448
	VI-D	0.460	0.449	0.440	0.429	0.428	0.422	0.416	0.411	0.408	0.426	0.443	0.459
		Isolet Database Considering $d = 25$											
RELEV index	MI-C	0.403	0.399	0.397	0.391	0.404	0.403	0.415	0.417	0.417	0.411	0.410	0.394
	MI-D	0.296	0.288	0.292	0.306	0.311	0.308	0.315	0.314	0.312	0.299	0.278	0.263
	VI-C	0.404	0.402	0.394	0.396	0.401	0.407	0.416	0.418	0.415	0.412	0.411	0.405
	VI-D	0.296	0.289	0.295	0.299	0.307	0.311	0.312	0.316	0.315	0.311	0.298	0.291
REDUN index	MI-C	0.433	0.429	0.446	0.452	0.427	0.422	0.425	0.425	0.426	0.433	0.461	0.472
	MI-D	0.514	0.506	0.507	0.502	0.495	0.499	0.491	0.490	0.486	0.497	0.512	0.521
	VI-C	0.437	0.433	0.429	0.461	0.452	0.437	0.429	0.424	0.428	0.455	0.463	0.470
	VI-D	0.497	0.502	0.506	0.499	0.476	0.469	0.466	0.461	0.467	0.475	0.471	0.489

TABLE 5.9 Comparative Performance Analysis of Different f-Information Measures on Letter Database for $d = 6$

Type of f	Value of α	C4.5 Error Fuzzy	C4.5 Error Crisp	K-NN Error Fuzzy	K-NN Error Crisp	Separability Fuzzy	Separability Crisp	Entropy E Fuzzy	Entropy E Crisp	Entropy H_R Fuzzy	Entropy H_R Crisp	RELEV Index Fuzzy	RELEV Index Crisp	REDUN Index Fuzzy	REDUN Index Crisp
I_α	0.2	21.8	25.2	11.6	16.2	1.159	1.608	0.889	0.911	3.208	3.082	0.170	0.164	0.237	0.301
	0.5	16.3	19.1	7.8	15.2	1.087	1.108	0.881	0.898	3.316	3.264	0.231	0.219	0.207	0.283
	0.8	21.8	25.2	11.6	16.2	1.159	1.608	0.889	0.911	3.208	3.082	0.170	0.164	0.237	0.301
	1.0	23.2	26.7	16.9	22.8	1.345	1.562	0.892	0.914	3.172	3.005	0.137	0.118	0.299	0.316
	1.5	21.8	25.2	11.6	16.2	1.159	1.608	0.889	0.911	3.208	3.082	0.170	0.164	0.237	0.301
	2.0	23.2	26.7	16.9	22.8	1.345	1.562	0.892	0.914	3.172	3.005	0.137	0.118	0.299	0.316
	3.0	23.2	26.7	16.9	22.8	1.345	1.562	0.892	0.914	3.172	3.005	0.137	0.118	0.299	0.316
M_α	0.2	21.8	25.2	11.6	16.2	1.159	1.608	0.889	0.911	3.208	3.082	0.170	0.164	0.237	0.301
	0.5	16.3	19.1	7.8	15.2	1.087	1.108	0.881	0.898	3.316	3.264	0.231	0.219	0.207	0.283
	0.8	21.8	25.2	11.6	16.2	1.159	1.608	0.889	0.911	3.208	3.082	0.170	0.164	0.237	0.301
	1.0	17.5	21.3	8.7	17.2	1.055	1.114	0.874	0.903	3.469	3.187	0.240	0.213	0.193	0.294
χ^α	1.5	21.8	25.2	11.6	16.2	1.159	1.608	0.889	0.911	3.208	3.082	0.170	0.164	0.237	0.301
	2.0	23.2	26.7	16.9	22.8	1.345	1.562	0.892	0.914	3.172	3.005	0.137	0.118	0.299	0.316
	3.0	23.2	26.7	16.9	22.8	1.345	1.562	0.892	0.914	3.172	3.005	0.137	0.118	0.299	0.316
R_α	0.2	21.8	25.2	11.6	16.2	1.159	1.608	0.889	0.911	3.208	3.082	0.170	0.164	0.237	0.301
	0.5	21.8	25.2	11.6	16.2	1.159	1.608	0.889	0.911	3.208	3.082	0.170	0.164	0.237	0.301
	0.8	21.8	25.2	11.6	16.2	1.159	1.608	0.889	0.911	3.208	3.082	0.170	0.164	0.237	0.301
	1.0	23.2	26.7	16.9	22.8	1.345	1.562	0.892	0.914	3.172	3.005	0.137	0.118	0.299	0.316
	1.5	21.8	25.2	11.6	16.2	1.159	1.608	0.889	0.911	3.208	3.082	0.170	0.164	0.237	0.301
	2.0	23.2	26.7	16.9	22.8	1.345	1.562	0.892	0.914	3.172	3.005	0.137	0.118	0.299	0.316
	3.0	23.2	26.7	16.9	22.8	1.345	1.562	0.892	0.914	3.172	3.005	0.137	0.118	0.299	0.316

TABLE 5.10 Comparative Performance Analysis of Different f-Information Measures on Satimage Database for $d = 10$

Type of f	Value of α	C4.5 Error		K-NN Error		Separability		Entropy E		Entropy H_R		RELEV Index		REDUN Index	
		Fuzzy	Crisp	Fuzzy	Crisp	Fuzzy	Crisp	Fuzzy	Crisp	Fuzzy	Crisp	Fuzzy	Crisp	Fuzzy	Crisp
I_α	0.2	18.0	17.9	17.2	18.0	0.435	0.465	0.828	0.827	3.298	3.257	0.419	0.341	0.349	0.349
	0.5	18.6	18.1	17.2	18.8	0.369	0.467	0.829	0.832	3.282	3.248	0.426	0.334	0.345	0.350
	0.8	17.3	18.1	17.3	18.8	0.367	0.467	0.828	0.832	3.263	3.263	0.427	0.334	0.346	0.350
	1.0	17.3	18.1	17.3	18.8	0.366	0.467	0.824	0.832	3.366	3.263	0.427	0.334	0.345	0.350
	1.5	17.2	18.6	14.8	16.1	0.360	0.478	0.829	0.829	3.364	3.254	0.427	0.329	0.347	0.357
	2.0	17.2	18.6	14.8	16.1	0.361	0.478	0.801	0.829	3.364	3.254	0.428	0.329	0.346	0.357
	3.0	17.1	18.0	17.1	18.0	0.361	0.461	0.827	0.827	3.366	3.260	0.428	0.330	0.341	0.407
	4.0	16.6	18.4	13.6	16.0	0.359	0.473	0.801	0.827	3.458	3.248	0.437	0.324	0.336	0.372
	5.0	26.1	18.4	19.5	16.1	0.447	0.473	0.831	0.827	3.197	3.248	0.411	0.324	0.425	0.372
M_α	0.2	19.8	17.7	16.6	18.1	0.418	0.462	0.824	0.828	3.117	3.351	0.422	0.339	0.379	0.413
	0.8	17.4	18.1	16.6	18.1	0.369	0.467	0.807	0.830	3.289	3.217	0.429	0.340	0.416	0.411
	1.0	17.4	18.1	16.6	18.1	0.366	0.467	0.802	0.830	3.420	3.217	0.434	0.340	0.419	0.411
χ^α	1.5	17.4	18.1	15.1	18.8	0.364	0.467	0.806	0.830	3.282	3.278	0.427	0.337	0.346	0.409
	3.0	16.6	18.6	13.6	16.1	0.360	0.478	0.797	0.829	3.455	3.254	0.434	0.329	0.389	0.357
	4.0	26.1	18.6	19.5	16.1	0.441	0.478	0.816	0.829	3.197	3.254	0.408	0.329	0.312	0.357
	5.0	30.9	18.6	23.7	16.1	0.486	0.478	0.823	0.829	3.068	3.254	0.401	0.329	0.317	0.357
R_α	0.2	18.0	17.9	17.2	18.0	0.435	0.465	0.828	0.827	3.298	3.298	0.427	0.341	0.349	0.349
	0.5	18.6	18.1	17.2	18.8	0.369	0.467	0.829	0.832	3.282	3.261	0.426	0.334	0.345	0.350
	0.8	17.3	18.1	17.3	18.8	0.367	0.467	0.828	0.832	3.263	3.263	0.427	0.334	0.346	0.350
	1.5	17.2	20.6	14.1	19.7	0.360	0.491	0.829	0.834	3.364	3.118	0.428	0.326	0.347	0.351
	2.0	17.2	20.6	14.1	23.6	0.361	0.491	0.826	0.834	3.364	3.118	0.428	0.326	0.347	0.351
	3.0	16.6	18.6	13.9	16.1	0.359	0.478	0.801	0.829	3.457	3.254	0.435	0.329	0.339	0.357
	4.0	16.4	18.6	13.6	16.1	0.355	0.478	0.794	0.831	3.478	3.254	0.441	0.329	0.338	0.357
	5.0	27.2	18.6	21.5	16.1	0.497	0.478	0.828	0.829	3.206	3.254	0.408	0.329	0.430	0.357

TABLE 5.11 Comparative Performance Analysis of Different f-Information Measures on Isolet Database for $d = 25$

Type of f	Value of α	C4.5 Error		K-NN Error		Separability		Entropy E		Entropy H_R		RELEV Index		REDUN Index	
		Fuzzy	Crisp	Fuzzy	Crisp	Fuzzy	Crisp	Fuzzy	Crisp	Fuzzy	Crisp	Fuzzy	Crisp	Fuzzy	Crisp
I_α	0.2	10.3	12.1	8.6	9.8	0.129	0.338	0.283	0.304	4.469	4.429	0.406	0.325	0.461	0.473
	0.5	10.1	11.4	8.6	9.8	0.154	0.416	0.276	0.303	4.639	4.408	0.417	0.316	0.426	0.463
	0.8	8.8	12.3	6.1	9.3	0.113	0.344	0.276	0.313	4.629	4.407	0.416	0.314	0.432	0.486
	1.0	8.8	12.3	6.1	9.3	0.113	0.344	0.276	0.313	4.629	4.407	0.417	0.314	0.425	0.490
	1.5	8.7	12.7	6.4	9.1	0.116	0.351	0.276	0.302	4.634	4.332	0.417	0.313	0.423	0.477
	2.0	9.1	11.9	7.6	9.5	0.120	0.351	0.288	0.308	4.588	4.449	0.412	0.315	0.448	0.457
	3.0	8.5	11.2	6.1	9.9	0.108	0.353	0.276	0.303	4.634	4.417	0.415	0.313	0.419	0.455
	4.0	8.3	11.2	5.8	9.9	0.089	0.353	0.276	0.303	4.627	4.417	0.401	0.313	0.412	0.455
	5.0	11.6	11.2	6.4	9.9	0.133	0.353	0.289	0.303	4.512	4.417	0.381	0.313	0.447	0.455
M_α	0.2	11.9	11.1	8.6	9.8	0.142	0.367	0.285	0.303	4.056	4.562	0.401	0.327	0.435	0.458
	0.8	9.5	11.4	7.8	9.3	0.128	0.389	0.276	0.303	4.641	4.445	0.417	0.316	0.426	0.463
	1.0	8.4	11.4	5.8	10.7	0.097	0.358	0.276	0.303	4.647	4.445	0.418	0.316	0.424	0.461
χ^α	1.5	9.6	11.3	7.6	10.7	0.124	0.347	0.288	0.301	4.487	4.471	0.401	0.348	0.448	0.452
	3.0	8.6	11.9	6.1	9.5	0.088	0.351	0.279	0.308	4.646	4.449	0.404	0.315	0.413	0.459
	4.0	14.7	11.9	9.1	9.5	0.164	0.351	0.285	0.308	4.413	4.449	0.374	0.315	0.466	0.459
	5.0	16.1	11.9	9.6	9.5	0.189	0.351	0.294	0.308	4.416	4.449	0.371	0.315	0.487	0.459
$\mathcal{R}_\alpha$	0.2	10.3	12.1	7.6	9.5	0.129	0.338	0.286	0.308	4.469	4.429	0.406	0.339	0.461	0.473
	0.5	10.8	12.3	7.6	9.3	0.117	0.344	0.283	0.313	4.525	4.407	0.401	0.314	0.465	0.486
	0.8	8.8	12.3	6.1	9.3	0.113	0.344	0.283	0.312	4.629	4.407	0.416	0.314	0.432	0.486
	1.5	8.7	12.3	6.1	9.3	0.116	0.344	0.276	0.313	4.634	4.407	0.417	0.311	0.423	0.473
	2.0	8.7	12.8	6.1	9.5	0.116	0.367	0.276	0.316	4.634	4.354	0.422	0.311	0.429	0.485
	3.0	8.4	13.1	5.8	9.9	0.108	0.419	0.276	0.329	4.642	4.327	0.418	0.311	0.411	0.492
	4.0	8.2	13.1	5.8	9.9	0.083	0.419	0.276	0.329	4.642	4.327	0.425	0.311	0.413	0.492
	5.0	15.7	13.1	8.8	9.9	0.194	0.419	0.291	0.329	4.413	4.327	0.388	0.311	0.471	0.492

indices and other measures are comparable to the mutual information. However, the lowest value of REDUN index for Satimage data is achieved using χ^{α}-information measure at $\alpha = 4.0$ and 5.0.

5.7.7 Comparative Performance of Different Algorithms

Finally, Tables 5.12 and 5.13 compare the best performance of different f-information measures used in the feature selection method that is based on minimum redundancy and maximum relevance. The results are presented on the basis of the minimum classification error of both C4.5 and K-NN for the data sets reported in Section 5.7.1. The values of β and η are considered as 0.5 and 1.5, respectively. The best performance of the QR algorithm, both in fuzzy [35] and crisp [21] approximation spaces, is also provided on the same data sets for the sake of comparison. It is seen that the f-information measures in fuzzy approximation spaces are more effective than that in crisp approximation spaces. The f-information-measure-based feature selection method selects a set of features having lowest classification error of both C4.5 and K-NN, class separability, entropy, and REDUN index values and highest representation entropy and RELEV index values for all the cases. Also, several f-information measures, although slightly more difficult to optimize, can potentially yield significantly better results than the mutual information, both in fuzzy and crisp approximation spaces. Moreover, the f-information-measure-based feature selection method outperforms the existing QR algorithm, both in fuzzy and crisp approximation spaces. However, the QR algorithm achieves the best RELEV index value for all the data sets as it selects only relevant features of a data set without considering the redundancy among them.

The better performance of the FEPM-based method is achieved because of the FEPM provides an efficient way to calculate different useful f-information measures on fuzzy approximation spaces. In effect, a reduced set of features having maximum relevance and minimum redundancy is being obtained using the FEPM-based method.

Table 5.14 reports the corresponding execution time (in milliseconds) of different algorithms. The results of I_{α}-information, M_{α}-information, χ^{α}-information, and $\mathcal{R}_{\alpha}$-information are the average execution time computed over possible values of α. From the results reported in Table 5.14, it is seen that the execution time required for the FEPM-based method in fuzzy approximation spaces is comparable to that in crisp approximation spaces, irrespective of the data sets used. However, as the computational complexity of the existing QR algorithm is exponential in nature, it requires significantly higher execution time compared to that of the FEPM-based method, both in crisp and fuzzy approximation spaces. The significantly lesser execution time of the FEPM-based method is achieved because of the low computational complexity of the algorithm with respect to the number of selected features and the total number of features and samples in the original data set.

TABLE 5.12 Comparative Performance Analysis of Different Methods Using Different Feature Evaluation Indices

Data Sets	Method	K-NN Error		C4.5 Error		Separability		RELEV Index		Entropy E		Entropy H_R		REDUN Index	
		Fuzzy	Crisp	Fuzzy	Crisp	Fuzzy	Crisp	Fuzzy	Crisp	Fuzzy	Crisp	Fuzzy	Crisp	Fuzzy	Crisp
	MI	7.3	7.3	7.9	7.9	0.181	0.181	0.443	0.443	0.743	0.743	0.997	0.997	0.413	0.443
	$I_{4.0}$	4.1	5.8	6.7	7.9	0.138	0.181	0.442	0.443	0.756	0.743	0.985	0.997	0.406	0.443
	$M_{0.8}$	3.9	5.8	2.8	4.5	0.108	0.154	0.445	0.434	0.741	0.745	0.998	0.997	0.417	0.430
Wine	VI	3.9	5.8	2.8	4.5	0.108	0.154	0.445	0.434	0.741	0.745	0.998	0.997	0.417	0.430
$d = 2$	$\chi^{1.5}$	3.9	5.8	2.8	4.5	0.108	0.154	0.445	0.434	0.741	0.745	0.998	0.997	0.417	0.430
	$\mathcal{R}_{4.0}$	4.1	5.8	6.7	7.9	0.138	0.181	0.442	0.443	0.756	0.743	0.985	0.997	0.406	0.443
	QR	3.9	7.0	2.8	9.2	0.108	0.196	0.445	0.442	0.741	0.752	0.998	0.998	0.417	0.459
	MI	16.9	22.8	23.2	26.7	1.345	1.562	0.137	0.118	0.892	0.914	3.172	3.005	0.299	0.316
	$I_{0.5}$	7.8	15.2	16.3	19.1	1.087	1.108	0.231	0.219	0.881	0.898	3.316	3.264	0.207	0.283
	$M_{0.5}$	7.8	15.2	16.3	19.1	1.087	1.108	0.231	0.219	0.881	0.898	3.316	3.264	0.207	0.283
Letter	VI	8.7	17.2	17.5	21.3	1.055	1.114	0.240	0.213	0.874	0.903	3.469	3.187	0.193	0.294
$d = 6$	$\chi^{1.5}$	11.6	16.2	21.8	25.2	1.159	1.608	0.170	0.164	0.889	0.911	3.208	3.082	0.237	0.301
	$\mathcal{R}_{1.5}$	11.6	16.2	21.8	25.2	1.159	1.608	0.170	0.164	0.889	0.911	3.208	3.082	0.237	0.301
	QR	21.8	29.5	38.3	39.7	8.809	9.627	0.285	0.271	0.891	0.906	3.314	3.001	0.330	0.385

TABLE 5.13 Comparative Performance Analysis of Different Methods Using Different Feature Evaluation Indices

Data Sets	Method	K-NN Error		C4.5 Error		Separability		RELEV Index		Entropy E		Entropy H_R		REDUN Index	
		Fuzzy	Crisp	Fuzzy	Crisp	Fuzzy	Crisp	Fuzzy	Crisp	Fuzzy	Crisp	Fuzzy	Crisp	Fuzzy	Crisp
Ionosphere $d=10$	MI	2.6	4.9	2.6	4.8	1.655	3.283	0.393	0.387	0.765	0.765	3.298	3.293	0.374	0.375
	$I_{1.5}$	3.7	4.8	4.6	4.8	2.297	3.283	0.404	0.387	0.765	0.765	3.298	3.293	0.391	0.375
	$M_{0.8}$	3.7	4.8	4.6	4.8	2.297	3.283	0.404	0.387	0.765	0.765	3.298	3.293	0.391	0.375
	VI	2.6	4.8	4.0	4.8	2.681	3.283	0.406	0.387	0.762	0.765	3.295	3.293	0.387	0.375
	$\chi^{1.5}$	3.7	4.8	4.6	4.8	2.297	3.283	0.404	0.387	0.765	0.765	3.298	3.293	0.391	0.375
	$\mathcal{R}_{1.5}$	3.7	4.8	4.6	4.8	2.297	3.283	0.404	0.387	0.765	0.765	3.298	3.293	0.391	0.375
	QR	6.7	9.2	8.5	11.7	5.877	9.361	0.262	0.257	0.754	0.755	3.299	3.291	0.513	0.558
Satimage $d=10$	MI	17.3	18.8	17.3	18.1	0.366	0.467	0.427	0.334	0.824	0.832	3.366	3.263	0.345	0.350
	$I_{4.0}$	13.6	16.0	16.6	18.4	0.359	0.473	0.437	0.324	0.801	0.827	3.458	3.248	0.336	0.372
	$M_{0.8}$	16.6	18.1	17.4	18.1	0.369	0.467	0.429	0.340	0.807	0.830	3.289	3.217	0.416	0.411
	VI	16.6	18.1	17.4	18.1	0.366	0.467	0.434	0.340	0.802	0.830	3.420	3.217	0.419	0.411
	$\chi^{3.0}$	13.6	16.1	16.6	18.6	0.360	0.478	0.434	0.329	0.797	0.829	3.455	3.254	0.389	0.357
	$\mathcal{R}_{4.0}$	13.6	16.1	16.4	18.6	0.355	0.478	0.441	0.329	0.794	0.831	3.478	3.254	0.338	0.357
	QR	19.2	24.8	21.6	24.8	0.892	0.996	0.458	0.357	0.795	0.834	3.118	3.006	0.513	0.529
Isolet $d=25$	MI	6.1	9.3	8.8	12.3	0.113	0.344	0.417	0.314	0.276	0.313	4.629	4.407	0.425	0.490
	$I_{4.0}$	5.8	9.9	8.3	11.2	0.089	0.353	0.401	0.313	0.276	0.303	4.627	4.417	0.412	0.455
	$M_{0.8}$	7.8	9.3	9.5	11.4	0.128	0.389	0.417	0.316	0.276	0.303	4.641	4.445	0.426	0.463
	VI	5.8	10.7	8.4	11.4	0.097	0.358	0.418	0.316	0.276	0.303	4.647	4.445	0.424	0.461
	$\chi^{3.0}$	6.1	9.5	8.6	11.9	0.088	0.351	0.404	0.315	0.279	0.308	4.646	4.449	0.413	0.459
	$\mathcal{R}_{4.0}$	5.8	9.9	8.2	13.1	0.083	0.419	0.425	0.311	0.276	0.329	4.642	4.327	0.413	0.492
	QR	9.5	15.2	12.8	15.2	1.362	1.594	0.449	0.361	0.278	0.341	4.517	4.211	0.507	0.539

TABLE 5.14 Comparative Execution Time (ms) Analysis of Different Methods on Different Data Sets

Methods/ Measures	Wine ($d = 2$, $\mathcal{D} = 13$, $n = 178$)		Letter ($d = 6$, $\mathcal{D} = 16$, $n = 15000$)		Ionosphere ($d = 10$, $\mathcal{D} = 34$, $n = 351$)		Satimage ($d = 10$, $\mathcal{D} = 36$, $n = 4435$)		Isolet ($d = 25$, $\mathcal{D} = 617$, $n = 7797$)	
	Fuzzy	Crisp	Fuzzy	Crisp	Fuzzy	Crisp	Fuzzy	Crisp	Fuzzy	Crisp
MI	8	7	2758	2699	163	144	2384	2273	143,467	141,973
I_α	7	7	2687	2685	165	147	2407	2239	143,460	142,157
M_α	8	8	2706	2776	168	153	2395	2192	143,481	141,996
VI	8	7	2694	2782	162	142	2383	2244	143,459	142,018
χ^α	8	8	2688	2695	167	146	2393	2248	143,478	141,980
$\mathcal{R}_\alpha$	7	8	2691	2795	167	150	2388	2267	143,501	141,769
QR	7	9	24,568	19473	2483	2107	39,179	37,982	71,412,733	7,048,8913

155

5.8 CONCLUSION AND DISCUSSION

The problem of feature selection is highly important, particularly given the explosive growth of available information. The present chapter deals with this task in rough-fuzzy computing framework. Using the concept of f-information measures on fuzzy approximation spaces, an efficient algorithm is described in this chapter for finding nonredundant and relevant features of real-valued data sets. This formulation is geared toward maximizing the utility of rough sets, fuzzy sets, and information measures with respect to knowledge discovery tasks. Several quantitative indices are reported on the basis of fuzzy-rough sets to evaluate the performance of different feature selection methods on fuzzy approximation spaces for real-life data sets. Finally, the superiority of the FEPM-based method over some other related algorithms is demonstrated on a set of real-life data sets. Through these investigations and experiments, the potential utility of the fuzzy-rough method for feature selection from various data sets, in general, is demonstrated. A real-life application of this methodology in bioinformatics is described in Chapter 8, where the problem of selecting discriminative genes from microarray gene expression data sets is addressed.

So far, we have described in Chapters 3, 4, and 5 different rough-fuzzy clustering, classification, and feature selection methodologies with extensive experimental results demonstrating their characteristic features. The next four chapters deal with some of the specific real-life problems in bioinformatics and medical imaging, namely, selection of a minimum set of basis strings with maximum information for amino acid sequence analysis, grouping functionally similar genes from microarray gene expression data through clustering, selection of relevant genes from high dimensional microarray data, and segmentation of brain magnetic resonance (MR) images.

REFERENCES

1. P. A. Devijver and J. Kittler. *Pattern Recognition: A Statistical Approach*. Prentice Hall, Englewood Cliffs, NJ, 1982.

2. D. Koller and M. Sahami. Toward Optimal Feature Selection In *Proceedings of the International Conference on Machine Learning*, Bari, Italy, pages 284–292, 1996.

3. R. Kohavi and G. H. John. Wrappers for Feature Subset Selection. *Artificial Intelligence*, 97(1–2):273–324, 1997.

4. I. Guyon and A. Elisseeff. An Introduction to Variable and Feature Selection. *Journal of Machine Learning Research*, 3:1157–1182, 2003.

5. J. R. Quinlan. Induction of Decision Trees. *Machine Learning*, 1:81–106, 1986.

6. J. R. Quinlan. *C4.5: Programs for Machine Learning*. Morgan Kaufmann, San Francisco, CA, 1993.

7. J. R. Quinlan. Improved Use of Continuous Attributes in C4.5. *Journal of Artificial Intelligence Research*, 4:77–90, 1996.

8. M. Dash and H. Liu. Consistency Based Search in Feature Selection. *Artificial Intelligence*, 151(1–2):155–176, 2003.

9. R. Battiti. Using Mutual Information for Selecting Features in Supervised Neural Net Learning. *IEEE Transactions on Neural Network*, 5(4):537–550, 1994.

10. D. Huang and T. W. S. Chow. Effective Feature Selection Scheme Using Mutual Information. *Neurocomputing*, 63:325–343, 2004.

11. I. Vajda. *Theory of Statistical Inference and Information*. Kluwer Academic, Dordrecht, The Netherlands, 1989.

12. J. P. W. Pluim, J. B. A. Maintz, and M. A. Viergever. f-Information Measures in Medical Image Registration. *IEEE Transactions on Medical Imaging*, 23(12):1508–1516, 2004.

13. P. Maji and S. K. Pal. Feature Selection Using f-Information Measures in Fuzzy Approximation Spaces. *IEEE Transactions on Knowledge and Data Engineering*, 22(6):854–867, 2010.

14. P. Maji. f-Information Measures for Efficient Selection of Discriminative Genes from Microarray Data. *IEEE Transactions on Biomedical Engineering*, 56(4):1063–1069, 2009.

15. P. Maji and S. K. Pal. Fuzzy-Rough Sets for Information Measures and Selection of Relevant Genes from Microarray Data. *IEEE Transactions on Systems Man and Cybernetics Part B-Cybernetics*, 40(3):741–752, 2010.

16. Z. Pawlak. *Rough Sets: Theoretical Aspects of Reasoning About Data*. Kluwer, Dordrecht, The Netherlands, 1991.

17. Q. Shen and A. Chouchoulas. Combining Rough Sets and Data-Driven Fuzzy Learning for Generation of Classification Rules. *Pattern Recognition*, 32(12):2073–2076, 1999.

18. A. Skowron, R. W. Swiniarski, and P. Synak. Approximation Spaces and Information Granulation. *LNCS Transactions on Rough Sets*, 3:175–189, 2005.

19. K. H. Chen, Z. W. Ras, and A. Skowron. Attributes and Rough Properties in Information Systems. *International Journal of Approximate Reasoning*, 2:365–376, 1988.

20. D. Yamaguchi. Attribute Dependency Functions Considering Data Efficiency. *International Journal of Approximate Reasoning*, 51:89–98, 2009.

21. A. Chouchoulas and Q. Shen. Rough Set-Aided Keyword Reduction for Text Categorisation. *Applied Artificial Intelligence*, 15(9):843–873, 2001.

22. C. Cornelis, R. Jensen, G. H. Martin, and D. Slezak. Attribute Selection with Fuzzy Decision Reducts. *Information Sciences*, 180:209–224, 2010.

23. J. Komorowski, Z. Pawlak, L. Polkowski, and A. Skowron. Rough Sets: A Tutorial. In S. K. Pal and A. Skowron, editors, *Rough-Fuzzy Hybridization: A New Trend in Decision Making*, pages 3–98. Springer-Verlag, Singapore, 1999.

24. A. Skowron and C. Rauszer. The Discernibility Matrices and Functions in Information Systems. In R. Slowinski, editor, *Intelligent Decision Support*, pages 331–362. Kluwer Academic, The Netherlands, Dordrecht, 1992.

25. J. Bazan, A. Skowron, and P. Synak. Dynamic Reducts as a Tool for Extracting Laws from Decision Tables. In Z. W. Ras and M. Zemankova, editors, *Proceedings of the 8th Symposium on Methodologies for Intelligent Systems*, volume 869, pages 346–355. Lecture Notes in Artificial Intelligence, Springer-Verlag, Charlotte, North Carolina, 1994.

26. N. Parthalain, Q. Shen, and R. Jensen. A Distance Measure Approach to Exploring the Rough Set Boundary Region for Attribute Reduction. *IEEE Transactions on Knowledge and Data Engineering*, 22(3):305–317, 2010.

27. M. Modrzejewski. Feature Selection Using Rough Sets Theory. In *Proceedings of the 11th International Conference on Machine Learning*, Amherst, MA, pages 213–226, 1993.

28. N. Zhong, J. Dong, and S. Ohsuga. Using Rough Sets with Heuristics for Feature Selection. *Journal of Intelligent Information Systems*, 16:199–214, 2001.

29. A. T. Bjorvand and J. Komorowski. Practical Applications of Genetic Algorithms for Efficient Reduct Computation. In *Proceedings of the 15th IMACS World Congress on Scientific Computation, Modeling and Applied Mathematics*, Berlin, volume 4, pages 601–606, 1997.

30. D. Slezak. Approximate Reducts in Decision Tables. In *Proceedings of the 6th International Conference on Information Processing and Management of Uncertainty in Knowledge-Based Systems*, Granada, pages 1159–1164, 1996.

31. J. Wroblewski. Finding Minimal Reducts Using Genetic Algorithms. In *Proceedings of the 2nd Annual Joint Conference on Information Sciences*, North Carolina, pages 186–189, 1995.

32. D. Dubois and H. Prade. Rough Fuzzy Sets and Fuzzy Rough Sets. *International Journal of General Systems*, 17:191–209, 1990.

33. D. Dubois and H. Prade. Putting Fuzzy Sets and Rough Sets Together. In R. Slowiniski, editor, *Intelligent Decision Support: Handbook of Applications and Advances of Rough Sets Theory*, pages 203–232. Kluwer, Norwell, MA, 1992.

34. R. Jensen and Q. Shen. Fuzzy-Rough Attribute Reduction with Application to Web Categorization. *Fuzzy Sets and Systems*, 141:469–485, 2004.

35. R. Jensen and Q. Shen. Semantics-Preserving Dimensionality Reduction: Rough and Fuzzy-Rough-Based Approach. *IEEE Transactions on Knowledge and Data Engineering*, 16(12):1457–1471, 2004.

36. R. Jensen and Q. Shen. New Approaches to Fuzzy-Rough Feature Selection. *IEEE Transactions on Fuzzy Systems*, 17(4):824–838, 2009.

37. E. C. C. Tsang, D. Chen, D. S. Yeung, X.-Z. Wang, and J. Lee. Attributes Reduction Using Fuzzy Rough Sets. *IEEE Transactions on Fuzzy Systems*, 16(5):1130–1141, 2008.

38. H. Wu, Y. Wu, and J. Luo. An Interval Type-2 Fuzzy Rough Set Model for Attribute Reduction. *IEEE Transactions on Fuzzy Systems*, 17(2):301–315, 2009.

39. Q. Hu, Z. Xie, and D. Yu. Hybrid Attribute Reduction Based on a Novel Fuzzy-Rough Model and Information Granulation. *Pattern Recognition*, 40:3577–3594, 2007.

40. R. Jensen and Q. Shen. Fuzzy-Rough Sets Assisted Attribute Selection. *IEEE Transactions on Fuzzy Systems*, 15:73–89, 2007.

41. Q. Hu, D. Yu, J. Liu, and C. Wu. Neighborhood Rough Set Based Heterogeneous Feature Subset Selection. *Information Sciences*, 178:3577–3594, 2008.

42. T. Y. Lin. Granulation and Nearest Neighborhoods: Rough Set Approach. In W. Pedrycz, editor, *Granular Computing: An Emerging Paradigm*, pages 125–142. Physica-Verlag, Heidelberg, Germany, 2001.

43. Q. Hu, D. Yu, Z. Xie, and J. Liu. Fuzzy Probabilistic Approximation Spaces and Their Information Measures. *IEEE Transactions on Fuzzy Systems*, 14(2):191–201, 2007.

44. S. K. Das. Feature Selection with a Linear Dependence Measure. *IEEE Transactions on Computers*, 20(9):1106–1109, 1971.

45. M. Dash and H. Liu. Unsupervised Feature Selection. In *Proceedings of the Pacific Asia Conference on Knowledge Discovery and Data Mining*, pages 110–121, 2000.

46. M. A. Hall. Correlation Based Feature Selection for Discrete and Numerical Class Machine Learning. In *Proceedings of the 17th International Conference on Machine Learning*, Stanford, CA, 2000.

47. R. P. Heydorn. Redundancy in Feature Extraction. *IEEE Transactions on Computers*, pages 1051–1054, 1971.

48. A. K. Jain and R. C. Dubes. *Algorithms for Clustering Data*. Prentice Hall, Englewood Cliffs, NJ, 1988.

49. B. B. Kiranagi, D. S. Guru, and M. Ichino. Exploitation of Multivalued Type Proximity for Symbolic Feature Selection. In *Proceedings of the International Conference on Computing: Theory and Applications*, Kolkata, India, 2007.

50. P. Mitra, C. A. Murthy, and S. K. Pal. Unsupervised Feature Selection Using Feature Similarity. *IEEE Transactions on Pattern Analysis and Machine Intelligence*, 24(3):301–312, 2002.

51. C. Shannon and W. Weaver. *The Mathematical Theory of Communication*. University Illinois Press, Champaign, IL, 1964.

52. S. K. Pal and S. Mitra. *Neuro-Fuzzy Pattern Recognition: Methods in Soft Computing*. John Wiley & Sons, Inc., New York, 1999.

53. M. Banerjee, S. Mitra, and S. K. Pal. Rough-Fuzzy MLP: Knowledge Encoding and Classification. *IEEE Transactions on Neural Networks*, 9(6):1203–1216, 1998.

54. S. K. Pal, S. Mitra, and P. Mitra. Rough-Fuzzy MLP: Modular Evolution, Rule Generation, and Evaluation. *IEEE Transactions on Knowledge and Data Engineering*, 15(1):14–25, 2003.

55. R. O. Duda, P. E. Hart, and D. G. Stork. *Pattern Classification and Scene Analysis*. John Wiley & Sons, Inc., New York, 1999.

6

ROUGH FUZZY c-MEDOIDS AND AMINO ACID SEQUENCE ANALYSIS

6.1 INTRODUCTION

Recent advancement and wide use of high throughput technology for biological research are producing enormous size of biological data. Data mining techniques and machine learning methods provide useful tools for analyzing these biological data. The successful analysis of biological sequences relies on the efficient coding of the biological information contained in sequences or subsequences. For example, to recognize functional sites within a biological sequence, the subsequences obtained through moving a fixed length sliding window are generally analyzed. The problem with using most pattern recognition algorithms to analyze these biological subsequences is that they cannot recognize nonnumerical features such as the biochemical codes of amino acids. Investigating a proper encoding process before modeling the amino acids is then critical.

The most commonly used method for coding a subsequence is distributed encoding, which encodes each of the 20 amino acids using a 20-bit binary vector [1]. However, in this method the input space is expanded unnecessarily. Also, this method may not be able to encode biological content in sequences efficiently. On the other hand, different distances for different amino acid pairs have been defined by various mutation matrices and validated [2–4]. But, they cannot be used directly for encoding an amino acid to a unique numerical value.

In this background, Yang and Thomson [5–7] proposed the concept of biobasis function for analyzing biological sequences. It uses a kernel function to

Rough-Fuzzy Pattern Recognition: Applications in Bioinformatics and Medical Imaging,
First Edition. Pradipta Maji and Sankar K. Pal.
© 2012 John Wiley & Sons, Inc. Published 2012 by John Wiley & Sons, Inc.

transform biological sequences to feature vectors directly. The bio-basis strings consist of sections of biological sequence that code for a feature of interest in the study, and are responsible for the transformation of biological data to high dimensional feature space. Transformation of input data to high dimensional feature space is performed on the basis of the similarity of an input sequence to a bio-basis string with reference to a biological similarity matrix. Hence, the biological content in the sequences can be maximally utilized for accurate modeling. The use of similarity matrices to map features allows the bio-basis function to analyze biological sequences without the need for encoding. The bio-basis function has been successfully applied to predict different functional sites in proteins [5–10].

The most important issue for bio-basis function is how to select a minimum set of bio-basis strings with maximum information. Berry et al. [5] used genetic algorithms for bio-basis string selection considering Fisher ratio as the fitness function. Yang and Thomson [7] proposed a method to select bio-basis strings using mutual information. In principle, the bio-basis strings in nonnumerical sequence space should be such that the degree of resemblance (DOR) between pairs of bio-basis strings would be as minimum as possible. Each of them would then represent a unique feature in numerical feature space. As this is a feature selection problem, clustering method can be used, which partitions the given biological sequences into subgroups around each bio-basis string, each of which should be as homogeneous or informative as possible. However, the methods proposed in Berry et al. [5] and Yang and Thomson [7] have not adequately addressed this problem. Also, much attention has not been paid to it earlier.

In biological sequences, the only available information is the numerical values that represent the degrees to which pairs of sequences in the data set are related. Algorithms that generate partitions of that type of relational data are usually referred to as *relational* or *pair-wise clustering algorithms*. The relational clustering algorithms can be used to cluster biological subsequences if one can come up with a similarity measure to quantify the DOR between pairs of subsequences. The pair-wise similarities are usually stored in the form of a matrix called the *similarity matrix*.

One of the most popular relational clustering algorithms is the sequential agglomerative hierarchical nonoverlapping model [11]. It is a bottom-up approach that generates crisp clusters by sequentially merging pairs of clusters that are closest to each other in each step. The model gives rise to single, complete, and average linkage algorithms depending on how closeness between clusters is defined. A variation of this algorithm can be found in Guha et al. [12]. Another well-known relational clustering algorithm is c-medoids or partitioning around medoids (PAM) due to Kaufman and Rousseeuw [13]. This algorithm is based on finding representative objects, also known as *medoids* [14], from the data set in such a way that the sum of the within-cluster dissimilarities is minimized. A modified version of the PAM called *CLARA* (clustering large applications) to handle large data sets was also proposed by Kaufman and Rousseeuw [13]. Ng and Han [15] proposed another variation of the CLARA called *CLARANS*. This algorithm tries to make the search for the c-representative objects or medoids more efficiently by considering candidate sets of c medoids in the neighborhood of the current set of c medoids. However, the CLARANS is not designed for relational data. Finally, it is also interesting to note that Fu [16] suggested a technique very similar to the c-medoid technique in the context of clustering string patterns generated by grammar in syntactic pattern recognition. Some of the more recent algorithms for relational clustering include Bajcsy and Ahuja [17], Gowda and Diday [18], Ramkumar and Swami [19], and Sonbaty and Ismail [20].

One of the main problems in biological subsequence analysis is uncertainty. Some of the sources of this uncertainty include incompleteness and vagueness in class definitions. In this background, the possibility concept introduced by the fuzzy sets theory and rough sets theory has gained popularity in modeling and propagating uncertainty. Both fuzzy sets and rough sets provide a mathematical framework to capture uncertainties associated with the data. However, the relational clustering algorithms reported above generate crisp clusters. When the clusters are not well defined, that is, when they are overlapping, one may desire fuzzy clusters. Two of the early fuzzy relational clustering algorithms are those due to Ruspini [21] and Diday [22]. Other notable algorithms include Roubens' fuzzy nonmetric model [23], Windham's association prototype model [24], Hathaway and Bezdek's relational fuzzy c-means (RFCM) [25], and Kaufman and Rousseeuw's fuzzy analysis (FANNY) [13]. The FANNY is in fact very closely related to the RFCM and is essentially equivalent to the RFCM when the fuzzifier is equal to two. Some improvements on this algorithm can also be found in Hathaway and Bezdek [26] and Sen and Dave [27]. The NERFCM model [26] extends the RFCM to ease the restrictions that the RFCM imposes on the dissimilarity matrix. More recently, Sen and Dave [27] have generalized this approach further, including an extension to handle data sets containing noise and outliers.

A more recent fuzzy relational clustering algorithm is fuzzy c-medoids [28]. It offers the opportunity to deal with the data that belong to more than one cluster at the same time. Also, it can handle the uncertainties arising from overlapping cluster boundaries. However, it is very sensitive to noise and outliers as the memberships of a pattern in fuzzy c-medoids are inversely related to the relative distances of the pattern to the cluster prototypes. The possibilistic c-medoids [28] is an extension of fuzzy c-medoids, which efficiently handle the data sets containing noise and outliers considering typicalities or compatibilities of the patterns with the cluster prototypes. But, it sometimes generates coincident clusters. Moreover, typicalities can be very sensitive to the choice of additional parameters needed by possibilistic c-medoids. Recently, the use of both fuzzy (probabilistic) and possibilistic memberships in a relational clustering algorithm has been proposed in Maji and Pal [29].

A relational clustering algorithm, termed as *rough-fuzzy c-medoids*, has been described by Maji and Pal [30], on the basis of the integration of rough and fuzzy sets to cluster relational objects. While the membership function of fuzzy sets enables efficient handling of overlapping partitions, the concept of lower and upper approximations of rough sets deals with uncertainty, vagueness, and incompleteness in class definition. Each partition is represented by a medoid, a crisp lower approximation, and a fuzzy boundary. The lower approximation influences the fuzziness of the final partition. The medoid depends on the weighting average of the crisp lower approximation and fuzzy boundary.

The rough-fuzzy c-medoids is used in this chapter to select a minimum set of most informative bio-basis strings. One of the important problems of the rough-fuzzy c-medoids algorithm is how to select initial prototypes for different clusters. The concept of DOR, based on nongapped pair-wise homology alignment score, circumvents the initialization and local minima problems of c-medoids and enables efficient selection of a minimum set of most informative bio-basis strings. Some quantitative measures are presented on the basis of mutual information and nongapped pair-wise homology alignment scores to evaluate the quality of selected bio-basis strings. The effectiveness of the rough-fuzzy c-medoids algorithm, along with a comparison with hard c-medoids [13], rough c-medoids [30], fuzzy c-medoids [28], possibilistic c-medoids [28], fuzzy-possibilistic c-medoids [29], and the methods proposed by Berry et al. [5] and Yang and Thompson [7], is demonstrated on different protein data sets.

The structure of the rest of this chapter is as follows. Section 6.2 briefly introduces necessary notions of bio-basis function and the bio-basis string selection methods proposed by Berry et al. [5] and Yang and Thomson [7]. Section 6.3 gives a brief description of hard c-medoids, fuzzy c-medoids, possibilistic c-medoids, and fuzzy-possibilistic c-medoids. In Section 6.4, the rough-fuzzy c-medoids algorithm is described, along with the rough c-medoids algorithm, for relational data clustering, while Section 6.5 gives an overview of the application of different c-medoids algorithms for bio-basis string selection, along with the initialization method for the c-medoids algorithm. Some quantitative measures are presented in Section 6.6 to select most informative bio-basis strings. A few case studies and a comparison among different methods are reported in Section 6.7. Concluding remarks are given in Section 6.8.

6.2 BIO-BASIS FUNCTION AND STRING SELECTION METHODS

In this section, the basic notion in the theory of bio-basis function is reported, along with the bio-basis string selection methods proposed by Yang and Thomson [7] and Berry et al. [5].

6.2.1 Bio-Basis Function

The most successful method of sequence analysis is homology alignment [31, 32]. In this method, the function of a sequence is annotated through aligning a novel sequence with known sequences. If the homology alignment between a novel sequence and a known sequence gives a very high similarity score, the novel sequence is believed to have the same or similar function as the known sequence. In homology alignment, an amino acid mutation matrix is commonly used. Each mutation matrix has 20 columns and 20 rows. A value at ith row and jth column is a probability or a likelihood value that the ith amino acid mutates into the jth amino acid after a particular evolutionary time [3, 4].

However, the principle of homology alignment cannot be used directly for subsequence analysis because a subsequence may not contain enough information for conventional homology alignment. A high homology alignment score between a novel subsequence and a known subsequence cannot assert that the two subsequences have the same function. However, it can be assumed that they may have the same function statistically.

The design of bio-basis function is based on the principle of conventional homology alignment used in biology. Using a table look-up technique, a homology alignment score as a similarity value can be obtained for a pair of subsequences. The nongapped homology alignment method is used to calculate this similarity value, where no deletion or insertion is used to align the two subsequences. The definition of bio-basis function is as follows [6, 7]:

$$f(x_j, v_i) = \exp\left\{\gamma \frac{h(x_j, v_i) - h(v_i, v_i)}{h(v_i, v_i)}\right\}, \qquad (6.1)$$

where $h(x_j, v_i)$ is the nongapped pair-wise homology alignment score between a subsequence x_j and a bio-basis string v_i calculated using an amino acid mutation matrix [3, 4], $h(v_i, v_i)$ denotes the maximum homology alignment score of the ith bio-basis string v_i, and γ is a constant. Let $\mathbb{A}$ be the set of 20 amino acids, $X = \{x_1, \ldots, x_j, \ldots, x_n\}$ be the set of n subsequences with m residues, and $V = \{v_1, \ldots, v_i, \ldots, v_c\} \subset X$ be the set of c bio-basis strings such that $v_{ik}, x_{jk} \in \mathbb{A}$, $\forall_{i=1}^c, \forall_{j=1}^n, \forall_{k=1}^m$. The nongapped pair-wise homology alignment score between x_j and v_i is then defined as

$$h(x_j, v_i) = \sum_{k=1}^m M(x_{jk}, v_{ik}), \qquad (6.2)$$

where $M(x_{jk}, v_{ik})$ can be obtained from an amino acid mutation matrix using the table look-up method. The function value is high if the two subsequences are similar or close to each other and one for two identical subsequences. The value is small if two subsequences are distinct.

The bio-basis function transforms various homology alignment scores to a real number as a similarity within the interval $[0, 1]$. Each bio-basis string is a feature dimension in a numerical feature space. It needs a subsequence as a support. A collection of c bio-basis strings formulates a numerical feature space $\Re^c$. After the mapping using bio-basis strings, a nonnumerical subsequence space $\mathbb{A}^m$ will be mapped to a c-dimensional numerical feature space $\Re^c$, that is, $\mathbb{A}^m \rightarrow \Re^c$.

The most important assumption about bio-basis function is that the distribution of the amino acids in sequences depends on the specificity of the sequences. If the distribution of amino acids is in random, the selection of bio-basis string will be very difficult. Fortunately, the biological experiments have shown that the distribution of amino acids at the specific subsites in sequences does depend on the specificity of the sequences.

6.2.2   Selection of Bio-Basis Strings Using Mutual Information

Yang and Thomson [7] proposed a method for bio-basis string selection using mutual information [33]. The necessity for a bio-basis string to be an independent and informative feature can be determined by the shared information between the bio-basis string and the rest as well as the shared information between the bio-basis string and class label [7].

The mutual information is quantified as the difference between the initial uncertainty and the conditional uncertainty. Let $\Phi = \{v_i\}$ be a set of selected bio-basis strings, $\Theta = \{v_k\}$ a set of candidate bio-basis strings, and $\Phi = \phi$ (empty) at the beginning. A prior probability of a bio-basis string v_k is referred to as p(v_k). The initial uncertainty of v_k is defined as

$$H(v_k) = -p(v_k) \ln p(v_k). \tag{6.3}$$

Similarly, the joint entropy of two bio-basis strings v_k and v_i is given by

$$H(v_k, v_i) = -p(v_k, v_i) \ln p(v_k, v_i), \tag{6.4}$$

where $v_i \in \Phi$ and $v_k \in \Theta$. The mutual information between v_k and v_i is, therefore, given by

$$\begin{aligned} I(v_k, v_i) &= H(v_k) + H(v_i) - H(v_k, v_i) \\ &= \{-p(v_k) \ln p(v_k) - p(v_i) \ln p(v_i) + p(v_k, v_i) \ln p(v_k, v_i)\}. \end{aligned} \tag{6.5}$$

However, the mutual information of v_k with respect to all the bio-basis strings in Φ is

$$I(v_k, \Phi) = \sum_{v_i \in \Phi} I(v_k, v_i). \tag{6.6}$$

Combining Equations (6.5) and (6.6), we get [7]

$$I(v_k, \Phi) = \sum_{v_i \in \Phi} p(v_k, v_i) \ln \left\{ \frac{p(v_k, v_i)}{p(v_k)p(v_i)} \right\}. \tag{6.7}$$

Replacing Φ with the class label $\Omega = \{\Omega_1, \ldots, \Omega_j, \ldots, \Omega_M\}$, the mutual information

$$I(v_k, \Omega) = \sum_{\Omega_j \in \Omega} p(v_k, \Omega_j) \ln \left\{ \frac{p(v_k, \Omega_j)}{p(v_k)p(\Omega_j)} \right\} \tag{6.8}$$

measures the mutual relationship between v_k and Ω. A bio-basis string whose $I(v_k, \Omega)$ value is the largest will be selected as v_k and will make the largest contribution to modeling (discrimination using Ω) among all the remaining bio-basis strings in Θ. Therefore, there are two mutual information measurements for v_k, the mutual information between v_k and Ω ($I(v_k, \Omega)$) and the mutual information between v_k and Φ ($I(v_k, \Phi)$). In this method, the following criterion is used for the selection of bio-basis strings [7, 34]

$$J(v_k) = \alpha_{YT} I(v_k, \Omega) - (1 - \alpha_{YT}) I(v_k, \Phi), \tag{6.9}$$

where α_{YT} is a constant. In the current study, the value of α_{YT} is set at 0.7 to give more weightage in discrimination [7, 34]. The major drawback of the method proposed by Yang and Thomson [7] is that a huge number of prior and joint probabilities are to be calculated, which makes the method computationally expensive.

6.2.3   Selection of Bio-Basis Strings Using Fisher Ratio

Berry et al. [5] proposed a method to select a set $V = \{v_1, \ldots, v_i, \ldots, v_c\}$ of c bio-basis strings from the whole set $X = \{x_1, \ldots, x_j, \ldots, x_n\}$ of n subsequences based on their discriminant capability. The discriminant capability of each subsequence is calculated using the Fisher ratio. The Fisher ratio is used to maximize the discriminant capability of a subsequence in terms of interclass separation (as large as possible) and intraclass spread between subsequences (as small as possible). The larger the Fisher ratio value, the larger the discriminant capability of the subsequence. On the basis of the values of the Fisher ratio, n subsequences of X can be ranked from the strongest discriminant capability to the weakest one. The method yields a set V of c subsequences from X as the bio-basis strings that possess good discriminant capability between two classes, having evolved from the original data set.

However, the n subsequences of X would have different compositions of amino acids. Hence, they should have different pair-wise alignment scores with the other subsequences of X. As the class properties of these training subsequences are known, these similarity values can be partitioned into two groups or classes, namely, functional and nonfunctional, which are denoted as $X_A \subset X$ and $X_B \subset X$, respectively. Denoting the similarity between two subsequences x_i and x_j as $h(x_j, x_i)$, the mean and standard deviation values for these two groups with respect to the subsequence x_i are as follows:

$$U_{A_i} = \frac{1}{n_A} \sum h(x_j, x_i); \qquad \forall x_j \in X_A \tag{6.10}$$

$$U_{B_i} = \frac{1}{n_B} \sum h(x_k, x_i); \qquad \forall x_k \in X_B \tag{6.11}$$

$$\sigma_{A_i}^2 = \frac{1}{n_A} \sum \{h(x_j, x_i) - U_{A_i}\}^2; \quad \forall x_j \in X_A \tag{6.12}$$

$$\sigma_{B_i}^2 = \frac{1}{n_B} \sum \{h(x_k, x_i) - U_{B_i}\}^2; \quad \forall x_k \in X_B, \tag{6.13}$$

where n_A and n_B are the number of similarity values in X_A and X_B, respectively. On the basis of these four quantities, the discriminant capability of each subsequence can be measured using the Fisher ratio

$$F(x_i) = \frac{|U_{A_i} - U_{B_i}|}{\sqrt{\sigma_{A_i}^2 + \sigma_{B_i}^2}}. \tag{6.14}$$

The basic steps of this method follows next:

1. Calculate the discriminant capabilities of all subsequences of X using the Fisher ratio as in Equation (6.14).
2. Rank all subsequences of X on the basis of the values of Fisher ratio in descending order.
3. Select first c subsequences from X as the set V of bio-basis strings.

However, the bio-basis strings in nonnumerical sequence space should be such that the similarity between pairs of bio-basis strings would be as minimum as possible. Each of them would then represent a unique feature in numerical feature space. The methods proposed by Berry et al. [5] and Yang and Thomson [7] have not adequately addressed this problem. Also, not much attention has been paid to it earlier.

6.3   FUZZY-POSSIBILISTIC c-MEDOIDS ALGORITHM

Three relational clustering algorithms, namely, hard c-medoids [13], fuzzy c-medoids [28], and possibilistic c-medoids [28], are described first. Next, a new relational clustering algorithm, termed as *fuzzy-possibilistic c-medoids* [29], is presented on the basis of these algorithms.

6.3.1   Hard c-Medoids

The hard c-medoids algorithm [13, 14] uses the most centrally located object in a cluster, which is termed as the *medoid*. A medoid is essentially one of the actual data points from the cluster, which is closest to the mean of the cluster. The objective of the hard c-medoids algorithm for clustering relational data sets is to assign n objects to c clusters. Each of the clusters β_i is represented by a medoid v_i, which is the cluster representative for that cluster. The process begins by randomly choosing c objects as the medoids. The objects are assigned to one of the c clusters on the basis of the similarity or dissimilarity between the object x_j and the medoid v_i. The pair-wise similarity or dissimilarity can be calculated from the similarity or dissimilarity matrix, respectively. After the assignment of all the objects to various clusters, the new medoids are calculated as follows:

$$v_i = x_q, \text{ where } q = \arg\min\{d(x_k, x_j)\}; \quad x_j \in \beta_i; \quad x_k \in \beta_i, \tag{6.15}$$

where $d(x_k, x_j)$ represents the dissimilarity value between two objects x_j and x_k that can be obtained from the dissimilarity matrix. For relational objects with only pair-wise similarity values, the new medoids can be calculated as

$$v_i = x_q, \text{ where } q = \arg\max\{s(x_k, x_j)\}; \quad x_j \in \beta_i; \quad x_k \in \beta_i, \tag{6.16}$$

where $s(x_k, x_j)$ represents the similarity value between two objects x_j and x_k that can be obtained from the similarity matrix. The basic steps of the hard c-medoids algorithm are outlined as follows:

1. Arbitrarily choose c objects as the initial medoids v_i, $i = 1, 2, \ldots, c$.
2. Assign each remaining objects to the cluster for the closest medoid.
3. Compute the new medoid as per Equation (6.15) or (6.16).
4. Repeat steps 2 and 3 until no more new assignments can be made.

6.3.2 Fuzzy c-Medoids

This provides a fuzzification of the hard c-medoids algorithm [28]. For clustering relational data sets, it minimizes

$$J_{\mathrm{F}} = \sum_{j=1}^{n} \sum_{i=1}^{c} (\mu_{ij})^{\acute{m}_1} d(x_j, v_i), \tag{6.17}$$

where $1 \le \acute{m}_1 < \infty$ is the fuzzifier, v_i the ith medoid, $\mu_{ij} \in [0, 1]$ the fuzzy membership of the object x_j to cluster β_i such that

$$\mu_{ij} = \sum_{l=1}^{c} \left\{ \frac{d(x_j, v_i)}{d(x_j, v_l)} \right\}^{-1/(\acute{m}_1 - 1)} \tag{6.18}$$

subject to

$$\sum_{i=1}^{c} \mu_{ij} = 1, \forall j, \text{ and } 0 < \sum_{j=1}^{n} \mu_{ij} < n, \quad \forall i. \tag{6.19}$$

The new medoids are calculated as follows:

$$v_i = x_q, \text{ where } q = \arg\min \sum_{k=1}^{n} (\mu_{ik})^{\acute{m}_1} d(x_k, x_j); \quad 1 \le j \le n. \tag{6.20}$$

The basic algorithm proceeds as follows:

1. Assign initial medoids v_i, $i = 1, 2, \ldots, c$.
2. Choose values for fuzzifier $\acute{m}_1$ and threshold ϵ_1 and set iteration counter $t = 1$.
3. Compute μ_{ij} using Equation (6.18) for c clusters and n objects.
4. Update medoid v_i using Equation (6.20).
5. Repeat steps 3–5, by incrementing t, until $|\mu_{ij}(t) - \mu_{ij}(t-1)| > \epsilon_1$.

6.3.3   Possibilistic c-Medoids

In possibilistic c-medoids [28], the objective function can be formulated as follows:

$$J_P = \sum_{i=1}^{c}\sum_{j=1}^{n}(v_{ij})^{\acute{m}_2}d(x_j, v_i) + \sum_{i=1}^{c}\eta_i\sum_{j=1}^{n}(1 - v_{ij})^{\acute{m}_2}, \tag{6.21}$$

where η_i represents the bandwidth or resolution or scale parameter. The membership matrix v generated by possibilistic c-medoids is not a partition matrix in the sense that it does not satisfy the constraint

$$\sum_{i=1}^{c}v_{ij} = 1. \tag{6.22}$$

The membership update equation in possibilistic c-medoids is

$$v_{ij} = \frac{1}{1 + D}, \quad \text{where } D = \left\{\frac{d(x_j, v_i)}{\eta_i}\right\}^{1/(\acute{m}_2 - 1)} \tag{6.23}$$

subject to

$$v_{ij} \in [0, 1], \forall i, j; 0 < \sum_{j=1}^{n}v_{ij} \leq n, \forall i; \text{ and } \max_{i} v_{ij} > 0, \forall j. \tag{6.24}$$

The scale parameter η_i represents the zone of influence or size of the cluster β_i and its update equation is

$$\eta_i = K \cdot \frac{P}{Q}, \tag{6.25}$$

where

$$P = \sum_{j=1}^{n}(v_{ij})^{\acute{m}_2}d(x_j, v_i) \tag{6.26}$$

and

$$Q = \sum_{j=1}^{n} (v_{ij})^{\acute{m}_2}. \tag{6.27}$$

Typically K is chosen to be 1. From the standpoint of compatibility with the medoid, the membership v_{ij} of an object x_j in a cluster β_i should be determined solely by how close it is to the medoid v_i of the class, and should not be coupled with its similarity with respect to other classes. Hence, in each iteration, the updated value of v_{ij} depends only on the similarity between the object x_j and the medoid v_i. The resulting partition of the relational data can be interpreted as a possibilistic partition, and the membership values may be interpreted as degrees of possibility of the objects belonging to the classes, that is, the compatibilities of the objects with the medoids. The updating of the medoids proceeds exactly the same way as in the case of the fuzzy c-medoids algorithm.

6.3.4 Fuzzy-Possibilistic c-Medoids

Incorporating both fuzzy and possibilistic membership functions into the hard c-medoids algorithm, the fuzzy-possibilistic c-medoids algorithm is introduced in Maji and Pal [29]. It avoids the noise sensitivity defect of fuzzy c-medoids and the coincident clusters problem of possibilistic c-medoids. For relational data clustering, it minimizes the following objective function

$$J_{\text{FP}} = \sum_{j=1}^{n} \sum_{i=1}^{c} \{a(\mu_{ij})^{\acute{m}_1} + b(v_{ij})^{\acute{m}_2}\} d(x_j, v_i) + \sum_{i=1}^{c} \eta_i \sum_{j=1}^{n} (1 - v_{ij})^{\acute{m}_2}. \tag{6.28}$$

The constants a and b, respectively, define the relative importance of fuzzy and possibilistic memberships in the objective function and $a + b = 1$. Note that μ_{ij} has the same meaning of membership as that in fuzzy c-medoids. Similarly, v_{ij} has the same interpretation of typicality as in possibilistic c-medoids and is given by

$$v_{ij} = \frac{1}{1 + D}, \tag{6.29}$$

where

$$D = \left\{ \frac{bd(x_j, v_i)}{\eta_i} \right\}^{1/(\acute{m}_2 - 1)}. \tag{6.30}$$

Hence, the constant b has a direct influence on the possibilistic membership v_{ij}. The scale parameter η_i has the same expression as that in Equation (6.25). The new medoids are calculated as follows:

$$v_i = x_q, \tag{6.31}$$

where

$$q = \arg\min \sum_{k=1}^{n} \{a(\mu_{ik})^{\acute{m}_1} + b(v_{ik})^{\acute{m}_2}\} d(x_k, x_j); \quad 1 \le j \le n. \qquad (6.32)$$

The main steps of the algorithm are as follows:

1. Assign initial medoids v_i, $i = 1, 2, \ldots, c$.
2. Choose values for a, b, $\acute{m}_1$, $\acute{m}_2$, and threshold ϵ_1, and set iteration counter $t = 1$.
3. Compute μ_{ij} and v_{ij} using Equations (6.18) and (6.29), respectively, for c clusters and n objects.
4. Estimate η_i using Equation (6.25).
5. Update medoid v_i using Equation (6.31).
6. Repeat steps 3–6, by incrementing t, until $|v_{ij}(t) - v_{ij}(t - 1)| > \epsilon_1$.

The fuzzy-possibilistic c-medoids algorithm boils down to fuzzy c-medoids for $a = 1$, while the algorithm reduces to the possibilistic c-medoids algorithm at $a = 0$. Hence, it is the generalization of the existing fuzzy c-medoids and possibilistic c-medoids algorithms.

6.4 ROUGH-FUZZY c-MEDOIDS ALGORITHM

In this section, two relational clustering algorithms, namely, rough c-medoids [29, 30] and rough-fuzzy c-medoids [30], are described incorporating the concept of lower and upper approximations of rough sets into hard c-medoids and fuzzy c-medoids, respectively.

6.4.1 Rough c-Medoids

Let $\underline{A}(\beta_i)$ and $\overline{A}(\beta_i)$ be the lower and upper approximations of cluster β_i, and $B(\beta_i) = \{\overline{A}(\beta_i) \setminus \underline{A}(\beta_i)\}$ denotes the boundary region of cluster β_i (Fig. 6.1). In the rough c-medoids algorithm [29, 30], the concept of the c-medoids algorithm is extended by viewing each cluster β_i as an interval or rough set. However, it is possible to define a pair of lower and upper bounds $[\underline{A}(\beta_i), \overline{A}(\beta_i)]$ or a rough set for every cluster $\beta_i \subseteq U$, U being the set of objects of concern. The family of upper and lower bounds is required to follow some of the basic rough set properties such as the following:

1. An object x_j can be part of at most one lower bound.
2. $x_j \in \underline{A}(\beta_i) \Rightarrow x_j \in \overline{A}(\beta_i)$.
3. An object x_j is not part of any lower bound $\Rightarrow x_j$ belongs to two or more upper bounds.

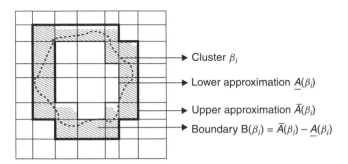

Cluster β_i

Lower approximation $\underline{A}(\beta_i)$

Upper approximation $\bar{A}(\beta_i)$

Boundary $B(\beta_i) = \bar{A}(\beta_i) - \underline{A}(\beta_i)$

Figure 6.1 Rough *c*-medoids: cluster β_i is represented by lower and upper bounds.

Incorporating rough sets into the *c*-medoids algorithm, the rough *c*-medoids algorithm [29, 30] is presented for clustering relational data sets. It adds the concept of lower and upper bounds of rough sets into hard *c*-medoids algorithm. It classifies the object space into two parts, namely, lower approximation and boundary region. The medoid is calculated on the basis of the weighting average of the lower bound and boundary region. All the objects in lower approximation take the same weight w, while all the objects in boundary take another weighting index $\tilde{w}$ uniformly. Calculation of the medoids is modified to include the effects of lower as well as upper bounds. The modified medoids calculation for rough *c*-medoids is given by

$$v_i = x_q, \tag{6.33}$$

where q is given by

$$q = \arg\min \begin{cases} w \times \mathcal{A} + \tilde{w} \times \mathcal{B} & \text{if } \underline{A}(\beta_i) \neq \emptyset, B(\beta_i) \neq \emptyset \\ \mathcal{A} & \text{if } \underline{A}(\beta_i) \neq \emptyset, B(\beta_i) = \emptyset \\ \mathcal{B} & \text{if } \underline{A}(\beta_i) = \emptyset, B(\beta_i) \neq \emptyset \end{cases} \tag{6.34}$$

where

$$\mathcal{A} = \sum_{x_k \in \underline{A}(\beta_i)} d(x_k, x_j) \tag{6.35}$$

and

$$\mathcal{B} = \sum_{x_k \in B(\beta_i)} d(x_k, x_j). \tag{6.36}$$

The parameters w and $\tilde{w}$ $(= 1 - w)$ correspond to the relative importance of lower bound and boundary region, respectively. Since the objects lying in lower approximation definitely belong to a cluster, they are assigned a higher weight w compared to $\tilde{w}$ of the objects lying in the boundary region. That is, $0 < \tilde{w} < w < 1$. The main steps of the rough c-medoids algorithm are as follows:

1. Assign initial medoids v_i, $i = 1, 2, \ldots, c$.
2. Choose values for w and threshold ϵ_2.
3. For each object x_j, calculate dissimilarity score $d(x_j, v_i)$ between itself and the medoid v_i of cluster β_i.
4. If $d(x_j, v_i)$ is minimum for $1 \leq i \leq c$ and $d(x_j, v_k) - d(x_j, v_i) \leq \epsilon_2$, then $x_j \in \overline{A}(\beta_i)$ and $x_j \in \overline{A}(\beta_k)$. Furthermore, x_j is not part of any lower bound.
5. Otherwise, $x_j \in \underline{A}(\beta_i)$ such that $d(x_j, v_i)$ is the minimum for $1 \leq i \leq c$. In addition, by properties of rough sets, $x_j \in \overline{A}(\beta_i)$.
6. Compute new medoid as per Equation (6.33).
7. Repeat steps 2–6 until no more new assignments can be made.

6.4.2   Rough-Fuzzy c-Medoids

Incorporating both fuzzy sets and rough sets, another version of the c-medoids algorithm termed as *rough-fuzzy c-medoids* [30], is reported. The rough-fuzzy c-medoids algorithm adds the concept of fuzzy membership of fuzzy sets, and lower and upper approximations of rough sets into c-medoids algorithm. While the lower and upper bounds of rough sets deal with uncertainty, vagueness, and incompleteness in class definition, the membership of fuzzy sets enables efficient handling of overlapping partitions.

In fuzzy c-medoids, the medoid depends on the fuzzy membership values of different objects, whereas in rough-fuzzy c-medoids, after computing the memberships for c clusters and n objects, the membership values of each object are sorted and the difference of two highest memberships is compared with a threshold value ϵ_2. Let μ_{ij} and μ_{kj} be the highest and second highest memberships of object x_j. If $(\mu_{ij} - \mu_{kj}) > \epsilon_2$, then $x_j \in \underline{A}(\beta_i)$ as well as $x_j \in \overline{A}(\beta_i)$ and $x_j \notin \underline{A}(\beta_k)$, otherwise $x_j \in B(\beta_i)$ and $x_j \in B(\beta_k)$. That is, the rough-fuzzy c-medoids algorithm first separates the core and overlapping portions of each cluster β_i based on the threshold value ϵ_2. The core portion of the cluster β_i is represented by its lower approximation $\underline{A}(\beta_i)$, while the boundary region $B(\beta_i)$ represents the overlapping portion. In effect, it minimizes the vagueness and incompleteness in cluster definition.

According to the definitions of lower approximations and boundary of rough sets, if an object $x_j \in \underline{A}(\beta_i)$, then $x_j \notin \underline{A}(\beta_k)$, $\forall k \neq i$, and $x_j \notin B(\beta_i)$, $\forall i$. That is, the object x_j is contained in β_i definitely. Hence, the weights of the objects in lower approximation of a cluster should be independent of other medoids and clusters, and should not be coupled with their similarity with respect to other medoids. Also, the objects in the lower approximation of a cluster should have similar influence on the corresponding medoid and cluster, whereas if $x_j \in B(\beta_i)$, then the object x_j possibly belongs to β_i and potentially belongs to another cluster. Hence, the objects in boundary regions should have different influences on the medoids and clusters.

So, in rough-fuzzy c-medoids, after assigning each object in lower approximations and boundary regions of different clusters based on ϵ_2, the memberships μ_{ij} of the objects are modified. The membership values of the objects in lower approximation are set to 1, while those in boundary regions remain unchanged. In other words, the rough-fuzzy c-medoids first partition the data into two classes, namely, lower approximation and boundary. The concept of fuzzy memberships is applied only to the objects in the boundary region, which enables the algorithm to handle overlapping clusters. Hence, in rough-fuzzy c-medoids, each cluster is represented by a medoid, a crisp lower approximation, and a fuzzy boundary (Fig. 6.2). The lower approximation influences the fuzziness of final partition. The fuzzy c-medoids can be reduced from rough-fuzzy c-medoids when $\underline{A}(\beta_i) = \emptyset, \forall i$. Hence, the rough-fuzzy c-medoids algorithm is the generalization of existing fuzzy c-medoids algorithm.

The new medoids are calculated on the basis of the weighting average of the crisp lower approximation and fuzzy boundary. Computation of the medoids is modified to include the effects of both fuzzy membership and lower and upper bounds. Since the objects lying in lower approximation definitely belong to a cluster, they are assigned a higher weight compared to that of the objects lying in boundary region. The modified medoids calculation for rough-fuzzy c-medoids is, therefore, given by

$$v_i = x_q, \tag{6.37}$$

where q is given by

$$q = \arg\min \begin{cases} w \times \mathcal{A} + \tilde{w} \times \mathcal{B} & \text{if } \underline{A}(\beta_i) \neq \emptyset, B(\beta_i) \neq \emptyset \\ \mathcal{A} & \text{if } \underline{A}(\beta_i) \neq \emptyset, B(\beta_i) = \emptyset , \\ \mathcal{B} & \text{if } \underline{A}(\beta_i) = \emptyset, B(\beta_i) \neq \emptyset \end{cases} \tag{6.38}$$

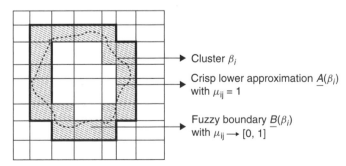

Cluster β_i

Crisp lower approximation $\underline{A}(\beta_i)$
with $\mu_{ij} = 1$

Fuzzy boundary $B(\beta_i)$
with $\mu_{ij} \rightarrow [0, 1]$

Figure 6.2 Rough-fuzzy c-medoids: cluster β_i is represented by crisp lower bound and fuzzy boundary.

where

$$\mathcal{A} = \sum_{x_k \in \underline{A}(\beta_i)} d(x_k, x_j) \tag{6.39}$$

and

$$\mathcal{B} = \sum_{x_k \in B(\beta_i)} (\mu_{ik})^{\acute{m}_1} d(x_k, x_j). \tag{6.40}$$

The main steps of this algorithm proceed as follows:

1. Assign initial medoids v_i, $i = 1, 2, \ldots, c$.
2. Choose values for fuzzifier $\acute{m}_1$ and thresholds ϵ_1 and ϵ_2, and set iteration counter $t = 1$.
3. Compute membership μ_{ij} using Equation (6.18) for c clusters and n objects.
4. If μ_{ij} is maximum for $1 \leq i \leq c$ and $(\mu_{ij} - \mu_{kj}) \leq \epsilon_2$, then $x_j \in \overline{A}(\beta_i)$ and $x_j \in \overline{A}(\beta_k)$. Furthermore, x_j is not part of any lower bound.
5. Otherwise, $x_j \in \underline{A}(\beta_i)$ such that μ_{ij} is the maximum for $1 \leq i \leq c$. In addition, by properties of rough sets, $x_j \in \overline{A}(\beta_i)$.
6. Compute new medoid as per Equation (6.37).
7. Repeat steps 2–6, by incrementing t, until $|\mu_{ij}(t) - \mu_{ij}(t-1)| > \epsilon_1$.

6.5 RELATIONAL CLUSTERING FOR BIO-BASIS STRING SELECTION

Different relational clustering algorithms, namely, hard c-medoids, fuzzy c-medoids, possibilistic c-medoids, fuzzy-possibilistic c-medoids, rough c-medoids, and rough-fuzzy c-medoids have been successfully used for the selection of bio-basis strings [29, 30]. The objective of different c-medoids algorithms for the selection of bio-basis strings is to assign n subsequences to c clusters. Each of the clusters β_i is represented by a bio-basis string v_i, which is the medoid for that cluster. The process begins by randomly choosing c subsequences as the bio-basis strings. The subsequences are assigned to one of the c clusters on the basis of the similarity or dissimilarity between the subsequence x_j and the bio-basis string v_i. The similarity is assessed through the nongapped pair-wise homology alignment score $h(x_j, v_i)$ between the subsequence x_j and the bio-basis string v_i, while the dissimilarity $d(x_j, v_i)$ is computed as follows:

$$d(x_j, v_i) = \{h(v_i, v_i) - h(x_j, v_i)\}. \tag{6.41}$$

The score $h(x_j, v_i)$ can be calculated as per Equation (6.2). After the assignment of all the subsequences to various clusters, the new bio-basis strings are calculated accordingly.

However, a limitation of the c-medoids algorithm is that it can only achieve a local optimum solution that depends on the initial choice of the bio-basis strings. Consequently, computing resources may be wasted in that some initial bio-basis strings get stuck in regions of the input space with a scarcity of data points and may therefore never have the chance to move to new locations where they are needed. To overcome this limitation of the c-medoids algorithm, another method is reported next to select initial bio-basis strings, which is based on a similarity measure using amino acid mutation matrix. It enables the algorithm to converge to an optimum or near-optimum solutions (bio-basis strings).

Before describing the initialization method of the c-medoids algorithm for selecting initial bio-basis strings, a quantitative measure is stated to evaluate the similarity between two subsequences in terms of nongapped pair-wise homology alignment score.

Degree of Resemblance The DOR between two subsequences x_i and x_j is defined as [29, 30]

$$\text{DOR}(x_j, x_i) = \frac{h(x_j, x_i)}{h(x_i, x_i)}. \tag{6.42}$$

It is the ratio between the nongapped pair-wise homology alignment scores of two input subsequences x_i and x_j based on an amino acid mutation matrix to the maximum homology alignment score of the subsequence x_i. It is used to quantify the similarity in terms of homology alignment score between pairs of subsequences. If the functions of two subsequences are different, the DOR between them is small. A high value of $\text{DOR}(x_i, x_j)$ between two subsequences x_i and x_j asserts that they may have the same function statistically. If two subsequences are the same, the DOR between them is maximum, that is, $\text{DOR}(x_i, x_i) = 1$. Hence, $0 < \text{DOR}(x_i, x_j) \leq 1$. Also, $\text{DOR}(x_i, x_j) \neq \text{DOR}(x_j, x_i)$.

On the basis of the concept of DOR, the method for selecting initial bio-basis strings is described next. The main steps of this method proceed as follows [30]:

1. For each subsequence x_i, calculate $\text{DOR}(x_j, x_i)$ between itself and the subsequence x_j, $\forall_{j=1}^{n}$.

2. Calculate the similarity score between subsequences x_i and x_j

$$S(x_j, x_i) = \begin{cases} 1 & \text{if } \text{DOR}(x_j, x_i) > \epsilon_3 \\ 0 & \text{otherwise} \end{cases}.$$

3. For each x_i, calculate the total number of similar subsequences of x_i as

$$N(x_i) = \sum_{j=1}^{n} S(x_j, x_i).$$

4. Sort n subsequences according to their values of $N(x_i)$ such that $N(x_1) > N(x_2) > \cdots > N(x_n)$.
5. If $N(x_i) > N(x_j)$ and $DOR(x_j, x_i) > \epsilon_3$, then x_j cannot be considered as a bio-basis string, resulting in a reduced set of subsequences to be considered for initial bio-basis strings.
6. Let there be $\acute{n}$ subsequences in the reduced set having $N(x_i)$ values such that $N(x_1) > N(x_2) > \cdots > N(x_{\acute{n}})$. A heuristic threshold function can be defined as

$$\text{Tr} = \frac{R}{\epsilon_4}, \quad \text{where } R = \sum_{i=1}^{\acute{n}} \frac{1}{N(x_i) - N(x_{i+1})},$$

where ϵ_4 is a constant ($= 0.5$, say), so that all subsequences in reduced set having $N(x_i)$ value higher than it are regarded as the initial bio-basis strings.

The value of Tr is high if most of the $N(x_i)$s are large and close to each other. The above condition occurs when a small number of large clusters are present. On the other hand, if the $N(x_i)$s have wide variation among them, then the number of clusters with smaller size increases. Accordingly, Tr attains a lower value automatically.

Note that the main motive of introducing this threshold function lies in reducing the number of bio-basis strings. It attempts to eliminate noisy bio-basis strings (subsequence representatives having lower values of $N(x_i)$) from the whole subsequences. The whole approach is, therefore, data dependent.

6.6   QUANTITATIVE MEASURES

In this section, some quantitative indices are reported to evaluate the quality of selected bio-basis strings incorporating the concepts of nongapped pair-wise homology alignment scores and mutual information.

6.6.1   Using Homology Alignment Score

On the basis of the nongapped pair-wise homology alignment scores, two indices, namely, β index and γ index, are presented next for evaluating quantitatively the quality of selected bio-basis strings.

6.6.1.1 β Index   It is defined as

$$\beta = \frac{1}{c} \sum_{i=1}^{c} \frac{1}{n_i} \sum_{x_j \in \beta_i} \frac{h(x_j, v_i)}{h(v_i, v_i)} \tag{6.43}$$

that is,

$$\beta = \frac{1}{c} \sum_{i=1}^{c} \frac{1}{n_i} \sum_{x_j \in \beta_i} \text{DOR}(x_j, v_i), \tag{6.44}$$

where n_i is the number of subsequences in the ith cluster β_i and $h(x_j, v_i)$ are the nongapped pair-wise homology alignment scores, obtained using an amino acid mutation matrix, between subsequence x_j and bio-basis string v_i. The β index is the average normalized homology alignment scores of input subsequences with respect to their corresponding bio-basis strings. A good clustering procedure for bio-basis string selection should make all input subsequences as similar to their bio-basis strings as possible. The β index increases with the increase in homology alignment scores within a cluster. Therefore, for a given data set and c value, the higher the homology alignment scores within the clusters, the higher would be the β value. The value of β also increases with c. In an extreme case when the number of clusters is maximum, that is, $c = n$, the total number of subsequences in the data set, we have $\beta = 1$. Hence, $0 < \beta \leq 1$.

6.6.1.2 γ Index   It can be defined as

$$\gamma = \max_{i,j} \frac{1}{2} \left\{ \frac{h(v_j, v_i)}{h(v_i, v_i)} + \frac{h(v_i, v_j)}{h(v_j, v_j)} \right\} \tag{6.45}$$

that is,

$$\gamma = \max_{i,j} \tfrac{1}{2} \left\{ \text{DOR}(v_j, v_i) + \text{DOR}(v_i, v_j) \right\} \tag{6.46}$$

$0 < \gamma < 1$. The γ index calculates the maximum normalized homology alignment score between bio-basis strings. A good clustering procedure for bio-basis string selection should make the homology alignment score between all bio-basis strings as low as possible. The γ index minimizes the between-cluster homology alignment score.

6.6.2 Using Mutual Information

Using the concept of mutual information, one can measure the within-cluster and between-cluster shared information. In principle, mutual information is regarded as a nonlinear correlation function and can be used to measure the mutual relation between a bio-basis string and the subsequences as well as the mutual relation between each pair of bio-basis strings. It is used to quantify the information shared by two objects. If two independent objects do not share much information, the mutual information value between them is small, while two highly nonlinearly correlated objects demonstrate a high mutual information value. In this case, the objects can be the bio-basis strings and the subsequences.

On the basis of the mutual information, the β index would be as follows:

$$\overline{\beta} = \frac{1}{c} \sum_{i=1}^{c} \frac{1}{n_i} \sum_{x_j \in \beta_i} \frac{\mathrm{MI}(x_j, v_i)}{\mathrm{MI}(v_i, v_i)}, \tag{6.47}$$

where $\mathrm{MI}(x_i, x_j)$ is the mutual information between subsequences x_i and x_j. The mutual information $\mathrm{MI}(x_i, x_j)$ is defined as

$$\mathrm{MI}(x_i, x_j) = H(x_i) + H(x_j) - H(x_i, x_j), \tag{6.48}$$

with $H(x_i)$ and $H(x_j)$ being the entropy of subsequences x_i and x_j, respectively, and $H(x_i, x_j)$ their joint entropy. $H(x_i)$ and $H(x_i, x_j)$ are defined as

$$H(x_i) = -p(x_i)\ln p(x_i) \tag{6.49}$$

$$H(x_i, x_j) = -p(x_i, x_j)\ln p(x_i, x_j) \tag{6.50}$$

$p(x_i)$ and $p(x_i, x_j)$ are the a priori probability of x_i and joint probability of x_i and x_j, respectively. The $\overline{\beta}$ index is the average normalized mutual information of input subsequences with respect to their corresponding bio-basis strings. A bio-basis string selection procedure should make the shared information between all input subsequences and their bio-basis strings as high as possible. The $\overline{\beta}$ index increases with the increase in mutual information within a cluster. Therefore, for a given data set and c value, the higher the mutual information within the clusters, the higher would be the $\overline{\beta}$ value. The value of $\overline{\beta}$ also increases with c. When $c = n$, $\overline{\beta} = 1$. Hence, $0 < \overline{\beta} \leq 1$.

Similarly, the γ index would be

$$\overline{\gamma} = \max_{i,j} \frac{1}{2} \left\{ \frac{\mathrm{MI}(v_i, v_j)}{\mathrm{MI}(v_i, v_i)} + \frac{\mathrm{MI}(v_i, v_j)}{\mathrm{MI}(v_j, v_j)} \right\}. \tag{6.51}$$

The $\overline{\gamma}$ index calculates the maximum normalized mutual information between bio-basis strings. A good clustering procedure for bio-basis string selection should make the shared information between all bio-basis strings as low as possible. The $\overline{\gamma}$ index minimizes the between-cluster mutual information.

6.7   EXPERIMENTAL RESULTS

The performance of rough-fuzzy c-medoids (RFCMdd) [30] is compared extensively with that of various other related ones. These involve different combinations of the individual components of the hybrid scheme as well as other related schemes. The algorithms compared are (i) hard c-medoids (HCMdd) [13], (ii) rough c-medoids (RCMdd) [29, 30], (iii) fuzzy c-medoids (FCMdd) [28], (iv) possibilistic c-medoids (PCMdd) [28], (v) fuzzy-possibilistic c-medoids (FPCMdd) [29], (vi) the method proposed by Yang and Thompson using mutual information (MI) [7], and (vii) the method proposed by Berry et al. using genetic algorithms and Fisher ratio (GAFR) [5]. All the algorithms are implemented in C language and run in LINUX environment having machine configuration Pentium IV, 3.2 GHz, 1 MB cache, and 1 GB RAM.

6.7.1   Description of Data Sets

To analyze the performance of different methods, several real data sets such as five whole HIV (human immunodeficiency virus) protein sequences, Cai–Chou HIV data set [35], and caspase cleavage protein sequences are used. The initial bio-basis strings for different c-medoids algorithms, which represent crude clusters in the nonnumerical sequence space, are generated by the methodology described in Section 6.5. The Dayhoff amino acid mutation matrix [2–4] is used to calculate the nongapped pair-wise homology alignment score between two subsequences.

6.7.1.1   *Five Whole HIV Protein Sequences*   The HIV protease belongs to the family of aspartyl proteases, which have been well characterized as proteolytic enzymes. The catalytic component is composed of carboxyl groups from two aspartyl residues located in both NH_2- and COOH-terminal halves of the enzyme molecule in the HIV protease [36]. They are strongly substrate selective and cleavage specific, demonstrating their capability of cleaving large, virus-specific polypeptides called *polyproteins* between a specific pair of amino acids. Miller et al. showed that the cleavage sites in the HIV polyprotein can extend to an octapeptide region [37]. The amino acid residues within this octapeptide region are represented by P_4-P_3-P_2-P_1-$P_{1'}$-$P_{2'}$-$P_{3'}$-$P_{4'}$, where P_4-P_3-P_2-P_1 is the NH_2-terminal half and $P_{1'}$-$P_{2'}$-$P_{3'}$-$P_{4'}$ the COOH-terminal half. Their counterparts in the HIV protease are represented by S_4-S_3-S_2-S_1-$S_{1'}$-$S_{2'}$-$S_{3'}$-$S_{4'}$ [38]. The HIV protease cleavage site is exactly between P_1 and $P_{1'}$.

The five whole HIV protein sequences have been downloaded from the National Center for Biotechnology Information (NCBI) (www.ncbi.nlm.nih.gov). The accession numbers are AAC82593, AAG42635, AAO40777, NP_057849, and NP_057850. Details of these five sequences are included in Table 6.1. Note that MA, CA, NC, TF, PR, RT, RH, and IN are matrix protein, capsid protein, nucleocapsid core protein, transframe

TABLE 6.1   Five Whole HIV Protein Sequences from NCBI

Accession No	Length	Cleavage Sites at P_1
AAC82593	500	132(MA/CA), 363(CA/p2), 377(p2/NC), 432(NC/p1), 448(p1/p6)
AAG42635	498	132(MA/CA), 363(CA/p2), 376(p2/NC), 430(NC/p1), 446(p1/p6)
AAO40777	500	132(MA/CA), 363(CA/p2), 377(p2/NC), 432(NC/p1), 448(p1/p6)
NP_057849	1435	488(TF/PR), 587(PR/RT), 1027(RT/RH), 1147(RH/IN)
NP_057850	500	132(MA/CA), 363(CA/p2), 377(p2/NC), 432(NC/p1), 448(p1/p6)

peptide, protease, reverse transcriptase, RNAse, and integrase, respectively. They are all cleavage products of the HIV protease; p1, p2, and p6 are also cleavage products [39]. For instance, 132 (MA/CA) means that the cleavage site is between the residues 132 (P_1) and 133 ($P_{1'}$), and the cleavage splits the polyprotein producing two functional proteins, the matrix protein and the capsid protein. The subsequences from each of five whole protein sequences are obtained through moving a sliding window with eight residues. Once a subsequence is produced, it is considered as functional if there is a cleavage site between P_1-$P_{1'}$; otherwise, it is labeled as nonfunctional. The total number of subsequences with eight residues in AAC82593, AAG42635, AAO40777, NP_057849, and NP_057850 are 493, 491, 493, 1428, and 493, respectively.

6.7.1.2   Cai–Chou HIV Data Set   Cai and Chou [35] have described a benchmark data set of the HIV. It consists of 114 positive oligopeptides and 248 negative oligopeptides, in total 362 subsequences with 8 residues. The data set has been collected from the University of Exeter, UK.

6.7.1.3   Caspase Cleavage Data Set   The programmed cell death, also known as *apoptosis*, is a gene-directed mechanism, which serves to regulate and control both cell death and tissue homeostasis during the development and the maturation of cells. The importance of apoptosis study is that many diseases such as cancer and ischemic damage result from apoptosis malfunction. A family of cysteine proteases called *caspases*, which are expressed initially in the cell as proenzymes, is the key to apoptosis [40]. As caspase cleavage is the key to programmed cell death, the study of caspase inhibitors could represent effective drugs against some diseases where blocking apoptosis is desirable. Without a careful study of caspase cleavage specificity, effective drug design could be difficult.

The 13 protein sequences containing various experimentally determined caspase cleavage sites have been downloaded from NCBI (www.ncbi.nih.gov). Table 6.2 represents the information contained in these sequences. Ci depicts the *i*th caspase. The total number of noncleaved subsequences is about 8340,

TABLE 6.2 Thirteen Caspase Cleavage Proteins from NCBI

Proteins	Gene	Length	Cleavage Sites
O00273	DFFA	331	117(C3), 224(C3)
Q07817	BCL2L1	233	61(C1)
P11862	GAS2	314	279(C1)
P08592	APP	770	672(C6)
P05067	APP	770	672(C6), 739(C3/C6/C8/C9)
Q9JJV8	BCL2	236	64(C3 and C9)
P10415	BCL2	239	34(C3)
O43903	GAS2	313	278(C)
Q12772	SREBF2	1141	468(C3 and C7)
Q13546	RIPK1	671	324(C8)
Q08378	GOLGA3	1498	59(C2), 139(C3), 311(C7)
O60216	RAD21	631	279(C3/C7)
O95155	UBE4B	1302	109(C3/C7), 123(C6)

while the number of cleaved subsequences is only 18. In total, there are 8358 subsequences with 8 residues.

6.7.2 Illustrative Example

Consider the data set NP_057849 with sequence length 1435. The number of subsequences obtained through moving a sliding window with 8 residues is 1428. The parameters generated in the DOR-based initialization method for bio-basis string selection are shown in Table 6.3 only for NP_057849 data, as an example. The values of different input parameters used are also presented in Table 6.3.

The similarity score of each subsequence in the original and reduced sets is shown in Fig. 6.3. The initial bio-basis strings for c-medoids algorithms are obtained from the reduced set using the threshold value of Tr. The initial bio-basis

TABLE 6.3 DOR-Based Initialization of Bio-Basis Strings for NP_057849

Sequence length = 1435; number of subsequences $n = 1428$
Value of $\epsilon_3 = 0.75$; number of subsequences in reduced set $\acute{n} = 223$
Value of $\epsilon_4 = 0.5$; value of Tr = 35.32; number of bio-basis strings $c = 36$

Values of fuzzifier $\acute{m}_1 = \acute{m}_2 = 2.0$; values of $w = 0.7$ and $\tilde{w} = 0.3$
Constants: $a = 0.5$ and $b = 0.5$; parameters: $\epsilon_1 = 0.001$ and $\epsilon_2 = 0.2$

Quantitative measures:
HCMdd: $\beta = 0.643$, $\gamma = 0.751$, $\bar{\beta} = 0.807$, $\bar{\gamma} = 1.000$
RCMdd: $\beta = 0.651$, $\gamma = 0.751$, $\bar{\beta} = 0.822$, $\bar{\gamma} = 1.000$
FCMdd: $\beta = 0.767$, $\gamma = 0.701$, $\bar{\beta} = 0.823$, $\bar{\gamma} = 0.956$
PCMdd: $\beta = 0.773$, $\gamma = 0.713$, $\bar{\beta} = 0.826$, $\bar{\gamma} = 0.953$
FPCMdd: $\beta = 0.782$, $\gamma = 0.703$, $\bar{\beta} = 0.825$, $\bar{\gamma} = 0.937$
RFCMdd: $\beta = 0.836$, $\gamma = 0.681$, $\bar{\beta} = 0.866$, $\bar{\gamma} = 0.913$

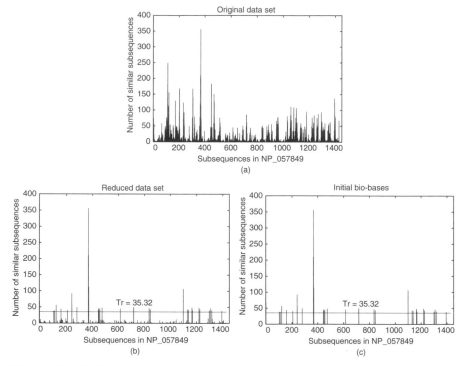

Figure 6.3 Similarity scores of subsequences of HIV protein NP_057849 considering $\epsilon_3 = 0.75$ and $\epsilon_4 = 0.50$. Similarity scores in (a) original data set and (b) reduced data set; (c) similarity scores of initial strings.

strings with similarity scores are also shown in Fig. 6.3. Each *c*-medoids algorithm is evolved using these initial bio-basis strings. The performance obtained by different *c*-medoids algorithms is shown in Table 6.3.

6.7.3 Performance Analysis

The experimental results on the data sets, reported in Section 6.7.1, are presented in Tables 6.4–6.12. Subsequent discussions analyze the results presented in these tables with respect to β, γ, $\overline{\beta}$, $\overline{\gamma}$, and execution time.

6.7.3.1 Optimum Value of Parameter ϵ_3 Table 6.4 reports the values of β, γ, $\overline{\beta}$, and $\overline{\gamma}$ of different algorithms for the data set NP_057849. Results are presented for different values of ϵ_3. Figure 6.4 shows the similarity scores of initial bio-basis strings as a function of ϵ_3. The parameters generated from the data set NP_057849 are shown in Table 6.4. The value of *c* is computed using the method described in Section 6.5. It may be noted that the optimal choice of *c* is a function of the value ϵ_3. From Fig. 6.4, it is seen that as the value of ϵ_3 increases, the initial bio-basis strings, which represent the crude clusters,

EXPERIMENTAL RESULTS

TABLE 6.4 Performance of Different Algorithms on NP_057849

Parameters	Algorithms	β	γ	$\overline{\beta}$	$\overline{\gamma}$
$\epsilon_3 = 0.60$ $\acute{n} = 15$ Tr = 4.05 $c = 13$	RFCMdd	0.736	0.914	0.817	1.000
	FPCMdd	0.722	0.914	0.809	1.000
	PCMdd	0.720	0.914	0.808	1.000
	FCMdd	0.719	0.914	0.805	1.000
	RCMdd	0.612	0.938	0.805	1.000
	HCMdd	0.607	0.938	0.801	1.000
$\epsilon_3 = 0.65$ $\acute{n} = 34$ Tr = 6.02 $c = 26$	RFCMdd	0.801	0.821	0.822	1.000
	FPCMdd	0.750	0.830	0.811	1.000
	PCMdd	0.748	0.833	0.811	1.000
	FCMdd	0.746	0.837	0.811	1.000
	RCMdd	0.632	0.836	0.807	1.000
	HCMdd	0.618	0.844	0.800	1.000
$\epsilon_3 = 0.70$ $\acute{n} = 84$ Tr = 16.58 $c = 27$	RFCMdd	0.801	0.819	0.822	0.982
	FPCMdd	0.752	0.819	0.816	0.988
	PCMdd	0.750	0.825	0.814	0.991
	FCMdd	0.746	0.828	0.811	0.996
	RCMdd	0.635	0.829	0.812	1.000
	HCMdd	0.621	0.827	0.803	1.000
$\epsilon_3 = 0.75$ $\acute{n} = 223$ Tr = 35.32 $c = 36$	RFCMdd	0.836	0.681	0.866	0.913
	FPCMdd	0.782	0.703	0.825	0.937
	PCMdd	0.773	0.713	0.826	0.953
	FCMdd	0.767	0.701	0.823	0.956
	RCMdd	0.651	0.751	0.822	1.000
	HCMdd	0.643	0.751	0.807	1.000
$\epsilon_3 = 0.80$ $\acute{n} = 594$ Tr = 28.05 $c = 6$	RFCMdd	0.682	0.937	0.809	1.000
	FPCMdd	0.674	0.938	0.810	1.000
	PCMdd	0.670	0.941	0.806	1.000
	FCMdd	0.667	0.941	0.805	1.000
	RCMdd	0.604	0.941	0.805	1.000
	HCMdd	0.605	0.938	0.807	1.000

are becoming more prominent. The best result is achieved at $\epsilon_3 = 0.75$. The subsequences selected as initial bio-basis strings at $\epsilon_3 = 0.75$ have higher values of $N(x_i)$.

It is seen from the results of Table 6.4 that the RFCMdd achieves consistently better performance than other algorithms with respect to the values of β, γ, $\overline{\beta}$, and $\overline{\gamma}$ for different values of ϵ_3. Also, the results reported in Table 6.4 establish the fact that as the value of ϵ_3 increases, the performance of the RFCMdd also increases. The best performance with respect to the values of β, γ, $\overline{\beta}$, and $\overline{\gamma}$, is achieved with $\epsilon_3 = 0.75$. At $\epsilon_3 = 0.75$, the values of $N(x_i)$ for most of the subsequences in reduced data set are large and close to each other. So, the

TABLE 6.5 Performance of HCMdd, RCMdd, and RFCMdd Algorithms

Data	Algorithms	Bio-Basis	β	γ	$\overline{\beta}$	$\overline{\gamma}$
	HCMdd	Random	0.615	0.817	0.809	1.000
		DOR	0.719	0.702	0.852	1.000
AAC8	RCMdd	Random	0.674	0.813	0.825	1.000
2593		DOR	0.815	0.677	0.872	0.983
	RFCMdd	Random	0.713	0.728	0.847	0.987
		DOR	0.874	0.633	0.913	0.916
	HCMdd	Random	0.657	0.799	0.803	1.000
		DOR	0.714	0.664	0.853	1.000
AAG4	RCMdd	Random	0.685	0.709	0.812	1.000
2635		DOR	0.768	0.681	0.882	1.000
	RFCMdd	Random	0.717	0.719	0.847	1.000
		DOR	0.831	0.611	0.912	0.957
	HCMdd	Random	0.651	0.864	0.837	1.000
		DOR	0.794	0.723	0.881	1.000
AAO4	RCMdd	Random	0.717	0.791	0.847	1.000
0777		DOR	0.809	0.633	0.879	0.977
	RFCMdd	Random	0.759	0.793	0.890	1.000
		DOR	0.856	0.613	0.930	0.947
	HCMdd	Random	0.601	0.882	0.801	1.000
		DOR	0.643	0.751	0.807	1.000
NP_05	RCMdd	Random	0.600	0.811	0.801	1.000
7849		DOR	0.651	0.751	0.822	1.000
	RFCMdd	Random	0.698	0.798	0.804	1.000
		DOR	0.836	0.681	0.866	0.913
	HCMdd	Random	0.611	0.913	0.792	1.000
		DOR	0.714	0.719	0.801	1.000
NP_05	RCMdd	Random	0.639	0.895	0.794	1.000
7850		DOR	0.758	0.702	0.826	0.993
	RFCMdd	Random	0.702	0.824	0.803	1.000
		DOR	0.851	0.629	0.911	0.928

threshold Tr attains a higher value compared to that of other values of ϵ_3. In effect, the subsequences selected as initial bio-basis strings with $\epsilon_3 = 0.75$, have higher values of $N(x_i)$. Hence, the quality of generated clusters using different *c*-medoids algorithms is better compared to other values of ϵ_3.

6.7.3.2 *Random versus DOR-Based Initialization*   Tables 6.5, 6.6, and 6.7 provide comparative results of different *c*-medoids algorithms with random initialization of bio-basis strings and the DOR-based initialization method

TABLE 6.6   Performance of FCMdd, PCMdd, and FPCMdd Algorithms

Data	Algorithms	Bio-Basis	β	γ	$\overline{\beta}$	$\overline{\gamma}$
	FCMdd	Random	0.655	0.791	0.821	1.000
		DOR	0.814	0.680	0.901	0.956
AAC8	PCMdd	Random	0.644	0.772	0.805	1.000
2593		DOR	0.815	0.677	0.904	0.949
	FPCMdd	Random	0.698	0.757	0.832	1.000
		DOR	0.821	0.677	0.909	0.952
	FCMdd	Random	0.698	0.706	0.818	1.000
		DOR	0.807	0.674	0.892	0.924
AAG4	PCMdd	Random	0.701	0.689	0.824	1.000
2635		DOR	0.811	0.672	0.897	0.937
	FPCMdd	Random	0.704	0.683	0.828	1.000
		DOR	0.811	0.659	0.894	0.928
	FCMdd	Random	0.718	0.804	0.842	1.000
		DOR	0.817	0.634	0.912	0.977
AAO4	PCMdd	Random	0.726	0.801	0.846	1.000
0777		DOR	0.821	0.630	0.911	0.962
	FPCMdd	Random	0.729	0.796	0.850	1.000
		DOR	0.824	0.629	0.914	0.972
	FCMdd	Random	0.606	0.802	0.811	1.000
		DOR	0.767	0.701	0.823	0.956
NP_05	PCMdd	Random	0.614	0.802	0.817	1.000
7849		DOR	0.773	0.713	0.826	0.953
	FPCMdd	Random	0.651	0.799	0.805	1.000
		DOR	0.782	0.703	0.825	0.937
	FCMdd	Random	0.648	0.881	0.796	1.000
		DOR	0.784	0.692	0.886	0.983
NP_05	PCMdd	Random	0.657	0.837	0.799	1.000
7850		DOR	0.801	0.692	0.889	0.983
	FPCMdd	Random	0.662	0.831	0.801	1.000
		DOR	0.807	0.688	0.890	0.971

described in Section 6.5 considering $\epsilon_3 = 0.75$. The DOR-based initialization is found to improve the performance in terms of β, γ, $\overline{\beta}$, and $\overline{\gamma}$ as well as reduce the time requirement of all c-medoids algorithms. It is also observed that the HCMdd with the DOR-based initialization performs similar to the RFCMdd with random initialization, although it is expected that the RFCMdd is superior to the HCMdd in partitioning subsequences. While in random initialization, different c-medoids algorithms get stuck in local optimums, the DOR-based scheme enables the algorithms to converge to an optimum or near-optimum solutions.

TABLE 6.7   Execution Time (milli second) of Different c-Medoids Algorithms

Methods	Bio-Basis	AAC8 2593	AAG4 2635	AAO4 0777	NP_05 7849	NP_05 7850
RFCMdd	Random	10326	17553	16218	316764	18038
	DOR	8981	12510	13698	251058	11749
FCMdd	Random	7349	16342	11079	293264	13217
	DOR	5898	11998	9131	240834	9174
PCMdd	Random	8217	13691	10983	295990	14372
	DOR	5982	10311	9618	241033	9713
FPCMdd	Random	9353	15892	12669	295874	15307
	DOR	6437	12133	12561	250963	10521
RCMdd	Random	6108	13816	8053	268199	10318
	DOR	5691	8015	5880	160563	5895
HCMdd	Random	2359	2574	2418	8728	2164
	DOR	535	534	532	4397	529

In effect, the execution time required for different c-medoids is lesser in the DOR-based initialization compared to random initialization.

6.7.3.3 Optimum Values of Parameters $\acute{m}_1$, w, and ϵ_2

The fuzzifier $\acute{m}_1$ has an influence on the clustering performance of both RFCMdd and FCMdd. Similarly, the performance of both RFCMdd and RCMdd depends on the parameter w and the threshold ϵ_2. Tables 6.8, 6.9 and 6.10 report the performance of different c-medoids algorithms for different values of $\acute{m}_1$, w and ϵ_2, respectively. The results and subsequent discussions are presented in these tables with respect to β, γ, $\overline{\beta}$, and $\overline{\gamma}$.

The fuzzifier $\acute{m}_1$ controls the extent of membership sharing between fuzzy clusters. From Table 6.8, it is seen that as the value of $\acute{m}_1$ increases, the values of β and $\overline{\beta}$ increase, while γ and $\overline{\gamma}$ decrease. The RFCMdd and FCMdd achieve their best performance with $\acute{m}_1 = 2.0$ for HIV protein NP_057849, $\acute{m}_1 = 1.9$ and 2.0 for Cai–Chou HIV data set, and $\acute{m}_1 = 2.0$ for caspase cleavage protein sequences, respectively. But, for $\acute{m}_1 > 2.0$, the performance of both algorithms decreases with the increase in $\acute{m}_1$. That is, the best performance of RFCMdd and FCMdd is achieved when the fuzzy membership value of a subsequence in a cluster is equal to its normalized homology alignment score with respect to all the bio-basis strings.

The parameter w has an influence on the performance of RFCMdd and RCMdd. Since the subsequences lying in lower approximation definitely belong to a cluster, they are assigned a higher weight w compared to $\tilde{w}$ of the subsequences lying in boundary regions. Hence, for both RFCMdd and RCMdd, $0 < \tilde{w} < w < 1$. Table 6.9 presents the performance of RFCMdd and RCMdd for different values w considering $\acute{m}_1 = 2.0$ and $\epsilon_2 = 0.20$. When the subsequences

TABLE 6.8 Performance of RFCMdd and FCMdd for Different Values of Fuzzifier

Value of Fuzzifier	Algorithms	HIV Protein NP_057849				Cai–Chou HIV Data Set				Caspase Cleavage Proteins			
		β	γ	$\overline{\beta}$	$\overline{\gamma}$	β	γ	$\overline{\beta}$	$\overline{\gamma}$	β	γ	$\overline{\beta}$	$\overline{\gamma}$
1.5	RFCMdd	0.759	0.744	0.832	0.997	0.755	0.714	0.867	0.992	0.748	0.662	0.882	0.989
	FCMdd	0.699	0.733	0.811	1.000	0.732	0.753	0.824	1.000	0.734	0.678	0.871	1.000
1.6	RFCMdd	0.762	0.717	0.839	0.966	0.781	0.692	0.878	0.979	0.773	0.658	0.899	0.985
	FCMdd	0.716	0.726	0.814	1.000	0.739	0.749	0.833	0.994	0.761	0.677	0.882	1.000
1.7	RFCMdd	0.799	0.702	0.843	0.956	0.794	0.677	0.895	0.950	0.785	0.647	0.907	0.977
	FCMdd	0.725	0.746	0.817	1.000	0.750	0.728	0.868	0.973	0.772	0.671	0.883	0.978
1.8	RFCMdd	0.814	0.695	0.852	0.947	0.818	0.639	0.907	0.932	0.803	0.628	0.923	0.972
	FCMdd	0.738	0.729	0.818	0.985	0.764	0.695	0.890	0.954	0.795	0.671	0.890	0.978
1.9	RFCMdd	0.831	0.681	0.858	0.913	0.829	0.618	0.911	0.927	0.814	0.611	0.937	0.965
	FCMdd	0.755	0.702	0.821	0.972	0.809	0.656	0.903	0.941	0.808	0.668	0.898	0.962
2.0	RFCMdd	0.836	0.681	0.866	0.913	0.829	0.618	0.911	0.927	0.839	0.608	0.942	0.944
	FCMdd	0.767	0.701	0.823	0.956	0.809	0.656	0.903	0.941	0.816	0.662	0.901	0.953
2.1	RFCMdd	0.835	0.684	0.861	0.927	0.811	0.622	0.908	0.945	0.826	0.617	0.935	0.949
	FCMdd	0.754	0.701	0.820	0.956	0.802	0.671	0.901	0.948	0.801	0.665	0.899	0.973
2.2	RFCMdd	0.817	0.699	0.847	0.931	0.802	0.640	0.903	0.958	0.817	0.639	0.928	0.954
	FCMdd	0.751	0.722	0.813	0.978	0.767	0.692	0.892	0.977	0.798	0.665	0.895	0.973
2.3	RFCMdd	0.816	0.712	0.847	0.959	0.791	0.658	0.882	0.961	0.801	0.641	0.901	0.961
	FCMdd	0.734	0.759	0.809	0.991	0.760	0.703	0.877	0.982	0.784	0.668	0.886	0.979
2.4	RFCMdd	0.802	0.712	0.835	0.959	0.774	0.699	0.878	0.967	0.792	0.642	0.894	0.961
	FCMdd	0.712	0.771	0.808	1.000	0.752	0.726	0.876	0.983	0.763	0.672	0.885	0.981
2.5	RFCMdd	0.795	0.751	0.829	0.984	0.767	0.711	0.863	0.967	0.785	0.657	0.887	0.979
	FCMdd	0.701	0.771	0.801	1.000	0.751	0.744	0.854	0.989	0.755	0.673	0.874	0.996

TABLE 6.9 Performance of RFCMdd and RCMdd for Different Values of Parameter $w(= 1 - \tilde{w})$

Value of w	Algorithms	HIV Protein NP_057849				Cai–Chou HIV Data Set				Caspase Cleavage Proteins			
		β	γ	$\bar{\beta}$	$\bar{\gamma}$	β	γ	$\bar{\beta}$	$\bar{\gamma}$	β	γ	$\bar{\beta}$	$\bar{\gamma}$
0.51	RFCMdd	0.632	0.941	0.742	1.000	0.684	0.827	0.806	1.000	0.683	0.714	0.808	1.000
	RCMdd	0.604	0.952	0.719	1.000	0.649	0.856	0.782	1.000	0.668	0.733	0.781	1.000
0.55	RFCMdd	0.637	0.839	0.756	1.000	0.739	0.743	0.837	1.000	0.727	0.688	0.833	1.000
	RCMdd	0.604	0.844	0.729	1.000	0.718	0.807	0.796	1.000	0.714	0.709	0.803	1.000
0.60	RFCMdd	0.648	0.750	0.795	0.983	0.788	0.708	0.883	0.991	0.779	0.649	0.883	0.983
	RCMdd	0.617	0.766	0.746	1.000	0.731	0.733	0.807	1.000	0.762	0.697	0.837	1.000
0.65	RFCMdd	0.817	0.708	0.839	0.934	0.811	0.633	0.902	0.958	0.821	0.622	0.927	0.957
	RCMdd	0.644	0.761	0.807	1.000	0.763	0.694	0.866	0.998	0.774	0.680	0.852	1.000
0.70	RFCMdd	0.836	0.681	0.866	0.913	0.829	0.618	0.911	0.927	0.839	0.608	0.942	0.944
	RCMdd	0.651	0.751	0.822	1.000	0.771	0.677	0.897	0.993	0.782	0.673	0.887	1.000
0.75	RFCMdd	0.819	0.713	0.844	0.940	0.807	0.627	0.899	0.951	0.838	0.611	0.939	0.953
	RCMdd	0.647	0.758	0.806	1.000	0.764	0.698	0.871	1.000	0.771	0.684	0.858	1.000
0.80	RFCMdd	0.766	0.784	0.813	0.992	0.793	0.651	0.874	0.978	0.817	0.622	0.914	0.964
	RCMdd	0.645	0.821	0.792	1.000	0.753	0.704	0.864	1.000	0.753	0.702	0.851	1.000
0.85	RFCMdd	0.713	0.839	0.802	1.000	0.781	0.692	0.853	0.991	0.804	0.647	0.887	1.000
	RCMdd	0.642	0.861	0.785	1.000	0.747	0.718	0.847	1.000	0.736	0.719	0.843	1.000
0.90	RFCMdd	0.648	0.841	0.788	1.000	0.748	0.711	0.829	1.000	0.761	0.682	0.825	1.000
	RCMdd	0.641	0.862	0.781	1.000	0.736	0.727	0.821	1.000	0.708	0.732	0.816	1.000
0.95	RFCMdd	0.639	0.862	0.759	1.000	0.702	0.774	0.818	1.000	0.728	0.711	0.814	1.000
	RCMdd	0.635	0.865	0.761	1.000	0.698	0.781	0.815	1.000	0.681	0.753	0.803	1.000
0.99	RFCMdd	0.602	0.968	0.736	1.000	0.671	0.813	0.802	1.000	0.675	0.762	0.798	1.000
	RCMdd	0.601	0.968	0.736	1.000	0.671	0.813	0.802	1.000	0.674	0.762	0.794	1.000

TABLE 6.10 Performance of RFCMdd and RCMdd for Different Values of Parameter ϵ_2

Value of ϵ_2	Algorithms	HIV Protein NP_057849				Cai–Chou HIV Data Set				Caspase Cleavage Proteins			
		β	γ	$\bar{\beta}$	$\bar{\gamma}$	β	γ	$\bar{\beta}$	$\bar{\gamma}$	β	γ	$\bar{\beta}$	$\bar{\gamma}$
0.00	RFCMdd	0.643	0.751	0.807	1.000	0.713	0.782	0.817	1.000	0.707	0.698	0.862	1.000
	RCMdd	0.643	0.751	0.807	1.000	0.713	0.782	0.817	1.000	0.707	0.698	0.862	1.000
0.05	RFCMdd	0.704	0.723	0.812	1.000	0.753	0.707	0.868	1.000	0.766	0.683	0.881	1.000
	RCMdd	0.644	0.751	0.810	1.000	0.716	0.754	0.839	1.000	0.723	0.687	0.863	1.000
0.10	RFCMdd	0.793	0.709	0.837	0.981	0.794	0.683	0.882	0.991	0.801	0.641	0.907	0.995
	RCMdd	0.647	0.751	0.814	1.000	0.738	0.726	0.841	1.000	0.738	0.681	0.872	1.000
0.15	RFCMdd	0.811	0.702	0.855	0.946	0.806	0.629	0.902	0.964	0.819	0.622	0.928	0.973
	RCMdd	0.651	0.751	0.819	1.000	0.744	0.694	0.856	1.000	0.764	0.678	0.879	1.000
0.20	RFCMdd	0.836	0.681	0.866	0.913	0.829	0.618	0.911	0.927	0.839	0.608	0.942	0.944
	RCMdd	0.651	0.751	0.822	1.000	0.771	0.677	0.897	0.993	0.782	0.673	0.887	1.000
0.25	RFCMdd	0.836	0.707	0.852	0.936	0.811	0.638	0.907	0.952	0.814	0.631	0.932	0.980
	RCMdd	0.651	0.792	0.819	1.000	0.759	0.698	0.881	1.000	0.767	0.692	0.855	1.000
0.30	RFCMdd	0.817	0.718	0.844	0.990	0.805	0.681	0.894	0.988	0.791	0.667	0.908	0.995
	RCMdd	0.648	0.828	0.801	1.000	0.738	0.731	0.878	1.000	0.741	0.723	0.839	1.000
0.35	RFCMdd	0.801	0.739	0.831	1.000	0.784	0.704	0.875	1.000	0.772	0.671	0.881	1.000
	RCMdd	0.631	0.857	0.779	1.000	0.706	0.757	0.849	1.000	0.728	0.756	0.814	1.000
0.40	RFCMdd	0.792	0.784	0.804	1.000	0.757	0.762	0.872	1.000	0.759	0.699	0.863	1.000
	RCMdd	0.629	0.914	0.758	1.000	0.681	0.796	0.826	1.000	0.719	0.779	0.794	1.000
0.45	RFCMdd	0.716	0.833	0.796	1.000	0.732	0.783	0.850	1.000	0.706	0.737	0.827	1.000
	RCMdd	0.617	0.957	0.792	1.000	0.667	0.817	0.803	1.000	0.678	0.793	0.779	1.000
0.50	RFCMdd	0.708	0.864	0.781	1.000	0.713	0.805	0.813	1.000	0.684	0.798	0.769	1.000
	RCMdd	0.617	0.962	0.781	1.000	0.659	0.836	0.793	1.000	0.659	0.822	0.746	1.000

TABLE 6.11　Comparative Performance of Different Methods

Data Set	Algorithms	β	γ	$\bar{\beta}$	$\bar{\gamma}$	Time
AAC8 2593	RFCMdd	0.847	0.633	0.913	0.916	8981
	FPCMdd	0.821	0.677	0.909	0.952	6437
	PCMdd	0.815	0.677	0.904	0.949	5982
	FCMdd	0.814	0.680	0.901	0.956	5898
	RCMdd	0.815	0.677	0.872	0.983	5691
	HCMdd	0.719	0.702	0.852	1.000	535
	MI	0.764	0.788	0.906	0.977	8617
	GAFR	0.736	0.814	0.826	1.000	12213
AAG4 2635	RFCMdd	0.831	0.611	0.912	0.957	12510
	FPCMdd	0.811	0.659	0.894	0.928	12133
	PCMdd	0.811	0.672	0.897	0.937	10311
	FCMdd	0.807	0.674	0.892	0.924	11998
	RCMdd	0.768	0.681	0.882	1.000	8015
	HCMdd	0.714	0.664	0.853	1.000	534
	MI	0.732	0.637	0.829	0.989	13082
	GAFR	0.707	0.713	0.801	1.000	12694
AAO4 0777	RFCMdd	0.856	0.613	0.930	0.947	13698
	FPCMdd	0.824	0.629	0.914	0.972	12561
	PCMdd	0.821	0.630	0.911	0.962	9618
	FCMdd	0.817	0.634	0.912	0.977	9131
	RCMdd	0.809	0.633	0.879	0.977	5880
	HCMdd	0.794	0.723	0.881	1.000	532
	MI	0.801	0.827	0.890	0.982	12974
	GAFR	0.773	0.912	0.863	1.000	11729
NP_05 7849	RFCMdd	0.836	0.681	0.866	0.913	251058
	FPCMdd	0.782	0.703	0.825	0.937	250963
	PCMdd	0.773	0.713	0.826	0.953	241033
	FCMdd	0.767	0.701	0.823	0.956	240834
	RCMdd	0.651	0.751	0.822	1.000	160563
	HCMdd	0.643	0.751	0.807	1.000	4397
	MI	0.637	0.854	0.802	1.000	250138
	GAFR	0.646	0.872	0.811	1.000	291413

of both lower approximation and boundary region are assigned approximately equal weights, the performance of RFCMdd and RCMdd is significantly poorer than the HCMdd. As the value of w increases, the values of β and $\bar{\beta}$ increase, while γ and $\bar{\gamma}$ decrease. The best performance of both algorithms is achieved with $w = 0.70$. The performance significantly reduces with $w \simeq 1.00$. In this case, since the clusters cannot see the subsequences of boundary regions, the mobility of the clusters and the bio-basis strings reduces. As a result, some bio-basis strings get stuck in local optimum. On the other hand, when the value

TABLE 6.12 Comparative Performance of Different Methods

Data Set	Algorithms	β	γ	$\overline{\beta}$	$\overline{\gamma}$	Time
NP_05 7850	RFCMdd	0.851	0.629	0.911	0.928	11749
	FPCMdd	0.807	0.688	0.890	0.971	10521
	PCMdd	0.801	0.692	0.889	0.983	9713
	FCMdd	0.784	0.692	0.886	0.983	9174
	RCMdd	0.758	0.702	0.826	0.993	5895
	HCMdd	0.714	0.719	0.801	1.000	529
	MI	0.736	0.829	0.833	1.000	9827
	GAFR	0.741	0.914	0.809	1.000	10873
Cai–Chou HIV data	RFCMdd	0.829	0.618	0.911	0.927	6217
	FPCMdd	0.816	0.640	0.904	0.935	6198
	PCMdd	0.811	0.649	0.903	0.941	4062
	FCMdd	0.809	0.656	0.903	0.941	4083
	RCMdd	0.771	0.677	0.897	0.993	3869
	HCMdd	0.713	0.782	0.817	1.000	718
	MI	0.764	0.774	0.890	1.000	6125
	GAFR	0.719	0.794	0.811	1.000	7016
Caspase cleavage	RFCMdd	0.839	0.608	0.942	0.944	513704
	FPCMdd	0.821	0.643	0.918	0.950	513406
	PCMdd	0.817	0.658	0.907	0.950	510802
	FCMdd	0.816	0.662	0.901	0.953	510961
	RCMdd	0.782	0.673	0.887	1.000	473380
	HCMdd	0.707	0.698	0.862	1.000	8326
	MI	0.732	0.728	0.869	1.000	511628
	GAFR	0.713	0.715	0.821	1.000	536571

of $w = 0.70$, the subsequences of lower approximations are assigned a higher weight compared to that of boundary regions as well as the clusters and the bio-basis strings have a greater degree of freedom to move. In effect, the quality of generated clusters is better compared to other values of w.

The performance of RFCMdd and RCMdd also depends on the value of ϵ_2, which determines the class labels of all the subsequences. In other words, the RFCMdd and RCMdd partition the data set of a cluster into two classes, namely, lower approximation and boundary, based on the value of ϵ_2. Table 6.10 presents the comparative performance of RFCMdd and RCMdd for different values of ϵ_2 considering $\acute{m}_1 = 2.0$ and $w = 0.70$. For $\epsilon_2 = 0.0$, all the subsequences will be in lower approximations of different clusters and $B(\beta_i) = \emptyset, \forall i$. In effect, both RFCMdd and RCMdd reduce to conventional HCMdd. On the other hand, for $\epsilon_2 = 1.0$, $\underline{A}(\beta_i) = \emptyset, \forall i$ and all the subsequences will be in the boundary regions of different clusters. That is, the RFCMdd boils down to the FCMdd. The best performance of RFCMdd and RCMdd with respect to $\beta, \overline{\beta}, \gamma$, and $\overline{\gamma}$ is achieved with $\epsilon_2 = 0.2$, which is approximately equal to the average difference

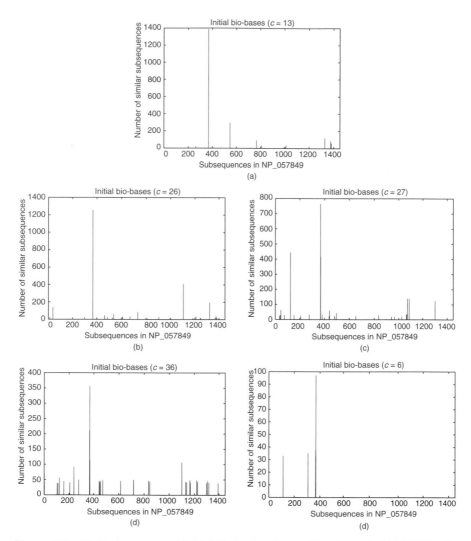

Figure 6.4 Similarity scores of initial bio-basis strings of HIV protein NP_057849 for different values of ϵ_3 considering $\epsilon_4 = 0.50$. (a) $\epsilon_3 = 0.60$ and Tr = 4.05, (b) $\epsilon_3 = 0.65$ and Tr = 6.02, (c) $\epsilon_3 = 0.70$ and Tr = 16.58, (d) $\epsilon_3 = 0.75$ and Tr = 35.32, (e) $\epsilon_3 = 0.80$ and Tr = 28.05.

of highest and second highest fuzzy membership values of all the subsequences. In practice, we find that both RFCMdd and RCMdd work well for $\epsilon_2 = \delta$, where

$$\delta = \frac{1}{n} \sum_{j=1}^{n} (\mu_{ij} - \mu_{kj}), \qquad (6.52)$$

where n is the total number of subsequences, μ_{ij} and μ_{kj} are the highest and second highest fuzzy membership values of the subsequence x_j, respectively. The values of δ for HIV protein NP_057849, Cai–Chou HIV data set, and caspase cleavage proteins are 0.197, 0.201, and 0.198, respectively.

6.7.3.4 Comparative Performance of Different Algorithms Finally, Tables 6.11 and 6.12 provide the comparative results of different algorithms for the protein sequences reported in Section 6.7.1. It is seen that the RFCMdd with the DOR-based initialization produces bio-basis strings having the highest β and $\overline{\beta}$ values and lowest γ and $\overline{\gamma}$ values for all the cases. Tables 6.11 and 6.12 also provide execution time (in millisecond) of different algorithms for all protein data sets. The execution time required for the RFCMdd is comparable to MI and GAFR. For the HCMdd, although the execution time is less, the performance is significantly poorer than that of RCMdd, FCMdd, PCMdd, FPCMdd, and RFCMdd.

The following conclusions can be drawn from the results reported in Tables 6.4–6.12:

1. It is observed that the RFCMdd is superior to the HCMdd both with random and the DOR-based initialization. However, the HCMdd requires considerably lesser time compared to the RFCMdd. But, the performance of the HCMdd is significantly poorer than the RFCMdd. The performance of FCMdd, PCMdd, FPCMdd, and RCMdd are intermediate between RFCMdd and HCMdd.
2. The DOR-based initialization is found to improve the values of β, γ, $\overline{\beta}$, and $\overline{\gamma}$ as well as reduce the time requirement substantially for all c-medoids algorithms.
3. Use of rough sets and fuzzy sets adds a small computational load to the HCMdd algorithm; however, the corresponding integrated methods (FCMdd, PCMdd, FPCMdd, RCMdd, and RFCMdd) show a definite increase in β and $\overline{\beta}$ values and decrease in γ and $\overline{\gamma}$ values.
4. Integration of three components, namely, rough sets, fuzzy sets, and c-medoids, in the RFCMdd algorithm produces a minimum set of most informative bio-basis strings in the least computation time compared to both MI and GAFR.
5. It is observed that the RFCMdd algorithm requires significantly lesser time compared to both MI and GAFR having comparable performance. Reduction in time is achieved because of the DOR-based initialization. The DOR-based initialization reduces the convergence time of the RFCMdd algorithm considerably compared to random initialization.

The best performance of the RFCMdd algorithm in terms of β, γ, $\overline{\beta}$, and $\overline{\gamma}$ is achieved because of the following three reasons:

1. The DOR-based initialization of bio-basis strings enables the algorithm to converge to optimum or near-optimum solutions;
2. Membership function of fuzzy sets handles efficiently overlapping partitions.
3. The concept of lower and upper bounds of rough sets deals with uncertainty, vagueness, and incompleteness in class definition.

In effect, the minimum set of bio-basis strings having maximum information is obtained using the RFCMdd algorithm.

6.8 CONCLUSION AND DISCUSSION

This chapter deals with the problem of relational clustering under uncertainty. Different relational clustering algorithms, including rough-fuzzy c-medoids, are described. The rough-fuzzy c-medoids comprise a judicious integration of the principles of rough sets, fuzzy sets, and c-medoids algorithm. While the membership function of fuzzy sets enables efficient handling of overlapping partitions, the concept of lower and upper bounds of rough sets deals with uncertainty, vagueness, and incompleteness in class definition. The concept of crisp lower bound and fuzzy boundary of a class, introduced in the rough-fuzzy c-medoids algorithm, enables efficient partitioning of relational objects.

The effectiveness of the rough-fuzzy c-medoids algorithm, along with a comparison with other c-medoids algorithms, is demonstrated for selection of the most informative bio-basis strings. The concept of DOR is found to be successful in effectively circumventing the initialization and local minima problems of iterative refinement clustering algorithms such as c-medoids. In addition, this concept enables efficient selection of a minimum set of most informative bio-basis strings compared to existing methods.

The methodology of integrating rough sets, fuzzy sets, and the c-medoids algorithm is general in the sense that it is applicable to situations in which the objects to be clustered cannot be represented by numerical features, rather, only represented with similarity or dissimilarity between pairs of objects. In effect, it can be used in several problems of web mining and bioinformatics if a similarity or dissimilarity measure can be defined to quantify the DOR between pairs of objects.

REFERENCES

1. N. Qian and T. J. Sejnowski. Predicting the Secondary Structure of Globular Proteins Using Neural Network Models. *Journal of Molecular Biology*, 202:865–884, 1988.
2. M. O. Dayhoff, R. M. Schwartz, and B. C. Orcutt. A Model of Evolutionary Change in Proteins. Matrices for Detecting Distant Relationships. *Atlas of Protein Sequence and Structure*, 5:345–358, 1978.

3. S. Henikoff and J. G. Henikoff. Amino Acid Substitution Matrices from Protein Blocks. *Proceedings of the National Academy of Sciences of the United States of America*, 89:10915–10919, 1992.

4. M. S. Johnson and J. P. Overington. A Structural Basis for Sequence Comparisons: An Evaluation of Scoring Methodologies. *Journal of Molecular Biology*, 233:716–738, 1993.

5. E. A. Berry, A. R. Dalby, and Z. R. Yang. Reduced Bio-Basis Function Neural Network for Identification of Protein Phosphorylation Sites: Comparison with Pattern Recognition Algorithms. *Computational Biology and Chemistry*, 28:75–85, 2004.

6. R. Thomson, C. Hodgman, Z. R. Yang, and A. K. Doyle. Characterising Proteolytic Cleavage Site Activity Using Bio-Basis Function Neural Network. *Bioinformatics*, 19:1741–1747, 2003.

7. Z. R. Yang and R. Thomson. Bio-Basis Function Neural Network for Prediction of Protease Cleavage Sites in Proteins. *IEEE Transactions on Neural Networks*, 16(1):263–274, 2005.

8. Z. R. Yang. Prediction of Caspase Cleavage Sites Using Bayesian Bio-Basis Function Neural Networks. *Bioinformatics*, 21:1831–1837, 2005.

9. Z. R. Yang and K. C. Chou. Predicting the O-Linkage Sites in Glycoproteins Using Bio-Basis Function Neural Networks. *Bioinformatics*, 20:903–908, 2004.

10. Z. R. Yang, R. Thomson, P. McNeil, and R. Esnouf. RONN: Use of the Bio-Basis Function Neural Network Technique for the Detection of Natively Disordered Regions in Proteins. *Bioinformatics*, 21:3369–3376, 2005.

11. P. H. A. Sneath and R. R. Sokal. *Numerical Taxonomy: The Principles and Practice of Numerical Classification*. W. H. Freeman, San Francisco, CA, 1973.

12. S. Guha, R. Rastogi, and K. Shim. CURE: An Efficient Clustering Algorithm for Large Databases. *Proceedings of the ACM SIGMOD International Conference on Management of Data*, pages 73–84, ACM Press, New York, 1998.

13. L. Kaufman and P. J. Rousseeuw. *Finding Groups in Data: An Introduction to Cluster Analysis*. John Wiley & Sons, Brussels, Belgium, 1990.

14. L. Kaufman and P. J. Rousseeuw. Clustering by Means of Medoids. In Y. Dodge, editor, *Statistical Data Analysis Based on the L_1 Norm*, pages 405–416. Elsevier, North Holland, Amsterdam, 1987.

15. R. T. Ng and J. Han. Efficient and Effective Clustering Methods for Spatial Data Mining. In *Proceedings of the 20th International Conference on Very Large Databases*, pages 144–155, Santiago, Chile, 1994.

16. K. S. Fu. *Syntactic Pattern Recognition and Applications*. Academic Press, London, 1982.

17. P. Bajcsy and N. Ahuja. Location- and Density-Based Hierarchical Clustering Using Similarity Analysis. *IEEE Transactions on Pattern Analysis and Machine Intelligence*, 20:1011–1015, 1998.

18. K. C. Gowda and E. Diday. Symbolic Clustering Using a New Similarity Measure. *IEEE Transactions on Systems Man and Cybernetics*, 20:368–377, 1992.

19. G. D. Ramkumar and A. Swami. Clustering Data without Distance Functions. *Data Engineering Bulletin*, 21(1):9–14, 1998.

20. Y. E. Sonbaty and M. A. Ismail. Fuzzy Clustering for Symbolic Data. *IEEE Transactions on Fuzzy Systems*, 6:195–204, 1998.

21. E. H. Ruspini. Numerical Methods for Fuzzy Clustering. *Information Sciences*, 2:319–350, 1970.

22. E. Diday. La Methode Des Nuees Dynamiques. *Revue de Statistique Appliquee*, 9(2):19–34, 1975.

23. M. Roubens. Pattern Classification Problems and Fuzzy Sets. *Fuzzy Sets and Systems*, 1:239–253, 1978.

24. M. P. Windham. Numerical Classification of Proximity Data with Assignment Measures. *Journal of Classification*, 2:157–172, 1985.

25. R. J. Hathaway, J. W. Devenport, and J. C. Bezdek. Relational Dual of the C-Means Clustering Algorithms. *Pattern Recognition*, 22(2):205–212, 1989.

26. R. J. Hathaway and J. C. Bezdek. NERF C-Means: Non-Euclidean Relational Fuzzy Clustering. *Pattern Recognition*, 27:429–437, 1994.

27. S. Sen and R. N. Dave. Clustering of Relational Data Containing Noise and Outliers. In *Proceedings of the IEEE International Conference on Fuzzy Systems*, pages 1411–1416, 1998.

28. R. Krishnapuram, A. Joshi, O. Nasraoui, and L. Yi. Low Complexity Fuzzy Relational Clustering Algorithms for Web Mining. *IEEE Transactions on Fuzzy System*, 9:595–607, 2001.

29. P. Maji and S. K. Pal. Protein Sequence Analysis Using Relational Soft Clustering Algorithms. *International Journal of Computer Mathematics*, 84(5):599–617, 2007.

30. P. Maji and S. K. Pal. Rough-Fuzzy C-Medoids Algorithm and Selection of Bio-Basis for Amino Acid Sequence Analysis. *IEEE Transactions on Knowledge and Data Engineering*, 19(6):859–872, 2007.

31. S. F. Altschul, M. S. Boguski, W. Gish, and J. C. Wootton. Issues in Searching Molecular Sequence Databases. *Nature Genetics*, 6:119–129, 1994.

32. S. F. Altschul, W. Gish, W. Miller, E. Myers, and D. J. Lipman. Basic Local Alignment Search Tool. *Journal of Molecular Biology*, 215:403–410, 1990.

33. C. Shannon and W. Weaver. *The Mathematical Theory of Communication*. University Illinois Press, Champaign, IL, 1964.

34. Z. R. Yang. Orthogonal Kernel Machine for the Prediction of Functional Sites in Proteins. *IEEE Transactions on Systems Man and Cybernetics Part B-Cybernetics*, 35(1):100–106, 2005.

35. Y. D. Cai and K. C. Chou. Artificial Neural Network Model for Predicting HIV Protease Cleavage Sites in Protein. *Advances in Engineering Software*, 29(2):119–128, 1998.

36. L. H. Pearl and W. R. Taylor. A Structural Model for the Retroviral Proteases. *Nature*, 329:351–354, 1987.

37. M. Miller, J. Schneider, B. K. Sathayanarayana, M. V. Toth, G. R. Marshall, L. Clawson, L. Selk, S. B. H. Kent, and A. Wlodawer. Structure of Complex of Synthetic HIV-1 Protease with Substrate-Based Inhibitor at 2.3 A° Resolution. *Science*, 246:1149–1152, 1989.

38. K. C. Chou. A Vectorised Sequence-Coupling Model for Predicting HIV Protease Cleavage Sites in Proteins. *Journal of Biological Chemistry*, 268:16938–16948, 1993.

39. K. C. Chou. Prediction of Human Immunodeficiency Virus Protease Cleavage Sites in Proteins. *Analytical Biochemistry*, 233:1–14, 1996.

40. T. T. Rohn, S. M. Cusack, S. R. Kessinger, and J. T. Oxford. Caspase Activation Independent of Cell Death is Required for Proper Cell Dispersal and Correct Morphology in PC12 Cells. *Experimental Cell Research*, 293:215–225, 2004.

7

CLUSTERING FUNCTIONALLY SIMILAR GENES FROM MICROARRAY DATA

7.1 INTRODUCTION

Microarray technology is one of the important biotechnological means that has made it possible to simultaneously monitor the expression levels of thousands of genes during important biological processes (BP) and across collections of related samples [1–3]. An important application of microarray data is to elucidate the patterns hidden in gene expression data for an enhanced understanding of functional genomics.

A microarray gene expression data set can be represented by an expression table, where each row corresponds to one particular gene, each column to a sample or time point, and each entry of the matrix is the measured expression level of a particular gene in a sample or time point, respectively [1–3]. However, the large number of genes and the complexity of biological networks greatly increase the challenges of comprehending and interpreting the resulting mass of data, which often consists of millions of measurements. A first step toward addressing this challenge is the use of clustering techniques, which is essential in the pattern recognition process to reveal natural structures and identify interesting patterns in the underlying data [4].

Cluster analysis is a technique to find natural groups present in the gene set. It divides a given gene set into a set of clusters in such a way that two genes from the same cluster are as similar as possible and the genes from different clusters are as dissimilar as possible [2, 3]. To understand gene function, gene regulation,

Rough-Fuzzy Pattern Recognition: Applications in Bioinformatics and Medical Imaging,
First Edition. Pradipta Maji and Sankar K. Pal.
© 2012 John Wiley & Sons, Inc. Published 2012 by John Wiley & Sons, Inc.

cellular processes, and subtypes of cells, clustering techniques have proved to be helpful. The co-expressed genes, that is, genes with similar expression patterns, can be clustered together with similar cellular functions. This approach may provide further understanding of the functions of many genes for which information has not been previously available [5, 6]. Furthermore, co-expressed genes in the same cluster are likely to be involved in the same cellular processes, and a strong correlation of expression patterns between those genes indicates co-regulation. Searching for common DNA sequences at the promoter regions of genes within the same cluster allows regulatory motifs specific to each gene cluster to be identified and *cis*-regulatory elements to be proposed [6, 7]. The inference of regulation through the gene expression data clustering also gives rise to hypotheses regarding the mechanism of the transcriptional regulatory network [8].

The purpose of gene clustering is to group together co-expressed genes that indicate co-function and co-regulation. Owing to the special characteristics of gene expression data, and the particular requirements from the biological domain, gene clustering presents several new challenges and is still an open problem. The cluster analysis is typically the first step in data mining and knowledge discovery. The purpose of clustering gene expression data is to reveal the natural data structures and gain some initial insights regarding data distribution. Therefore, a good clustering algorithm should depend as little as possible on prior knowledge, which is usually not available before cluster analysis. A clustering algorithm that can accurately estimate the true number of clusters in the data set would be more favored than one requiring the predetermined number of clusters.

However, gene expression data often contains a huge amount of noise due to the complex procedures of microarray experiments. Hence, clustering algorithms for gene expression data should be capable of extracting useful information from a high level of background noise. Also, the empirical study has demonstrated that gene expression data are often highly connected [9], and clusters may be highly overlapping with each other or even embedded one in another [10]. Therefore, gene clustering algorithms should be able to effectively handle this situation.

This chapter deals with the application of different rough-fuzzy clustering algorithms for clustering functionally similar genes from microarray gene expression data sets. Details of these algorithms are reported in Chapter 3 and in Maji and Pal [11, 12]. The effectiveness of the algorithms, along with a comparison with other related gene clustering algorithms, is demonstrated on a set of microarray gene expression data sets using some standard validity indices.

The rest of this chapter is organized as follows: Section 7.2 reports a brief overview of different gene clustering algorithms. Several quantitative and qualitative performance measures are reported in Section 7.3 to evaluate the quality of gene clusters. A brief description of different microarray gene expression data sets is presented in Section 7.4. Implementation details, experimental results, and a comparison among different algorithms are presented in Section 7.5. Concluding remarks are given in Section 7.6.

7.2 CLUSTERING GENE EXPRESSION DATA

Clustering is one of the major tasks in gene expression data analysis. When applied to gene expression data analysis, clustering algorithms can be applied on both gene and sample dimensions [4, 13]. The conventional clustering methods group a subset of genes that are interdependent or correlated with each other. In other words, genes in a cluster are more correlated with each other, whereas genes in different clusters are less correlated [13]. After clustering genes, a reduced set of genes can be selected for further analysis. The conventional gene clustering methods allow genes with similar expression patterns, that is, co-expressed genes, to be identified [4]. Different clustering techniques such as hierarchical clustering [14], k-means algorithm [15], self-organizing map (SOM)[16], principal component analysis [17], graph-theoretical approaches [18–21], model-based clustering [22–25], density-based approach [10], and fuzzy clustering algorithms [26, 27] have been widely applied to find groups of co-expressed genes from microarray data. This section presents a brief overview of different clustering algorithms that have been applied to group genes. A comprehensive survey of various gene clustering algorithms can also be found in Jiang et al. [4].

7.2.1 k-Means Algorithm

The k-means or hard c-means (HCM) algorithm [15] is a typical partition-based clustering method. A brief description of this algorithm is reported in Chapter 3. Given a prespecified number k, the algorithm partitions the data set into k disjoint subsets. However, it has several drawbacks as a gene clustering algorithm. As the number of gene clusters in a gene expression data set is usually unknown in advance, the users usually run the algorithms repeatedly with different values of k and compare the clustering results to detect the optimal number of clusters. This extensive parameter fine-tuning process may not be practical for a large gene expression data set containing thousands of genes. Also, as the k-means algorithm forces each gene into a cluster, it may cause the algorithm to be sensitive to noise present in the gene expression data sets. Recently, several new clustering algorithms have been proposed to overcome the drawbacks of the k-means algorithm. These algorithms typically use some global parameters such as the maximal radius of a cluster and/or the minimal distance between clusters to control the quality of resulting clusters. Clustering is the process of extracting all of the qualified clusters from the data set. In this way, the number of clusters can be automatically determined and those data objects that do not belong to any qualified clusters are regarded as outliers. However, the qualities of clusters in gene expression data sets may vary widely. Hence, it is often a difficult problem to choose the appropriate globally constraining parameters [4].

7.2.2 Self-Organizing Map

In an SOM [16], a prespecified number and an initial spatial structure of clusters are required. However, this may be hard to come up with in real problems.

Furthermore, if the data set is abundant with irrelevant data points such as genes with invariant patterns, the SOM may produce an output in which this type of data will populate the vast majority of clusters. In this case, the SOM is not effective because most of the interesting patterns may be merged into only one or two clusters and cannot be identified [4].

7.2.3 Hierarchical Clustering

The partition-based clustering such as k-means algorithm directly decomposes the data set into a set of disjointed clusters. In contrast, hierarchical clustering generates a hierarchical series of nested clusters that can be graphically represented by a tree, called *dendrogram*. The branches of a dendrogram not only record the formation of the clusters but also indicate the similarity between the clusters. By cutting the dendrogram at some level, one can obtain a specified number of clusters. By reordering the objects such that the branches of the corresponding dendrogram do not cross, the data set can be arranged with similar objects placed together [4]. The hierarchical clustering identifies sets of correlated genes with similar behavior across the samples, but yields thousands of clusters in a treelike structure, which makes the identification of functional groups very difficult [14].

7.2.4 Graph-Theoretical Approach

Graph-theoretical clustering techniques are explicitly presented in terms of a graph, where each gene corresponds to a vertex. For some clustering methods, each pair of genes is connected by an edge with weight assigned according to the proximity value between the genes [20, 21]. For other methods, proximity is mapped only to either 0 or 1 on the basis of some threshold [18, 19]. Hence, this approach converts the problem of clustering a gene set into graph-theoretical problems as finding minimum cut or maximal cliques in the proximity graph.

The CLICK (cluster identification via connectivity kernels) [20] seeks to identify highly connected components in the proximity graph as clusters. It makes the probabilistic assumption that after standardization, pairwise similarity values between elements are normally distributed, no matter whether they are in the same cluster or not. Under this assumption, the weight of an edge between two vertices is defined as the probability that two vertices are in the same cluster. The clustering process of the CLICK iteratively finds the minimum cut in the proximity graph and recursively splits the gene set into a set of connected components from the minimum cut. The CLICK also takes two postpruning steps, namely, adoption and merging, to refine the clustering results. The adoption step handles the remaining singletons and updates the current clusters, whereas the merging step iteratively merges two clusters with similarity exceeding a predefined threshold.

In Shamir and Sharan [20], the clustering results of the CLICK on two public gene expression data sets are compared with those of the gene cluster

[16], a SOM approach, and Eisen's hierarchical approach [5], respectively. In both cases, clusters obtained by CLICK demonstrated better quality in terms of homogeneity and separation. However, CLICK has little guarantee of not going astray and generating highly unbalanced partitions. Furthermore, in gene expression data, two clusters of co-expressed genes may be highly overlapping with each other. In such situations, the CLICK is not able to split them and reports as one highly connected component.

Ben-Dor et al. [18] have presented both a theoretical algorithm and a practical heuristic called CAST (cluster affinity search technique). The CAST takes as input a real, symmetric similarity matrix and an affinity threshold. The algorithm searches the clusters one at a time. It alternates between adding high affinity elements to the current cluster and removing low affinity elements from it. When the process stabilizes, the current cluster is considered as a complete cluster, and this process continues with each new cluster until all elements have been assigned to a cluster. The affinity threshold of the CAST algorithm is actually the average of pairwise similarities within a cluster. The CAST specifies the desired cluster quality through the affinity threshold and applies a heuristic searching process to identify qualified clusters one at a time. Therefore, the CAST does not depend on a user-defined number of clusters and deals with outliers effectively. However, the CAST has the usual difficulty in determining a good value for the global affinity threshold [4].

7.2.5 Model-Based Clustering

The model-based clustering approaches [22–25] provide a statistical framework to model the cluster structure of gene expression data. The data set is assumed to come from a finite mixture of underlying probability distributions, with each component corresponding to a different cluster. The goal is to estimate the parameters that maximize the likelihood. Usually, the parameters are estimated by the expectation maximization (EM) algorithm. The EM algorithm iterates between expectation (E) and maximization (M) steps. In the E step, hidden parameters are conditionally estimated from the data with the current estimated model parameter. In the M step, model parameters are estimated so as to maximize the likelihood of complete data given the estimated hidden parameters. When the EM algorithm converges, each gene is assigned to the component or cluster with the maximum conditional probability [4].

An important advantage of model-based approaches is that they provide an estimated probability about the belongingness of a gene to a cluster. As gene expression data are typically highly connected, there may be instances in which a single gene has a high correlation with two different clusters. Hence, the probabilistic feature of model-based clustering is particularly suitable for gene expression data. However, model-based clustering relies on the assumption that the gene set fits a specific distribution. This may not be true in many cases. Several kinds of commonly used data transformations have been studied by Yeung et al. [25] to fit gene expression data sets.

7.2.6   Density-Based Hierarchical Approach

A density-based hierarchical clustering method, called the DHC, is proposed in Jiang et al. [10] to identify the co-expressed gene groups from gene expression data. The DHC is developed on the basis of the notions of density and attraction of data objects. The basic idea is to consider a cluster as a high dimensional dense area, where data objects are attracted to each other. At the core part of the dense area, objects are crowded closely with each other and thus have high density. Objects at the peripheral area of the cluster are relatively sparsely distributed and are attracted to the core part of the dense area. Once the density and attraction of data objects are defined, the DHC organizes the cluster structure of the data set in two-level hierarchical structures [4].

As a density-based approach, the DHC effectively detects the co-expressed genes from noise, and thus is robust in the noisy environment. Furthermore, the DHC is particularly suitable for the high connectivity characteristic of gene expression data because it first captures the core part of the cluster and then divides the borders of clusters on the basis of the attraction between the data objects. The two-level hierarchical representation of the data set not only discloses the relationship between the clusters but also organizes the relationship between data objects within the same cluster. However, to compute the density of data objects, the DHC calculates the distance between each pair of data objects in the data set. Also, two global parameters are used in the DHC to control the splitting process of dense areas. Therefore, the DHC does not escape from the typical difficulty to determine the appropriate value of parameters [4].

7.2.7   Fuzzy Clustering

A fuzzy logic method introduced to microarray data analysis [26–29] has been shown to reveal additional information concerning gene co-expression. In particular, information regarding overlapping clusters and overlapping cellular pathways has been identified from fuzzy clustering results [27]. The method of choice in all up-to-date applications has been the fuzzy c-means (FCM). The FCM is the fuzzy logic extension of the k-means heuristic used for crisp clustering. The FCM method searches for the membership degrees and centroids until there is no further improvement in the objective function value, thereby risking the possibility of remaining in a local minimum of a poor value. A brief description of this algorithm is reported in Chapter 3. An alternative fuzzy clustering method called *fuzzy J-means* has been reported in Belacel et al. [30], which is embedded into the variable neighborhood search meta-heuristic for appropriate gene cluster arrangements.

7.2.8   Rough-Fuzzy Clustering

Different rough and rough-fuzzy clustering algorithms such as rough c-means [31], rough-fuzzy c-means (RFCM) [11], rough-possibilistic c-means (RPCM), and rough-fuzzy-possibilistic c-means (RFPCM) [12] can be used for clustering functionally similar genes from microarray gene expression data sets. Details of these algorithms are reported in Chapter 3 and in Maji and Pal [11, 12]. The rough c-means algorithm [31] has been applied successfully [32] to discover value-coherent overlapping gene biclusters. Recently, fuzzy-rough supervised attribute clustering algorithm is proposed in Maji [33] to find groups of co-regulated genes whose collective expression is strongly associated with sample categories. This chapter presents the application of different rough-fuzzy clustering algorithms for clustering functionally similar genes and compares their performance with that of some existing gene clustering algorithms.

7.3 QUANTITATIVE AND QUALITATIVE ANALYSIS

The following quantitative and qualitative indices are generally used, along with other measures [4], to evaluate the performance of different gene clustering algorithms for grouping functionally similar genes from microarray gene expression data sets.

7.3.1 Silhouette Index

To assess the quality of clusters, the Silhouette measure proposed by Rousseeuw [34] can be used. For computing the Silhouette value of a gene x_i, two scalars $a(x_i)$ and $b(x_i)$ are first estimated. Let us note β_r the cluster to which gene x_i belongs. The scalar $a(x_i)$ is the average distance between gene x_i and all other genes of β_r. For any other cluster $\beta_s \neq \beta_r$, let $d(x_i, \beta_s)$ denote the average distance of gene x_i to all genes of β_s. The scalar $b(x_i)$ is the smallest of these $d(x_i, \beta_s), r \neq s = 1, \ldots, c$. The Silhouette $s(x_i)$ of gene x_i is then defined as

$$s(x_i) = \frac{b(x_i) - a(x_i)}{\max\{a(x_i), b(x_i)\}}. \tag{7.1}$$

The Silhouette value lies between -1 and 1. When its value is less than zero, the corresponding gene is poorly classified.

7.3.2 Eisen and Cluster Profile Plots

In Eisen plot [5], the expression value of a gene at a specific time point is represented by coloring the corresponding cell of the data matrix with a color similar to the original color of its spot on the microarray. The shades of red color represent higher expression level, the shades of green color represent low expression level and the colors toward black represent absence of differential expression values. In the present representation, the genes are ordered before plotting so that the genes that belong to the same cluster are placed one after another. The cluster boundaries are identified by white colored blank rows. On the other hand, the cluster profile plot shows for each gene cluster the normalized gene expression values of the genes of that cluster with respect to the samples or time points.

7.3.3 Z Score

The Z score [35, 36] is calculated by observing the relation between a clustering result and the functional annotation of the genes in the cluster. Here, an attribute database is used to create an $n \times m$ gene-attribute table for n genes and m attributes in which a "1" in position (i, j) indicates that the gene i is known to possess attribute A_j, and a "0" indicates the lack of knowledge about whether gene i possesses attribute A_j or not. With this gene-attribute table, a contingency table for each cluster-attribute pair is constructed, from which the entropies H_{A_jC} for each cluster-attribute pair is computed [35, 36], H_C for clustering result independent of attributes, and H_{A_j} for each of the N_A attributes in the table independent of clusters. Using the definition of mutual information between two variables and assuming both absolute and conditional independence of attributes, the total mutual information is expanded as a sum of mutual information between clusters and each individual attribute. The total mutual information between the clustering result C and all the attributes A_j is computed as [35, 36]

$$\mathrm{MI}(C, A_1, \ldots, A_{N_A}) = \sum_t \mathrm{MI}(C, A_t) = N_A H_C + \sum_t H_{A_t} - \sum_t H_{A_t C}, \quad (7.2)$$

where summation is taken over all attributes A_j. Hence, the Z score is defined as [35, 36]

$$Z = \frac{\mathrm{MI}_{\mathrm{real}} - \mathrm{MI}_{\mathrm{random}}}{s_{\mathrm{random}}}, \quad (7.3)$$

where $\mathrm{MI}_{\mathrm{real}}$ is the computed mutual information for the clustered data using the attribute database. $\mathrm{MI}_{\mathrm{random}}$ is computed by computing mutual information again, for a clustering obtained by randomly assigning genes to clusters of uniform size and repeating until a distribution of values is obtained. The mean of these mutual information values, computed for a randomly obtained cluster is $\mathrm{MI}_{\mathrm{random}}$ and standard deviation of these mutual information values is s_{random}. A higher value of Z indicates that the genes are better clustered by function, indicating a biologically relevant clustering result [35, 36].

7.3.4 Gene-Ontology-Based Analysis

To interpret the biological significance of the gene clusters, the Gene Ontology (GO) Term Finder can be used [37, 38]. It finds the most significantly enriched GO terms associated with the genes belonging to a cluster. The GO project aims to build tree structures, controlled vocabularies, also called *ontologies*, that describe gene products in terms of their associated BP, molecular functions (MF) or cellular components (CC). The GO Term Finder determines whether any GO term annotates a specified list of genes at a frequency greater than that would be expected by chance, calculating the associated p value by using the hypergeometric distribution and the Bonferroni multiple-hypothesis correction [37, 38]. The closer the p value is to zero, the more significant the particular GO term associated with the group of genes is, that is, the less likely the observed annotation of the particular GO term to a group of genes occurs by chance. On the other hand, the false discovery rate is a multiple-hypothesis testing error measure, indicating the expected proportion of false positives among the set of significant results.

7.4 DESCRIPTION OF DATA SETS

In this chapter, several publicly available microarray gene expression data sets are used to compare the performance of different gene clustering methods.

7.4.1 Fifteen Yeast Data

This section gives a brief description of the following 15 microarray gene expression data sets, which are downloaded from *Gene Expression Omnibus* (http://www.ncbi.nlm.nih.gov/geo/).

7.4.1.1 GDS608 It is a temporal analysis of wild-type diploid cells shifted from yeast-form growth in SHAD liquid (plentiful glucose and ammonium) to filamentous-form growth on SLAD agar (low ammonium). The filamentous-form cells were collected hourly for 10 h. The number of genes and time points of this data are 6303 and 10, respectively.

7.4.1.2 GDS759 This data set is related to analysis of gene expression in temperature-sensitive pre-mRNA splicing factor mutants prp17 null, prp17-1, and prp22-1 at various time points following a shift from the permissive temperature of 23°C to the restrictive temperature of 37°C. The number of genes and time points of this data are 6350 and 24, respectively.

7.4.1.3 GDS922 It contains the comparison of total transcription profiles for temperature-sensitive TOR2 mutant strain SH121 to its isogenic wild-type counterpart SH100. The number of genes and time points of this data are 9275 and 12, respectively.

7.4.1.4 GDS1013 It contains the analysis of overexpression of essential ribosomal protein activator IFH1. The cells engineered to express IFH1 form a galactose inducible promoter. The expression was examined at various time points up to 60 min following galactose addition. The data set contains 9275 genes with 24 time points.

7.4.1.5   GDS1550   This represents the analysis of cells depleted of Paf1, an RNA polymerase II-associated protein. The cells were examined at various time points up to 8 h following the inactivation of Paf1 expression from a tetracycline-regulated promoter using doxycyline. The data set contains 9275 genes and six time points.

7.4.1.6   GDS1611   It is the analysis of wild-type and mutant Upf1 strains up to 60 min after treatment with 10 μg/ml thiolutin, a global transcription inhibitor. The Upf1 is an RNA helicase required for nonsense-mediated mRNA decay. The data set contains 9275 genes with 96 time points.

7.4.1.7   GDS2002   It contains the analysis of catabolite-repressed (glucose) or derepressed (galactose) wild-type JM43 cells shifted from aerobiosis to anaerobiosis (two generations). The number of genes in this data set is 5617, whereas the number of time points is 30.

7.4.1.8   GDS2003   It is the analysis of catabolite-derepressed (galactose) wild-type JM43 and isogenic msn2/4 mutant KKY8 cells shifted to short-term anaerobiosis (two generations). The Msn2 and Msn4 are key stress factors. The number of genes and time points are 5617 and 30, respectively.

7.4.1.9   GDS2196   It contains the analysis of yeast cells for up to 3 days after treatment with the saponins alpha-tomatine or tomatidine or the sterol biosynthesis inhibitor fenpropimorph. The saponins are glycoside compounds accumulated by many plant species, and possess antimicrobial activity and various pharmacological properties. The number of genes and time points are 9275 and 12, respectively.

7.4.1.10   GDS2267   It contains the analysis of nutrient-limited continuous-culture cells at twelve 25-min intervals for three cycles. The cells grown under such conditions exhibit robust, periodic cycles in the form of respiratory bursts. The number of genes and time points are 9275 and 36, respectively.

7.4.1.11   GDS2318   It is the analysis of yox1 yhp double mutants across two cell cycles, a length of 2 h after synchronization with alpha factor. The number of genes in the data set is 6216, whereas the number of time points is 13.

7.4.1.12   GDS2347   It contains the analysis of wild-type W303 cells across two cell cycles, a length of 2 h after synchronization with the alpha factor. The number of genes and time points are 6228 and 13, respectively.

7.4.1.13 GDS2712 It represents the analysis of *Saccharomyces cerevisiae* BY4743 cells subjected to controlled air drying and subsequent rehydration (I) for up to 360 min. The data contain 9275 genes and 21 time points.

7.4.1.14 GDS2713 It represents the analysis of *S. cerevisiae* BY4743 cells subjected to controlled air drying and subsequent rehydration (II) for up to 360 min. The data contain 9275 genes and 21 time points.

7.4.1.15 GDS2715 It represents the analysis of *S. cerevisiae* BY4743 cells subjected to controlled air drying and subsequent rehydration (III) for up to 360 min. The data contain 9275 genes and 21 time points.

7.4.2 Yeast Sporulation

This data set consists of 6118 genes measured across seven time points (0, 0.5, 2, 5, 7, 9, and 11.5 h) during the sporulation process of budding yeast [39]. The data are then log-transformed. The sporulation data set is publicly available at http://cmgm.stanford.edu/pbrown/sporulation. Among the 6118 genes, the genes whose expression levels did not change significantly during the harvesting have been ignored from further analysis. This is determined with a threshold level of 1.6 for the root mean squares of the log2-transformed ratios. The resulting set consists of 474 genes.

7.4.3 Auble Data

Auble data is in the form of a tab-delimited file that contains the log2 ratios from four microarray hybridizations [40]. The first three columns were mot1-14mut#1, #2 and #3 and the log2 ratios are for the mot1-14 temperature-sensitive mutant/wild type. The fourth column contains the log2 ratios for a single hybridization of wild-type versus a different allele named mot1-4. There are 6226 genes measured across four conditions of temperature sensitivity. The Mot1 microarray data set in text format is available at http://dir.niehs.nih.gov/microarray/datasets/auble_data.txt.

7.4.4 Cho et al. Data

This data set contains gene expressions of 6457 genes at 10-min intervals following release from cell cycle arrest at a restrictive temperature [41]. The time points were taken up to 160 min. Data up to time points 40 reflects both temperature-induced and cell cycle effects. This data set can be downloaded from http://yfgdb.princeton.edu/download/yeast_datasets/.

7.4.5 Reduced Cell Cycle Data

It contains gene expression of 384 genes across 17 conditions. The data set can be downloaded from http://www.math.unipa.it/lobosco/genclust/.

7.5   EXPERIMENTAL RESULTS

In this section, the performance of different rough-fuzzy clustering algorithms is presented for clustering functionally similar genes from microarray data sets. The experimentation is done in two parts. In the first part, the results of the RFCM algorithm are presented on 15 yeast microarray data sets obtained from the *Gene Expression Omnibus*. The performance of the RFCM is also compared extensively with that of different algorithms, namely, HCM [42], FCM [43], SOM [16], and CLICK [20]. In the second part, the performance of RFPCM is reported on the other four data sets, namely, yeast sporulation [39], Auble [40], Cho et al. [41], and reduced cell cycle data sets. The results are compared with the performance of the FCM [43], possibilistic c-means (PCM) [44], fuzzy-possibilistic c-means (FPCM) [45], RFCM, and RPCM algorithms.

The CLICK [20] algorithm, which combines graph-theoretic and statistical techniques for automatic identification of clusters in a microarray gene expression data set, is used to arrive at an estimation of the number of clusters. The final centroids of the HCM are used to initialize all c-means algorithms. In all the experiments, the values of both probabilistic and possibilistic fuzzifiers are set at 2.0, which are held constant across all runs. The values of threshold δ for the RFCM, RPCM, and RFPCM are calculated using the procedure reported in Chapter 3. The major metrics for evaluating the performance of different methods are the Silhouette index [34], Z score [35, 36], Eisen plot [5], and GO Term Finder [37].

7.5.1   Performance Analysis of Rough-Fuzzy c-Means

This section presents the performance of the RFCM algorithm on the data sets reported in Section 7.4.1. The results are presented in Tables 7.1 and 7.2 with respect to the Silhouette index and Z score, respectively, for different values of weight parameter w as the parameter w has an influence on the performance of the RFCM. From the results reported in Tables 7.1 and 7.2, it can be seen that the RFCM algorithm performs better for $0.50 < w < 1.00$. That is, the quality of generated clusters is better when the genes of lower approximations are assigned a higher weight compared to that of boundary regions.

7.5.2   Comparative Analysis of Different c-Means

Table 7.3 compares the performance of the RFCM algorithm with that of both the HCM and FCM. The results are presented in this table with respect to both the Silhouette index and the Z score. From the results reported in Table 7.3, it is seen that the performance of the RFCM is significantly better than that of the HCM and FCM in most of the cases. Among 15 data sets, the RFCM algorithm provides better Silhouette index and Z score values in 12 and 11 cases, respectively. Only the FCM algorithm for GDS1013 and GDS1550 data sets, and the HCM algorithm for GDS2713 data set attain higher Silhouette index values. For microarray data sets such as GDS1013, GDS2318, and GDS2715, the

TABLE 7.1 Variation in Silhouette Index for Different Values of Parameter w

Data Sets	No. of Clusters	Different Values of Parameter w										
		0.510	0.550	0.600	0.650	0.700	0.750	0.800	0.850	0.900	0.950	0.990
GDS608	26	0.019	0.020	0.024	0.050	0.057	0.065	0.076	0.083	0.090	0.104	0.110
GDS759	25	0.058	0.061	0.067	0.071	0.075	0.078	0.086	0.086	0.098	0.119	0.121
GDS922	48	0.140	0.137	0.150	0.152	0.147	0.154	0.157	0.168	0.170	0.170	0.180
GDS1013	18	0.198	0.199	0.198	0.200	0.193	0.194	0.202	0.203	0.205	0.212	0.213
GDS1550	21	0.221	0.224	0.229	0.231	0.228	0.225	0.229	0.236	0.239	0.243	0.242
GDS1611	26	0.129	0.135	0.142	0.146	0.149	0.152	0.159	0.163	0.166	0.168	0.177
GDS2002	25	0.030	0.040	0.050	0.057	0.064	0.068	0.079	0.085	0.092	0.099	0.109
GDS2003	23	0.000	0.007	0.030	0.046	0.059	0.074	0.081	0.089	0.095	0.102	0.128
GDS2196	24	0.292	0.295	0.302	0.304	0.309	0.317	0.316	0.305	0.304	0.308	0.312
GDS2267	14	0.223	0.225	0.225	0.226	0.226	0.231	0.232	0.233	0.233	0.233	0.233
GDS2318	21	0.056	0.084	0.121	0.136	0.151	0.159	0.165	0.180	0.206	0.217	0.220
GDS2347	18	0.051	0.076	0.086	0.104	0.111	0.118	0.125	0.132	0.132	0.145	0.160
GDS2712	15	0.245	0.247	0.247	0.248	0.248	0.249	0.249	0.251	0.250	0.250	0.251
GDS2713	14	0.211	0.213	0.213	0.215	0.216	0.218	0.219	0.219	0.220	0.220	0.221
GDS2715	16	0.213	0.215	0.216	0.215	0.217	0.218	0.220	0.220	0.222	0.224	0.224

TABLE 7.2 Variation in Z Score for Different Values of Parameter w

Data Sets	No. of Clusters	Different Values of Parameter w										
		0.51	0.55	0.60	0.65	0.70	0.75	0.80	0.85	0.90	0.95	0.99
GDS608	26	1.51	1.21	2.69	1.7	3.04	1.98	2.47	2.79	2.61	1.26	2.00
GDS759	25	1.20	-0.08	0.21	0.32	0.57	0.52	-0.48	-0.67	-1.67	-1.07	-0.98
GDS922	48	0.19	0.04	0.11	0.38	0.15	0.11	-0.29	-0.30	0.16	-0.07	0.00
GDS1013	18	0.64	0.74	0.56	0.57	0.52	0.50	0.66	0.64	0.72	0.68	0.74
GDS1550	21	2.04	2.38	2.58	2.18	2.12	2.12	2.27	2.36	2.05	2.84	2.12
GDS1611	26	5.19	5.82	5.15	4.52	4.27	3.89	4.47	4.69	4.07	3.84	4.06
GDS2002	25	0.22	0.46	0.46	-0.13	0.44	-0.28	-0.18	-0.20	-0.38	-0.04	-0.05
GDS2003	23	2.10	1.90	1.23	1.50	1.53	1.82	1.23	1.15	0.94	1.06	0.76
GDS2196	24	2.43	2.25	2.38	2.44	2.52	2.55	2.78	2.25	2.35	2.79	2.81
GDS2267	14	0.35	0.22	0.18	0.40	0.41	0.84	0.77	0.90	0.79	1.17	1.18
GDS2318	21	-1.56	-1.61	-1.24	-1.16	-1.17	-0.79	-0.91	-1.04	-0.37	-0.37	-0.55
GDS2347	18	1.39	1.46	1.59	1.33	-0.45	-0.06	-0.21	0.08	0.36	-0.70	-0.22
GDS2712	15	-0.37	-0.51	-0.59	-0.60	-0.57	-0.42	-0.38	-0.44	-0.59	-0.56	-0.40
GDS2713	14	-0.14	-0.05	0.08	0.09	0.09	0.02	0.07	0.07	0.14	0.26	0.31
GDS2715	16	-0.08	-0.15	-0.07	-0.06	0.00	-0.13	-0.36	-0.37	-0.40	-0.25	-0.23

TABLE 7.3 **Performance Analysis of Different c-Means Algorithm**

Data Sets	Silhouette Index			Z Score		
	HCM	FCM	RFCM	HCM	FCM	RFCM
GDS608	0.078	0.005	**0.110**	2.44	2.48	**3.04**
GDS759	0.082	0.017	**0.121**	−0.54	0.58	**1.20**
GDS922	0.173	0.124	**0.180**	0.31	−0.02	**0.38**
GDS1013	0.220	**0.249**	0.213	0.54	**0.78**	0.74
GDS1550	0.245	**0.259**	0.243	2.18	2.22	**2.84**
GDS1611	0.158	0.088	**0.177**	4.79	4.11	**5.82**
GDS2002	0.079	*	**0.109**	−0.80	0.44	**0.46**
GDS2003	0.082	0.014	**0.128**	1.28	0.36	**2.10**
GDS2196	0.309	0.300	**0.317**	2.64	2.59	**2.81**
GDS2267	0.230	0.197	**0.233**	1.11	1.17	**1.18**
GDS2318	0.153	0.065	**0.220**	−0.75	**0.14**	−0.37
GDS2347	0.134	0.031	**0.160**	−0.72	−1.35	**1.59**
GDS2712	0.250	0.208	**0.251**	−0.6	−0.66	**−0.37**
GDS2713	**0.223**	0.201	0.221	**0.34**	0.13	0.31
GDS2715	0.222	0.177	**0.224**	−0.02	**0.23**	0.00

Bold value signifies highest value and * indicates missing value.

FCM algorithm achieves better Z score values than that of the RFCM, whereas the HCM attains a higher value for the GDS2713 data set only.

7.5.3 Biological Significance Analysis

The GO Term Finder is used to determine the statistical significance of the association of a particular GO term with the genes of different clusters produced by the RFCM algorithm. It is used to compute the p value for all the GO terms from the BP, MF, and CC ontology and the most significant term, that is, the one with the lowest p value, is chosen to represent the set of genes of the best cluster. Table 7.4 presents the p values for the BP, MF, and CC on different data sets. The results corresponding to the best clusters of the FCM algorithm are also provided on the same data sets for the sake of comparison. The "*" in Table 7.4 represents that no significant shared term is found considering p-value cutoff as 0.05. From the results reported in Table 7.4, it is seen that the best cluster generated by the RFCM algorithm can be assigned to the GO BP, MF, and CC with high reliability in terms of p value. That is, the RFCM algorithm describes accurately the known classification, the one given by the GO, and thus reliable for extracting new biological insights.

7.5.4 Comparative Analysis of Different Algorithms

Finally, Table 7.5 compares the performance of the RFCM algorithm with that of the SOM [16] and CLICK [20]. All the results are reported in this table with respect to the Silhouette index and Z score values. From all the results reported

TABLE 7.4   **Biological Significance Analysis Using Gene Ontology**

	Biological Processes		Molecular Functions		Cellular Components	
Data Set	RFCM	FCM	RFCM	FCM	RFCM	FCM
GDS608	1.0E-079	4.6E-079	1.3E-039	1.8E-038	1.3E-080	4.1E-080
GDS759	3.7E-048	2.3E-041	7.3E-103	1.4E-087	8.1E-120	1.0E-105
GDS922	4.9E-017	1.0E-008	1.5E-027	2.8E-016	4.3E-031	1.1E-017
GDS1013	6.2E-036	2.2E-026	4.4E-064	8.6E-042	8.5E-074	1.9E-048
GDS1550	4.3E-027	6.5E-012	2.5E-034	7.4E-019	3.4E-033	2.6E-021
GDS1611	6.6E-038	1.2E-021	1.9E-058	9.2E-036	7.8E-076	1.1E-046
GDS2002	3.7E-087	*	9.5E-048	*	3.0E-102	*
GDS2003	5.6E-093	1.1E-068	3.5E-100	1.8E-029	9.7E-134	3.0E-074
GDS2196	9.5E-013	1.2E-010	5.9E-026	1.2E-017	4.0E-031	2.0E-020
GDS2267	3.5E-032	7.8E-029	9.0E-062	4.2E-055	1.5E-073	8.6E-065
GDS2318	6.6E-003	*	1.5E-04	2.9E-002	2.5E-002	1.9E-002
GDS2347	2.7E-076	3.0E-066	2.2E-022	6.9E-016	1.8E-082	1.1E-069
GDS2712	1.1E-034	6.7E-039	3.8E-072	2.3E-065	4.2E-086	2.3E-083
GDS2713	1.9E-029	1.4E-032	1.9E-052	1.4E-050	1.9E-067	8.3E-066
GDS2715	1.0E-029	4.3E-037	7.9E-053	9.7E-072	3.6E-066	1.5E-091

TABLE 7.5   **Comparative Analysis of Different Algorithms**

	Silhouette Index			Z Score		
Data Sets	RFCM	CLICK	SOM	RFCM	CLICK	SOM
GDS608	**0.110**	−0.042	−0.025	**3.04**	−1.11	0.74
GDS759	**0.121**	−0.079	−0.020	1.20	**2.28**	1.70
GDS922	**0.180**	−0.577	0.099	**0.38**	−0.28	0.21
GDS1013	**0.210**	−0.523	0.064	**0.74**	−0.31	0.33
GDS1550	**0.240**	−0.490	0.145	**2.84**	−0.67	1.56
GDS1611	**0.177**	−0.274	0.051	5.82	**7.20**	2.76
GDS2002	**0.109**	−0.117	−0.051	0.46	−1.07	**0.79**
GDS2003	**0.128**	−0.093	−0.063	**2.10**	−0.80	−0.03
GDS2196	**0.317**	−0.529	0.166	**2.81**	0.45	1.41
GDS2267	**0.233**	−0.418	0.022	1.18	−0.51	**0.79**
GDS2318	**0.220**	−0.131	−0.112	**−0.37**	−1.42	−0.79
GDS2347	**0.160**	−0.112	−0.126	**1.59**	0.39	−0.35
GDS2712	**0.251**	−0.420	0.065	−0.37	**0.21**	0.16
GDS2713	**0.220**	−0.394	0.065	0.31	1.39	**1.54**
GDS2715	**0.224**	−0.407	0.080	0.00	−0.40	**0.33**

in Table 7.5, it can be seen that the RFCM algorithm achieves highest Silhouette index values for all microarray gene expression data sets, while the RFCM attains the highest Z score in 8 cases out of the total 15 data sets. The CLICK [20] and SOM [16] achieve best Z score values for three (namely, GDS759, GDS1611, and

GDS2712), and four (namely, GDS2002, GDS2267, GDS2713, and GDS2715) data sets, respectively.

7.5.5 Performance Analysis of Rough-Fuzzy-Possibilistic c-Means

In this section, the performance of the RFPCM algorithm is presented on yeast sporulation [39], Auble [40], Cho et al. [41], and reduced cell cycle data sets. The value of weight parameter w for the RFCM, RPCM, and RFPCM is set at 0.95, while details of the experimental setup and objective of the experiments are the same as those in the previous sections. The performance of the RFPCM is also compared with that of the FCM, PCM, FPCM, RFCM, and RPCM algorithms with respect to the Silhouette index, Z score, p value, and Eisen plot.

7.5.5.1 Qualitative Analysis The Eisen plot gives a visual representation of the clustering result. The clustering results produced by the FCM, RFCM, and RFPCM algorithms on yeast sporulation data are visualized by TreeView software, which is available at http://rana.lbl.gov/EisenSoftware and reported in Fig. 7.1. From the Eisen plots presented in Fig. 7.1, it is evident that the expression profiles of the genes in a cluster are similar to each other and they produce similar color patterns, whereas the genes from different clusters differ in color patterns. Also, the results obtained by both RFCM and RFPCM algorithms are more promising than that by the FCM algorithm.

7.5.5.2 Quantitative Analysis Tables 7.6 and 7.7 compare the performance of different c-means algorithm on yeast sporulation [39], Auble [40], Cho et al. [41], and reduced cell cycle data sets with respect to the Silhouette index and Z score, respectively. From the results reported in Tables 7.6 and 7.7, it can be seen that for yeast sporulation and Auble data, highest values of the Silhouette index are obtained for the RFPCM and RPCM, respectively. However, for Cho et al. data, the best result is found for FPCM, while the FCM, PCM, and RPCM achieve very good values of the Silhouette index on reduced cell cycle data set. On the other hand, a very high value of Z score is observed for the RFPCM for the yeast sporulation data set. Similarly, for Auble data and Cho et al. data, the best Z score is obtained for the RFPCM. However, the FPCM attains the best score for reduced cell cycle data set.

Finally, Table 7.8 presents the p value of BP, along with the significant shared GO term, on yeast sporulation data for both FCM and RFPCM algorithms. From the results reported in Table 7.8, it can be seen that the RFPCM algorithm generates more functionally enriched gene clusters than the FCM algorithm.

7.6 CONCLUSION AND DISCUSSION

This chapter presents the results of clustering functionally similar genes from microarray data sets using different rough-fuzzy clustering algorithms. Several

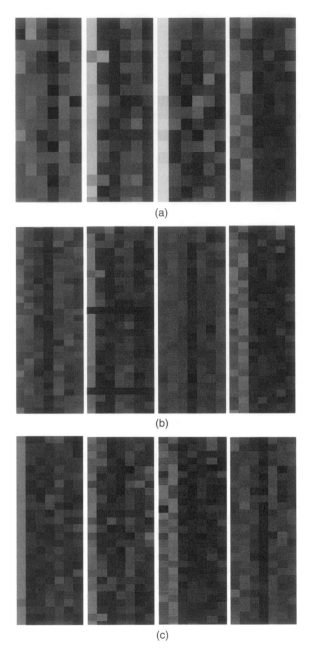

Figure 7.1 Eisen plots of four clusters of yeast sporulation data. (a) Fuzzy c-means, (b) rough-fuzzy c-means, (c) rough-fuzzy-possibilistic c-means.

TABLE 7.6 Silhouette Index Values of RFPCM Algorithm

Data Set	FCM	PCM	FPCM	RFCM	RPCM	RFPCM
Yeast	−8.8E-04	−2.1E-03	−1.8E-03	−3.3E-03	−1.7E-03	3.8E-04
Reduced cell	4.6E-04	5.6E-04	−1.4E-03	−1.9E-03	1.8E-04	−1.1E-03
Auble	−1.5E-04	−1.4E-04	−1.5E-04	−1.6E-04	5.5E-05	−1.5E-04
Cho et al.	−1.2E-04	−8.3E-05	−7.6E-05	−1.5E-04	−8.4E-05	−1.4E-04

TABLE 7.7 Z Score Values of RFPCM Algorithm

Data Set	FCM	PCM	FPCM	RFCM	RPCM	RFPCM
Yeast	31.7	−0.995	13.1	19.4	19.9	36.3
Reduced cell	0.699	0.748	4.17	2.15	2.89	1.4
Auble	0.623	−0.0396	0.0753	0.766	0.698	1.3
Cho et al.	5.66	1.43	0.358	6.53	4.72	7.34

TABLE 7.8 p Value of FCM and RFPCM on Yeast Sporulation Data

Algorithm	Cluster	p Value	Biological Processes
	0	2.81E-41	Sporulation resulting in the formation of cellular spore
FCM	1	7.24E-24	Meiosis I
	2	1.84E-16	Ribosome biogenesis
	3	7.19E-10	Glycosis
	0	2.39E-42	Sporulation resulting in the formation of cellular spore
RFPCM	1	1.02E-24	Meiosis I
	2	5.21E-21	Ribosome biogenesis
	3	1.66E-11	Pyruvate Metabolic process

quantitative and qualitative indices are reported for evaluating the quality of generated gene clusters. The effectiveness of different rough-fuzzy clustering algorithms, along with a comparison with other gene clustering algorithms, has been demonstrated both qualitatively and quantitatively on a set of yeast microarray gene expression data sets.

The algorithms described in this chapter group genes according to similarity measures computed from the gene expressions, without using any information about the response variables. The information of response variables may be incorporated in gene clustering to find groups of co-regulated genes with strong association to the response variables [46]. In this background, some supervised gene clustering algorithms such as supervised gene clustering [46], gene shaving

[47], tree harvesting [48], partial least square procedure [49], mutual-information-based supervised attribute clustering [50], and fuzzy-rough supervised attribute clustering [33] have been reported recently to reveal various groups of co-regulated genes with strong association to the response variables. The supervised attribute clustering is defined as the grouping of genes or attributes, controlled by the information of response variables.

In the next chapter, another important task of bioinformatics, namely, selection of discriminative genes from microarray gene expression, is handled in rough-fuzzy framework. Here the fuzzy equivalence partition matrix, described in Chapter 5, is used to deal with the problem.

REFERENCES

1. H. Causton, J. Quackenbush, and A. Brazma. *Microarray Gene Expression Data Analysis: A Beginner's Guide*. Wiley-Blackwell, MA, USA, 2003.

2. E. Domany. Cluster Analysis of Gene Expression Data. *Journal of Statistical Physics*, 110:1117–1139, 2003.

3. T. R. Golub, D. K. Slonim, P. Tamayo, C. Huard, M. Gaasenbeek, J. P. Mesirov, H. Coller, M. L. Loh, J. R. Downing, M. A. Caligiuri, C. D. Bloomfield, and E. S. Lander. Molecular Classification of Cancer: Class Discovery and Class Prediction by Gene Expression Monitoring. *Science*, 286:531–537, 1999.

4. D. Jiang, C. Tang, and A. Zhang. Cluster Analysis for Gene Expression Data: A Survey. *IEEE Transactions on Knowledge and Data Engineering*, 16(11):1370–1386, 2004.

5. M. B. Eisen, P. T. Spellman, O. Patrick, and D. Botstein. Cluster Analysis and Display of Genome-Wide Expression Patterns. *Proceedings of the National Academy of Sciences of the United States of America*, 95(25):14863–14868, 1998.

6. S. Tavazoie, D. Hughes, M. J. Campbell, R. J. Cho, and G. M. Church. Systematic Determination of Genetic Network Architecture. *Nature Genetics*, 22:281–285, 1999.

7. A. Brazma and J. Vilo. Minireview: Gene Expression Data Analysis. *Federation of European Biochemical Societies*, 480:17–24, 2000.

8. P. D'haeseleer, X. Wen, S. Fuhrman, and R. Somogyi. Mining the Gene Expression Matrix: Inferring Gene Relationships from Large Scale Gene Expression Data. In R. C. Paton and M. Holcombe, editors, *Information Processing in Cells and Tissues*, pages 203–212, Plenum Press, New York, 1998.

9. D. Jiang, J. Pei, and A. Zhang. Interactive Exploration of Coherent Patterns in Time-Series Gene Expression Data. In *Proceedings of the 9th ACM SIGKDD International Conference on Knowledge Discovery and Data Mining*, Washington, DC, pages 24–27, 2003.

10. D. Jiang, J. Pei, and A. Zhang. DHC: A Density-Based Hierarchical Clustering Method for Time-Series Gene Expression Data. In *Proceedings of the 3rd IEEE International Symposium on Bioinformatics and Bioengineering*, Bethesda, MD, 2003.

11. P. Maji and S. K. Pal. RFCM: A Hybrid Clustering Algorithm Using Rough and Fuzzy Sets. *Fundamenta Informaticae*, 80(4):475–496, 2007.

12. P. Maji and S. K. Pal. Rough Set Based Generalized Fuzzy C-Means Algorithm and Quantitative Indices. *IEEE Transactions on Systems Man and Cybernetics Part B-Cybernetics*, 37(6):1529–1540, 2007.

13. W.-H. Au, K. C. C. Chan, A. K. C. Wong, and Y. Wang. Attribute Clustering for Grouping, Selection, and Classification of Gene Expression Data. *IEEE/ACM Transactions on Computational Biology and Bioinformatics*, 2(2):83–101, 2005.

14. J. Herrero, A. Valencia, and J. Dopazo. A Hierarchical Unsupervised Growing Neural Network for Clustering Gene Expression Patterns. *Bioinformatics*, 17:126–136, 2001.

15. L. J. Heyer, S. Kruglyak, and S. Yooseph. Exploring Expression Data: Identification and Analysis of Coexpressed Genes. *Genome Research*, 9:1106–1115, 1999.

16. P. Tamayo, D. Slonim, J. Mesirov, Q. Zhu, S. Kitareewan, E. Dmitrovsky, E. S. Lander, and T. R. Golub. Interpreting Patterns of Gene Expression with Self-Organizing Maps: Methods and Application to Hematopoietic Differentiation. *Proceedings of the National Academy of Sciences of the United States of America*, 96(6):2907–2912, 1999.

17. G. J. McLachlan, K.-A. Do, and C. Ambroise. *Analyzing Microarray Gene Expression Data*. John Wiley & Sons, Inc., Hoboken, NJ, 2004.

18. A. Ben-Dor, R. Shamir, and Z. Yakhini. Clustering Gene Expression Patterns. *Journal of Computational Biology*, 6(3–4):281–297, 1999.

19. E. Hartuv and R. Shamir. A Clustering Algorithm Based on Graph Connectivity. *Information Processing Letters*, 76(4–6):175–181, 2000.

20. R. Shamir and R. Sharan. CLICK: A Clustering Algorithm for Gene Expression Analysis. In *Proceedings of the 8th International Conference on Intelligent Systems for Molecular Biology*, San Diego, CA, 2000.

21. E. P. Xing and R. M. Karp. CLIFF: Clustering of High-Dimensional Microarray Data via Iterative Feature Filtering Using Normalized Cuts. *Bioinformatics*, 17(1):306–315, 2001.

22. C. Fraley and A. E. Raftery. How Many Clusters? Which Clustering Method? Answers Via Model-Based Cluster Analysis. *Computer Journal*, 41(8):578–588, 1998.

23. D. Ghosh and A. M. Chinnaiyan. Mixture Modelling of Gene Expression Data from Microarray Experiments. *Bioinformatics*, 18:275–286, 2002.

24. G. J. McLachlan, R. W. Bean, and D. Peel. A Mixture Model-Based Approach to the Clustering of Microarray Expression Data. *Bioinformatics*, 18:413–422, 2002.

25. K. Y. Yeung, C. Fraley, A. Murua, A. E. Raftery, and W. L. Ruzz. Model-Based Clustering and Data Transformations for Gene Expression Data. *Bioinformatics*, 17:977–987, 2001.

26. D. Dembele and P. Kastner. Fuzzy C-Means Method for Clustering Microarray Data. *Bioinformatics*, 19(8):973–980, 2003.

27. A. P. Gasch and M. B. Eisen. Exploring the Conditional Coregulation of Yeast Gene Expression Through Fuzzy K-Means Clustering. *Genome Biology*, 3(11):1–22, 2002.

28. E. R. Dougherty, J. Barrera, M. Brun, S. Kim, R. M. Cesar, Y. Chen, M. Bittner, and J. M. Trent. Inference from Clustering with Application to Gene-Expression Microarrays. *Journal of Computational Biology*, 9:105–126, 2002.

29. P. J. Woolf and Y. Wang. A Fuzzy Logic Approach to Analyzing Gene Expression Data. *Physiological Genomics*, 3:9–15, 2000.

30. N. Belacel, M. Cuperlovic-Culf, M. Laflamme, and R. Ouellette. Fuzzy J-Means and VNS Methods for Clustering Genes from Microarray Data. *Bioinformatics*, 20(11):1690–1701, 2004.

31. P. Lingras and C. West. Interval Set Clustering of Web Users with Rough K-Means. *Journal of Intelligent Information Systems*, 23(1):5–16, 2004.

32. R. Wang, D. Miao, G. Li, and H. Zhang. Rough Overlapping Biclustering of Gene Expression Data. In *Proceedings of the 7th IEEE International Conference on Bioinformatics and Bioengineering*, Boston, MA, pages 828–834, 2007.

33. P. Maji. Fuzzy-Rough Supervised Attribute Clustering Algorithm and Classification of Microarray Data. *IEEE Transactions on Systems Man and Cybernetics Part B-Cybernetics*, 41(1):222–233, 2011.

34. J. P. Rousseeuw. Silhouettes: A Graphical Aid to the Interpretation and Validation of Cluster Analysis. *Journal of Computational and Applied Mathematics*, 20:53–65, 1987.

35. F. Gibbons and F. Roth. Judging the Quality of Gene Expression Based Clustering Methods Using Gene Annotation. *Genome Research*, 12:1574–1581, 2002.

36. W. H. Press, S. A. Teukolsky, W. T. Vetterling, and B. P. Flannery. *Numerical Recipes: the Art of Scientific Computing*. Cambridge University Press, Cambridge, 2003.

37. E. I. Boyle, S. Weng, J. Gollub, H. Jin, D. Botstein, J. M. Cherry, and G. Sherlock. GO :: Term Finder Open Source Software for Accessing Gene Ontology Information and Finding Significantly Enriched Gene Ontology Terms Associated with a List of Genes. *Bioinformatics*, 20:3710–3715, 2004.

38. J. L. Sevilla, V. Segura, A. Podhorski, E. Guruceaga, J. M. Mato, L. A. Martinez-Cruz, F. J. Corrales, and A. Rubio. Correlation Between Gene Expression and GO Semantic Similarity. *IEEE/ACM Transactions on Computational Biology and Bioinformatics*, 2(4):330–338, 2005.

39. S. Chu, J. DeRisi, M. Eisen, J. Mulholland, D. Botstein, P. O Brown, and I. Herskowitz. The Transcriptional Program of Sporulation in Budding Yeast. *Science*, 282(5389):699–705, 1998.

40. A. Dasgupta, R. P. Darst, K. J. Martin, C. A. Afshari, and D. T. Auble. Mot1 Activates and Represses Transcription by Direct, ATPase-Dependent Mechanisms. *Proceedings of the National Academy of Sciences of the United States of America*, 99:2666–2671, 2002.

41. R. J. Cho, M. J. Campbell, E. A. Winzeler, L. Steinmetz, A. Conway, L. Wodicka, T. G. Wolfsberg, A. E. Gabrielian, D. Landsman, D. J. Lockhart, and R. W. Davis. A Genome-Wide Transcriptional Analysis of the Mitotic Cell Cycle. *Molecular Cell*, 2(1):65–73, 1998.

42. A. K. Jain and R. C. Dubes. *Algorithms for Clustering Data*. Prentice Hall, Englewood Cliffs, NJ, 1988.

43. J. C. Bezdek. *Pattern Recognition with Fuzzy Objective Function Algorithm*. Plenum, New York, 1981.

44. R. Krishnapuram and J. M. Keller. A Possibilistic Approach to Clustering. *IEEE Transactions on Fuzzy Systems*, 1(2):98–110, 1993.

45. N. R. Pal, K. Pal, J. M. Keller, and J. C. Bezdek. A Possibilistic Fuzzy C-Means Clustering Algorithm. *IEEE Transactions on Fuzzy Systems*, 13(4):517–530, 2005.

46. M. Dettling and P. Buhlmann. Supervised Clustering of Genes. *Genome Biology*, 3(12):1–15, 2002.

47. T. Hastie, R. Tibshirani, M. B. Eisen, A. Alizadeh, R. Levy, L. Staudt, W. C. Chan, D. Botstein, and P. Brown. 'Gene Shaving' as a Method for Identifying Distinct Sets of Genes with Similar Expression Patterns. *Genome Biology*, 1(2):1–21, 2000.

48. T. Hastie, R. Tibshirani, D. Botstein, and P. Brown. Supervised Harvesting of Expression Trees. *Genome Biology*, 1:1–12, 2001.

49. D. Nguyen and D. Rocke. Tumor Classification by Partial Least Squares Using Microarray Gene Expression Data. *Bioinformatics*, 18:39–50, 2002.

50. P. Maji. Mutual Information Based Supervised Attribute Clustering for Microarray Sample Classification. *IEEE Transactions on Knowledge and Data Engineering*. DOI:10.1109/TKDE.2010.210.

8

SELECTION OF DISCRIMINATIVE GENES FROM MICROARRAY DATA

8.1 INTRODUCTION

Microarray technology allows us to record the expression levels of thousands of genes simultaneously within a number of different samples. A microarray gene expression data set can be represented by an expression table, $\mathcal{T} = \{w_{ij} | i = 1, \ldots, m, j = 1, \ldots, n\}$, where $w_{ij} \in \Re$ is the measured expression level of gene $\mathbb{G}_i$ in the jth sample, m and n represent the total number of genes and samples, respectively. Each row in the expression table corresponds to one particular gene and each column to a sample [1–3].

The wide use of high throughput technology produces an explosion in using gene expression phenotype for the identification and classification in a variety of diagnostic areas. An important application of gene expression data in functional genomics is to classify samples according to their gene expression profiles such as to classify cancer versus normal samples or to classify different types or subtypes of cancer [1–3]. However, for most gene expression data, the number of training samples is still very small compared to the large number of genes involved in the experiments. For example, the colon cancer data set consists of 62 samples and 2000 genes, and the leukemia data set contains 72 samples and 7129 genes. The number of samples is likely to remain small for many areas of investigation, especially for human data, because of the difficulty in collecting and processing microarray samples [2]. When the number of genes is significantly greater than

Rough-Fuzzy Pattern Recognition: Applications in Bioinformatics and Medical Imaging,
First Edition. Pradipta Maji and Sankar K. Pal.
© 2012 John Wiley & Sons, Inc. Published 2012 by John Wiley & Sons, Inc.

the number of samples, it is possible to find biologically relevant correlations of gene behavior with sample categories.

However, among the large amount of genes, only a small fraction is effective for performing a certain task. Also, a small subset of genes is desirable in developing gene-expression-based diagnostic tools for delivering precise, reliable, and interpretable results. With the gene selection results, the cost of biological experiment and decision can be greatly reduced by analyzing only the marker genes. Hence, identifying a reduced set of most relevant genes is the goal of gene selection. The small number of training samples and the large number of genes make gene selection a more relevant and challenging problem in gene-expression-based classification. This is an important problem in machine learning and is referred to as *feature selection* [4].

In the gene selection process, an optimal gene subset is always relative to a certain criterion. In general, different criteria may lead to different optimal gene subsets. However, every criterion tries to measure the discriminating ability of a gene or a subset of genes to distinguish the different class labels. To measure the gene–class relevance, different statistical measures such as the F-test and t-test [5, 6] are widely used. While the distance measure is a traditional discrimination or divergence measure [7], the dependence or correlation measure is mainly utilized to find the correlation between two genes or a gene and class labels [8]. As these measures depend on the actual values of the training data, they are very much sensitive to the noise or outlier of the data set [5, 7–9]. On the other hand, information measures such as the entropy, mutual information [5, 9–11], and f-information [12] compute the amount of information or the uncertainty of a gene for classification. As the information measure depends only on the probability distribution of a random variable rather than on its actual values, it has been widely used in gene selection [5, 9–14].

However, for real-valued gene expression data, the estimation of different information measures is a difficult task as it requires the knowledge of the underlying probability density functions of the data and the integration of these functions. In general, the continuous expression values are divided into several discrete partitions, and the information measures are calculated using the definitions for discrete cases [5, 12]. The inherent error that exists in the discretization process is of major concern in the computation of information measures of continuous gene expression values. In Battiti [10] and Kwak and Choi [15], histograms are used to estimate the true density functions, and the computational difficulty in performing integration can be circumvented in an efficient way. However, the histogram-based approaches are only applicable to a relatively low dimensional data as the sparse data distribution encountered in a high dimensional data set may greatly degrade the reliability of histograms [16, 17].

The rough set theory is a new paradigm to deal with uncertainty, vagueness, and incompleteness. It has been applied successfully to analyze microarray gene expression data [18–24]. Hu et al. [25] have used the concept of crisp equivalence relation of rough sets to compute entropy and mutual information in crisp approximation spaces that can be used for feature selection of discrete-valued data sets. However, there are usually real-valued data and fuzzy information in real-world applications [26]. Judicious integration of fuzzy sets and rough sets provides an important direction in reasoning with uncertainty for real-valued data sets as they are complementary in some aspects. The generalized theories of rough-fuzzy sets and fuzzy-rough sets have been applied successfully to feature selection of real-valued data. Several feature selection methodologies based on rough sets and fuzzy-rough sets have been mentioned earlier in Chapter 5, which can be used for gene selection from microarray data sets. Hu et al. [25] have also used the concept of fuzzy equivalence relation matrix of fuzzy-rough sets to compute entropy and mutual information in fuzzy approximation spaces, which can be used for feature selection from real-valued data sets. However, many useful information measures such as several f-information measures cannot be computed from the fuzzy equivalence relation matrix [25] as it does not provide a way to compute marginal and joint distributions directly.

In this regard, it may be noted that the concept of fuzzy equivalence partition matrix (FEPM) [27] is reported in Chapter 5 for computing different information measures on fuzzy approximation spaces. Each row of the matrix presents a fuzzy equivalence partition that offers an efficient way to estimate true density functions of continuous-valued gene expression data required for computing different information measures. Hence, the relevance and redundancy of the genes can be calculated using several information measures on fuzzy approximation spaces on the basis of the concept of FEPM [28]. This chapter establishes the effectiveness of the FEPM for the problem of gene selection from microarray data and compares its performance with some existing methods on a set of microarray gene expression data sets.

The structure of the rest of this chapter is as follows: Section 8.2 briefly introduces various evaluation criteria used for computing both the relevance and redundancy of the genes, while Section 8.3 presents several approaches to approximate the true probability density function for continuous-valued gene expression data. The problem of gene selection from microarray data sets using information theoretic approaches is described in Section 8.4. A few case studies and a comparison among different approximation methods are reported in Section 8.5. Concluding remarks are given in Section 8.6.

8.2 EVALUATION CRITERIA FOR GENE SELECTION

The F-test value [5, 6], information gain, mutual information [5, 9], normalized mutual information [13], and f-information [12] are typically used to measure the gene−class relevance $\hat{f}(\mathbb{G}_i, \mathbb{C})$ of the ith gene $\mathbb{G}_i$ with respect to the class labels $\mathbb{C}$ and the same or a different metric such as mutual information, f-information [12], the L_1 distance, Euclidean distance, and Pearson's correlation coefficient [5, 7, 9] is employed to calculate the gene−gene redundancy $\tilde{f}(\mathbb{G}_i, \mathbb{G}_j)$.

8.2.1   Statistical Tests

To measure the relevance of a gene, the t-value is widely used in the literature. Assuming that there are two classes of samples in a gene expression data set, the t-value $t(\mathbb{G}_i)$ for gene $\mathbb{G}_i$ is given by

$$t(\mathbb{G}_i) = \frac{\mu_1 - \mu_2}{\sqrt{\sigma_1^2/n_1 + \sigma_2^2/n_2}}, \tag{8.1}$$

where μ_c and σ_c are the mean and the standard deviation of the expression levels of gene $\mathbb{G}_i$ for class c, respectively, and n_c the number of samples in class c for $c = 1, 2$. When there are multiple classes of samples, the t-value is typically computed for one class versus all the other classes.

For multiple classes of samples, an F-statistic between a gene and the class label can be used to calculate the relevance score of that gene. The F-statistic value of gene $\mathbb{G}_i$ in K classes denoted by $\mathbb{C}$ is defined as follows:

$$F(\mathbb{G}_i, \mathbb{C}) = \frac{1}{\sigma^2} \left[\sum_{c=1}^{K} \frac{n_c(\overline{w}_{ic} - \overline{w}_i)^2}{(K - 1)}, \right] \tag{8.2}$$

where $\overline{w}_i$ is the mean of w_{ij} in all samples, $\overline{w}_{ic}$ the mean of w_{ij} in the cth class, K the number of classes, and

$$\sigma^2 = \sum_{c} \frac{(n_c - 1)\sigma_c^2}{(n - c)} \tag{8.3}$$

is the pooled variance, and n_c and σ_c are the size and the variance of the cth class, respectively. Hence, the F-test reduces to the t-test for the two-class problem with the relation $F = t^2$.

8.2.2   Euclidean Distance

Given two genes $\mathbb{G}_i$ and $\mathbb{G}_j$, $i, j \in \{1, \ldots, m\}, i \neq j$, the Euclidean distance between $\mathbb{G}_i$ and $\mathbb{G}_j$ is given by

$$d_E(\mathbb{G}_i, \mathbb{G}_j) = \sqrt{\sum_{k=1}^{n}(w_{ik} - w_{jk})^2}, \tag{8.4}$$

where $w_{ik}, w_{jk} \in \Re$ is the measured expression level. The d_E measures the difference in the individual magnitudes of each gene. However, the genes regarded as similar by the Euclidean distance may be very dissimilar in terms of their shapes. Similarly, the Euclidean distance between two genes having an identical shape may be large if they differ from each other by a large scaling factor. But, the overall shapes of genes are of primary interest for gene expression data. Hence, the Euclidean distance may not be able to yield a good proximity measurement of genes [7].

8.2.3 Pearson's Correlation

The Pearson's correlation coefficient between genes $\mathbb{G}_i$ and $\mathbb{G}_j$ is defined as

$$d_C(\mathbb{G}_i, \mathbb{G}_j) = \frac{\sum_{k=1}^{n}(w_{ik} - \overline{w}_i)(w_{jk} - \overline{w}_j)}{\sqrt{\sum_{k=1}^{n}(w_{ik} - \overline{w}_i)^2}\sqrt{\sum_{k=1}^{n}(w_{jk} - \overline{w}_j)^2}}, \tag{8.5}$$

where $\overline{w}_i$ and $\overline{w}_j$ are the means of w_{ik} and w_{jk}, respectively. It considers each gene as a random variable with n observations and measures the similarity between the two genes by calculating the linear relationship between the distributions of the two corresponding random variables. An empirical study has shown that Pearson's correlation coefficient is not robust to outliers and it may assign a high similarity score to a pair of dissimilar genes [8].

However, as the F-test value, Euclidean distance, and Pearson's correlation depend on the actual gene expression values of the microarray data, they are very much sensitive to the noise or outlier of the data set. On the other hand, as the information theoretic measures such as entropy, mutual information, and f-information [12] depend only on the probability distribution of a random variable rather than on its actual values, they are more effective to evaluate the gene–class relevance as well as the gene–gene redundancy [5, 9].

8.2.4 Mutual Information

In principle, the mutual information is used to quantify the information shared by two objects. If two independent objects do not share much information, the mutual information value between them is small, while two highly correlated objects demonstrate a high mutual information value [29]. The objects can be the class label and the genes. The necessity for a gene to be independent and informative can, therefore, be determined by the shared information between the gene and the rest as well as the shared information between the gene and class label [5, 9]. If a gene has expression values randomly or uniformly distributed in different classes, its mutual information with these classes is zero. If a gene is strongly and differentially expressed for different classes, it should have large mutual information. Hence, the mutual information can be used as a measure of the relevance of genes. Similarly, the mutual information may be used to measure the level of similarity between the genes. The idea of minimum redundancy is to select the genes such that they are mutually and maximally dissimilar. Minimal redundancy will make the gene set a better representation of the entire data set.

The entropy is a measure of uncertainty of random variables. If a discrete random variable X has $\mathcal{X}$ alphabets and the probability density function is $p(x) = \Pr\{X = x\}, x \in \mathcal{X}$, the entropy of X is defined as follows:

$$H(X) = -\sum_{x \in \mathcal{X}} p(x) \log p(x). \tag{8.6}$$

Similarly, the joint entropy of two random variables X with $\mathcal{X}$ alphabets and Y with $\mathcal{Y}$ alphabets is given by

$$H(X, Y) = -\sum_{x \in \mathcal{X}} \sum_{y \in \mathcal{Y}} p(x, y) \log p(x, y), \tag{8.7}$$

where $p(x, y)$ is the joint probability density function. The mutual information between X and Y is, therefore, given by

$$I(X, Y) = H(X) + H(Y) - H(X, Y). \tag{8.8}$$

However, for continuous gene expression values (random variables), the differential entropy and joint entropy can be defined as

$$H(X) = -\int p(x) \log p(x) dx; \tag{8.9}$$

$$H(X, Y) = -\int p(x, y) \log p(x, y) dx dy. \tag{8.10}$$

8.2.5 f-Information Measures

The extent to which two probability distributions differ can be expressed by a so-called measure of divergence. Such a measure will reach a minimum value when the two probability distributions are identical and the value increases with increasing disparity between the two distributions. A specific class of divergence measures is the set of f-divergence measures [12, 30, 31].

A special case of f-divergence measures are the f-information measures. These are defined similar to f-divergence measures, but apply only to specific probability distributions, namely, the joint probability of two variables and their marginal probabilities' product. Hence, the f-information is a measure of dependence: it measures the distance between a given joint probability and the joint probability when the variables are independent [12, 30, 31]. The f-information measures have been applied successfully in Maji [12] for the selection of relevant and nonredundant genes from microarray gene expression data. Details of different f-information measures are reported in Chapter 5.

8.3 APPROXIMATION OF DENSITY FUNCTION

In microarray gene expression data, the class labels of samples are represented by discrete symbols, while the expression values of genes are continuous. Hence, to measure both gene–class relevance and gene–gene redundancy using information theoretic measures such as entropy, mutual information, and f-information measures [12], the true density functions of continuous-valued genes have to be approximated. In this regard, some density approximation approaches are reported next.

8.3.1 Discretization

The following two discretization methods can be used to approximate the true marginal and joint distributions of continuous gene expression values.

8.3.1.1 Mean and Standard Deviation The discretization method reported in Maji [12] can be employed to discretize the continuous gene expression values for computing marginal and joint distributions of real-valued gene expression data. The continuous expression values of a gene are discretized using mean μ and standard deviation σ computed over n values of that gene: any value larger than $(\mu + \frac{\sigma}{2})$ is transformed to state 1; any value between $(\mu - \frac{\sigma}{2})$ and $(\mu + \frac{\sigma}{2})$ is transformed to state 0; any value smaller than $(\mu - \frac{\sigma}{2})$ is transformed to state -1 [12]. These three states correspond to overexpression, baseline, and underexpression of continuous-valued genes, respectively. The marginal and joint distributions are then computed to calculate both the relevance and redundancy of genes using the information theoretic measures.

8.3.1.2 Equal Frequency Binning Equal frequency binning can also be applied for discretizing the continuous gene expression values to approximate the true density functions of genes [21]. The continuous gene expression values are discretized by fixing the number of intervals c and examining the histogram of each gene, $(c - 1)$ cuts are determined so that approximately the same number of objects fall into each of the c intervals. The $(c - 1)$ cuts are introduced such that the area between two neighboring cuts in the normalized histogram is as close to $1/c$ as possible.

However, the inherent error that exists in the discretization process is of major concern in the computation of relevance and redundancy of continuous-valued genes [32]. To address this problem, the Parzen window density estimator has been used [9, 15, 33] to approximate the true marginal distributions and joint distributions of continuous variables.

8.3.2 Parzen Window Density Estimator

The Parzen window density estimator can be used to approximate the probability density $p(x)$ of a vector of continuous random variables X [34]. It involves the superposition of a normalized window function centered on a set of random samples. Given a set of n d-dimensional samples of a variable X, the approximate density function $\hat{p}(x)$ has the following form:

$$\hat{p}(x) = \frac{1}{n} \sum_{i=1}^{n} \delta(x - x_i, h), \tag{8.11}$$

where $\delta(\cdot)$ is the Parzen window function, x_i the ith sample, and h the window width parameter. Parzen has proved that the estimation $\hat{p}(x)$ converges to the true density $p(x)$ if $\delta(\cdot)$ and h are selected properly [34]. The window function is required to be a finite-valued nonnegative density function where

$$\int \delta(z, h)\, dz = 1 \tag{8.12}$$

and the window width parameter h is required to be a function of n such that

$$\lim_{n \to \infty} h(n) = 0 \quad \text{and} \quad \lim_{n \to \infty} nh^d(n) = \infty. \tag{8.13}$$

Usually, the window function $\delta(\cdot)$ is chosen as the Gaussian window that is given by

$$\delta(z, h) = \frac{1}{(2\pi)^{d/2} h^d |\sum|^{1/2}} \exp\left(-\frac{z^T \sum^{-1} z}{2h^2},\right) \tag{8.14}$$

where $z = (x - x_i)$, d is the dimension of the sample x, and $\sum$ the covariance matrix of a d-dimensional vector of random variable z. When $d = 1$, Equation (8.14) returns the estimated marginal density; when $d = 2$, Equation (8.14) can be used to estimate the density of the bivariate variable (x, y), $p(x, y)$, which is actually the joint density of x and y.

Hence, the approximate density function for a gene $\mathbb{G}_i$ with continuous expression values will be as follows:

$$\hat{p}(\mathbb{G}_i) = \frac{1}{n\sqrt{2\pi h^2 |\sum_i|}} \sum_{k=1}^{n} \exp\left\{-\frac{(w_i - w_{ik})^2}{2h^2 \sum_i},\right\} \tag{8.15}$$

where $w_{ik} \in \Re$ is the measured expression level of gene $\mathbb{G}_i$ in the kth sample and $\sum_i$ represents the variance of continuous expression values of gene $\mathbb{G}_i$. Similarly, the approximate joint density function between two genes $\mathbb{G}_i$ and $\mathbb{G}_j$ is given by

$$\hat{p}(\mathbb{G}_i, \mathbb{G}_j) = \frac{1}{2\pi n h^2 \sqrt{|\sum_i| + |\sum_j|}} \times$$

$$\sum_{k=1}^{n} \exp\left\{-\frac{1}{2h^2}\left\{\frac{(w_i - w_{ik})^2}{\sum_i} + \frac{(w_j - w_{jk})^2}{\sum_j}\right\}\right\}. \tag{8.16}$$

However, in a gene selection problem, the sample categories or class labels are discrete values while the gene expression values are usually continuous values. In this case, the approximate joint density function $\hat{p}(\mathbb{G}_i, c)$ between a gene $\mathbb{G}_i$

and the sample category c would be as follows:

$$\hat{p}(\mathbb{G}_i, c) = \frac{1}{n_c\sqrt{2\pi h_c^2 |\sum_c|}} \sum_{w_{ik}\in c} \exp\left\{-\frac{(w_i - w_{ik})^2}{2h_c^2\sum_c},\right\} \tag{8.17}$$

where n_c is the number of the training examples belonging to class c, h_c the class specific window width, and Σ_c represents the variance of expression values of gene $\mathbb{G}_i$ that belong to class c. The value of h is generally determined in a way developed in Silverman [35], that is,

$$h = \left\{\frac{4}{d+2}\right\}^{1/(d+4)} \times n^{-1/(d+4)}, \tag{8.18}$$

where $d = 1$ and 2 for the estimation of marginal density and joint density, respectively. Hence, to compute the relevance of a gene with respect to sample categories and the redundancy between two genes using different information theoretic measures, Equations (8.15)–(8.17) can be used to approximate the required marginal and joint distributions.

8.3.3 Fuzzy Equivalence Partition Matrix

The FEPM, introduced in Maji and Pal [27] and reported in Chapter 5, can be used to compute different information measures on fuzzy approximation spaces. Given a finite set $\mathbb{U}$, $\mathbb{A}$ is a fuzzy attribute set in $\mathbb{U}$, which generates a fuzzy equivalence partition on $\mathbb{U}$. If c denotes the number of fuzzy equivalence classes generated by the fuzzy equivalence relation and n is the number of objects in $\mathbb{U}$, then c-partitions of $\mathbb{U}$ can be conveniently arrayed as a $(c \times n)$ FEPM $\mathbb{M}_{\mathbb{A}} = [m_{ij}^{\mathbb{A}}]$, where

$$\mathbb{M}_{\mathbb{A}} = \begin{pmatrix} m_{11}^{\mathbb{A}} & m_{12}^{\mathbb{A}} & \cdots & m_{1n}^{\mathbb{A}} \\ m_{21}^{\mathbb{A}} & m_{22}^{\mathbb{A}} & \cdots & m_{2n}^{\mathbb{A}} \\ \cdots & \cdots & \cdots & \cdots \\ m_{c1}^{\mathbb{A}} & m_{c2}^{\mathbb{A}} & \cdots & m_{cn}^{\mathbb{A}} \end{pmatrix} \tag{8.19}$$

where $m_{ij}^{\mathbb{A}} \in [0, 1]$ represents the membership of object x_j in the ith fuzzy equivalence partition or class F_i. The fuzzy relative frequency corresponding to fuzzy equivalence partition F_i is then defined as follows:

$$\lambda_{F_i} = \frac{1}{n}\sum_{j=1}^{n} m_{ij}^{\mathbb{A}}, \tag{8.20}$$

which appears to be a natural generalization of the crisp set.

Given $<\mathbb{U}, \mathbb{A}>$, $\mathbb{P}$ and $\mathbb{Q}$ are two subsets of $\mathbb{A}$. Let p and q be the numbers of fuzzy equivalence partitions or classes generated by the fuzzy attribute sets $\mathbb{P}$ and $\mathbb{Q}$, respectively, and P_i and Q_j represent the corresponding ith and jth fuzzy equivalence partitions, respectively. Then, the joint frequency of P_i and Q_j is given by

$$\lambda_{P_i Q_j} = \frac{1}{n} \sum_{k=1}^{n} (m_{ik}^{\mathbb{P}} \cap m_{jk}^{\mathbb{Q}}). \tag{8.21}$$

Hence, to compute the relevance of a gene with respect to sample categories and the redundancy between two genes using different information theoretic measures, Equations (8.20) and (8.21) can be used to approximate the required marginal and joint distributions.

In this context, it should be noted that Hu et al. have introduced the concept of $n \times n$ fuzzy equivalence relation matrix in Hu et al. [25] to compute entropy and mutual information in fuzzy approximation spaces. However, this matrix does not provide a way to compute marginal and joint distributions directly. In effect, many useful information measures cannot be computed from the FEPM [25]. Also, the complexity of this approach is $\mathcal{O}(n^2)$, which is higher than $\mathcal{O}(cn)$ of the FEPM-based approach as $c \ll n$, where c is the number of fuzzy equivalence classes or partitions.

8.4   GENE SELECTION USING INFORMATION MEASURES

In microarray data analysis, the data set may contain a number of redundant genes with low relevance to the classes. The presence of such redundant and nonrelevant genes leads to a reduction in the useful information. Ideally, the selected genes should have high relevance with the classes, while the redundancy among them would be as low as possible. The genes with high relevance are expected to be able to predict the classes of the samples. However, the prediction capability is reduced if many redundant genes are selected. In contrast, a microarray data set that contains genes not only with high relevance with respect to the classes but with low mutual redundancy is more effective in its prediction capability. Hence, to assess the effectiveness of the genes, both relevance and redundancy need to be measured quantitatively. An information-measure-based gene selection method is presented next to address this problem.

Let $\mathbb{G} = \{\mathbb{G}_1, \ldots, \mathbb{G}_i, \ldots, \mathbb{G}_j, \ldots, \mathbb{G}_m\}$ denote the set of genes or fuzzy condition attributes of a given microarray data set and $\mathbb{S}$ be the set of selected genes. Define $\tilde{f}(\mathbb{G}_i, \mathbb{D})$ as the relevance of the gene $\mathbb{G}_i$ with respect to the class label $\mathbb{D}$, while $\tilde{f}(\mathbb{G}_i, \mathbb{G}_j)$ as the redundancy between two genes $\mathbb{G}_i$ and $\mathbb{G}_j$. The total relevance of all selected genes is, therefore, given by

$$\mathcal{J}_{\text{relev}} = \sum_{\mathbb{G}_i \in \mathbb{S}} \tilde{f}(\mathbb{G}_i, \mathbb{D}), \tag{8.22}$$

while the total redundancy among the selected genes is defined as follows:

$$\mathcal{J}_{\text{redun}} = \sum_{\mathbb{G}_i, \mathbb{G}_j \in \mathbb{S}} \tilde{f}(\mathbb{G}_i, \mathbb{G}_j). \tag{8.23}$$

Therefore, the problem of selecting a set $\mathbb{S}$ of nonredundant and relevant genes from the whole set of genes $\mathbb{G}$ is equivalent to maximize $\mathcal{J}_{\text{relev}}$ and minimize $\mathcal{J}_{\text{redun}}$, that is, to maximize the objective function $\mathcal{J}$, where

$$\mathcal{J} = \mathcal{J}_{\text{relev}} - \mathcal{J}_{\text{redun}} = \sum_{\mathbb{G}_i \in \mathbb{S}} \tilde{f}(\mathbb{G}_i, \mathbb{D}) - \sum_{\mathbb{G}_i, \mathbb{G}_j \in \mathbb{S}} \tilde{f}(\mathbb{G}_i, \mathbb{G}_j). \tag{8.24}$$

The gene selection problem reported above can be solved using the greedy algorithm reported in Chapter 5. Both the relevance $\tilde{f}(\mathbb{G}_i, \mathbb{D})$ of a gene $\mathbb{G}_i$ with respect to the class labels $\mathbb{D}$ and the redundancy $\tilde{f}(\mathbb{G}_i, \mathbb{G}_j)$ between two genes $\mathbb{G}_i$ and $\mathbb{G}_j$ can be calculated using any one of the information measures. The true marginal and joint distributions of continuous-valued genes can be approximated using any one of the density approximation methods reported earlier.

8.5 EXPERIMENTAL RESULTS

The performance of the FEPM-based density approximation approach is extensively compared with that of two existing methods, namely, the discretization-based approach (discrete) [5, 12] and the Parzen-window-based approach (Parzen) [15]. Results are reported with respect to three widely used information measures such as mutual information, V-information, and χ^2-information. All these measures are applied to calculate both gene-class relevance and gene–gene redundancy. To analyze the performance of different methods, the experimentation is done on five microarray gene expression data sets. The major metrics for evaluating the performance of different methods are the class separability index [4], which is reported in Chapter 5, and the classification accuracy of support vector machine (SVM) [36], which is reported next.

8.5.1 Support Vector Machine

The SVM [36] is a new and promising classification method. It is a margin classifier that draws an optimal hyperplane in the feature vector space; this defines a boundary that maximizes the margin between data samples in different classes, thereby leading to good generalization properties. A key factor in the SVM is to use kernels to construct nonlinear decision boundary. In this work, linear kernels are used. The source code of the SVM is downloaded from http://www.csie.ntu.edu.tw/~cjlin/libsvm.

8.5.2 Gene Expression Data Sets

In this chapter, three publicly available cancer and two arthritis data sets are used. Since binary classification is a typical and fundamental issue in diagnostic and prognostic prediction of cancer and arthritis, different methods are compared using the following five binary class data sets.

8.5.2.1 Breast Cancer   The breast cancer data set contains expression levels of 7129 genes in 49 breast tumor samples [37]. The samples are classified according to their estrogen receptor (ER) status: 25 samples are ER positive whereas the other 24 samples are ER negative.

8.5.2.2 Leukemia   It is an affymetrix high density oligonucleotide array that contains 7070 genes and 72 samples from two classes of leukemia: 47 acute lymphoblastic leukemia and 25 acute myeloid leukemia [2].

8.5.2.3 Colon Cancer   The colon cancer data set contains the expression levels of 40 tumor and 22 normal colon tissues. Only the 2000 genes with the highest minimal intensity were selected by Alon et al. [38].

8.5.2.4 Rheumatoid Arthritis versus Osteoarthritis (RAOA)   The rheumatoid arthritis versus osteoarthritis (RAOA) data set consists of gene expression profiles of 30 patients: 21 with the rheumatoid arthritis (RA) and 9 with the osteoarthritis (OA) [39]. The Cy5-labeled experimental cDNA and the Cy3-labeled common reference sample were pooled and hybridized to the lymphochips containing ~18,000 cDNA spots representing genes of relevance in immunology [39].

8.5.2.5 Rheumatoid Arthritis versus Healthy Controls (RAHC)   The rheumatoid arthritis versus healthy controls (RAHC) data set consists of gene expression profiling of peripheral blood cells from 32 patients with RA, 3 patients with probable RA, and 15 age- and sex-matched healthy controls performed on microarray with a complexity of ~26K unique genes (43K elements) [40].

8.5.3 Performance Analysis of the FEPM

To select genes from the microarray data set using the FEPM-based method, the π function in the one-dimensional form is used to assign membership values to different fuzzy equivalence classes for the input genes. Each input real-valued gene in quantitative form is assigned to different fuzzy equivalence classes in terms of membership values using the π fuzzy set with appropriate parameters. The parameters of the π functions along each gene axis are determined automatically from the distribution of the training patterns. The parameters of each π fuzzy set are computed according to the procedure reported in Section 5.5.3 of Chapter 5.

The η is a multiplicative parameter controlling the extent of overlapping between fuzzy sets low and medium or medium and high. There is insignificant overlapping between the π functions low and medium or medium and high as η decreases. This implies that certain regions along the ith gene axis $\mathbb{G}_i$ go underrepresented such that three membership values corresponding to three fuzzy sets low, medium, and high attain small values. On the other hand, as η is increased, the amount of overlapping between the π functions increases.

8.5.3.1 *Class Separability Analysis*   Figures 8.1–8.5 depict the performance of the FEPM-based method for the five microarray data sets in terms of the within-class scatter matrix, between-class scatter matrix, and class separability index. The results are presented for 30 top ranked genes selected by the FEPM-based method for the three information measures and the five microarray data sets. Each data set is preprocessed by standardizing each sample to zero mean and unit variance. From the results reported in Figs. 8.1–8.5, it can be seen that as the value of multiplicative parameter η increases, the values of within-class scatter matrix and class separability index decrease, while the between-class

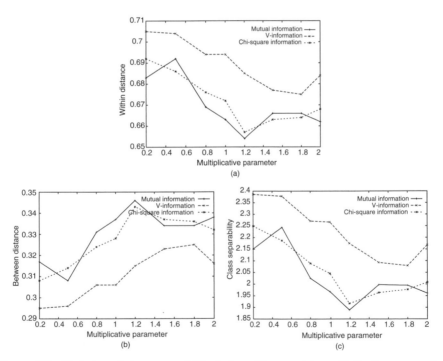

Figure 8.1   Variation of class separability index with respect to multiplicative parameter η for breast cancer data set. (a) Within class scatter matrix, (b) between class scatter matrix, and (c) class separability index.

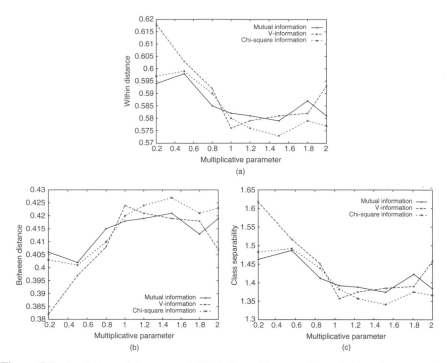

Figure 8.2   Variation of class separability index with respect to multiplicative parameter η for leukemia data set. (a) Within class scatter matrix, (b) between class scatter matrix, and (c) class separability index.

scatter matrix increases, irrespective of the data sets and information measures used. The best performance of the FEPM-based method is achieved for $1.0 \leq \eta \leq 1.8$. For $\eta > 1.8$, the performance of the FEPM-based method decreases with the increase in η.

Table 8.1 presents the best performance achieved by the FEPM-based method for different data sets and information measures used in terms of the within-class scatter matrix (S_w), between class scatter matrix (S_b), and class separability index (S), along with the corresponding η value. The FEPM-based method achieves best performance with $\eta = 1.0$ for leukemia data using V-information, colon cancer data using χ^2-information, RAOA data using mutual information and χ^2-information, and RAHC data using mutual information, respectively. Similarly, the best performance is achieved with $\eta = 1.2$ for breast cancer data using mutual information and χ^2-information, colon cancer data using mutual information, RAOA data using V-information, and RAHC data using χ^2-information, respectively. At $\eta = 1.5$, mutual information and χ^2-information provide best result for leukemia data, while V-information gives best performance for RAHC data. On the other hand, V-information provides the best performance for both

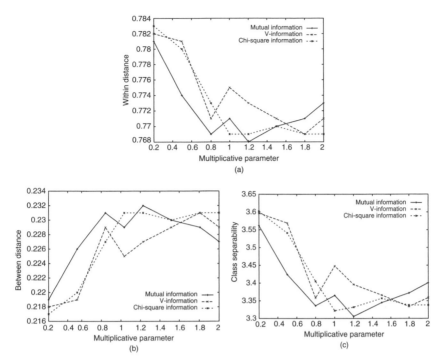

Figure 8.3 Variation of class separability index with respect to multiplicative parameter η for colon cancer data set. (a) Within class scatter matrix, (b) between class scatter matrix, and (c) class separability index.

breast and colon cancer data with $\eta = 1.8$. However, for $\eta > 1.8$, the performance of the FEPM-based method decreases with the increase in η for the three measures and five microarray gene expression data sets used.

8.5.3.2 Classification Accuracy Analysis Tables 8.2–8.16 represent the performance of the FEPM-based method in terms of the classification accuracy of the SVM for different values of η. Results are presented for the five microarray data sets considering three widely used information measures such as mutual information, V-information, and χ^2-information. To compute the prediction accuracy of the SVM, the leave-one-out cross-validation is performed on each gene expression data set. The values of η investigated are 0.2, 0.5, 0.8, 1.0, 1.2, 1.5, 1.8, and 2.0. The number of genes selected ranges from 1 to 30; however, the results are reported only for 20 top-ranked genes, and each data set is preprocessed by standardizing each sample to zero mean and unit variance.

Tables 8.2, 8.3, and 8.4 depict the results for the breast cancer data set with respect to the three information measures. The 100% classification accuracy of the SVM is obtained for mutual information and χ^2-information considering 8 and 7 top-ranked genes, respectively, with both $\eta = 0.5$ and 0.8, while in

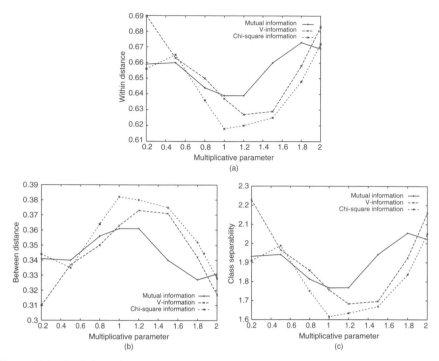

Figure 8.4 Variation of class separability index with respect to multiplicative parameter η for RAOA data set. (a) Within class scatter matrix, (b) between class scatter matrix, and (c) class separability index.

case of V-information, 5 top-ranked genes are required to achieve this accuracy with $\eta = 0.5$. Similarly, maximum 98.6% accuracy in case of leukemia data is obtained for both mutual information and χ^2-information using 9 genes with $\eta = 1.0$ and 13 genes with $\eta = 0.2$ and 0.5, respectively. On the other hand, V-information provides 100% accuracy using only 3 genes with $\eta = 0.5$. The results corresponding to leukemia data are reported in Tables 8.5–8.7. The results reported in Tables 8.8–8.10 are based on the predictive accuracy of the SVM on the colon cancer data. While both V- and χ^2-information attain maximum 91.9% accuracy with 10 genes using $\eta = 0.5$ and 20 genes using $\eta = 1.0$, respectively, mutual information provides maximum 90.3% accuracy using 7 genes for $\eta = 1.0$ to 1.8.

Tables 8.11–8.16 present the results of two RA data sets, namely, RAOA and RAHC. For the RAOA data set, mutual information, V-information, and χ^2-information attain 100% accuracy using 3 genes for $\eta = 0.2$, 0.5, and 1.2, 2 genes with $\eta = 1.5$ and 3 genes with $\eta = 0.2$ and 2.0, respectively. Similarly, 100% accuracy is obtained for the RAHC data set in case of these three measures using 7 genes with $\eta = 0.5$, 6 genes with $\eta = 0.2$, and 16 genes with $\eta = 1.2$, respectively. All the results reported in Tables 8.2–8.16 establish the fact that the

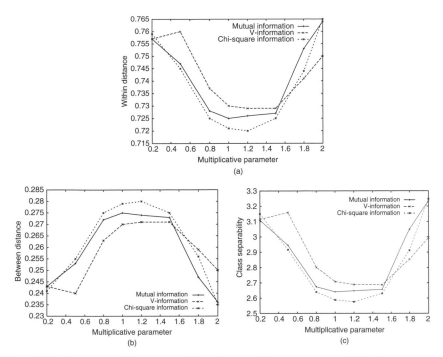

Figure 8.5 Variation of class separability index with respect to multiplicative parameter η for RAHC data set. (a) Within class scatter matrix, (b) between class scatter matrix, and (c) class separability index.

TABLE 8.1   Class Separability Analysis

Data Set	Measure	η	S_w	S_b	S
Breast	I	1.2	0.654	0.346	1.889
	V	1.5	0.675	0.325	2.079
	χ^2	1.2	0.657	0.343	1.917
Leukemia	I	1.5	0.579	0.421	1.374
	V	1.0	0.576	0.424	1.357
	χ^2	1.5	0.573	0.427	1.341
Colon	I	1.2	0.768	0.232	3.306
	V	1.8	0.769	0.231	3.334
	χ^2	1.0	0.769	0.231	3.322
RAOA	I	1.0	0.639	0.361	1.767
	V	1.2	0.627	0.373	1.683
	χ^2	1.0	0.618	0.382	1.616
RAHC	I	1.0	0.725	0.275	2.641
	V	1.5	0.729	0.271	2.687
	χ^2	1.2	0.720	0.280	2.576

TABLE 8.2   Mutual Information on Breast Cancer Data

Gene/η	0.2	0.5	0.8	1.0	1.2	1.5	1.8	2.0
1	85.7	85.7	85.7	85.7	85.7	85.7	85.7	85.7
2	89.8	89.8	89.8	89.8	89.8	91.8	91.8	95.9
3	91.8	91.8	91.8	91.8	91.8	91.8	91.8	93.9
4	87.8	95.9	95.9	95.9	95.9	91.8	91.8	93.9
5	89.8	95.9	95.9	95.9	95.9	87.8	93.9	93.9
6	87.8	93.9	93.9	93.9	93.9	83.7	91.8	91.8
7	93.9	93.9	91.8	93.9	93.9	85.7	93.9	91.8
8	93.9	100	100	91.8	93.9	91.8	89.8	89.8
9	93.9	98.0	100	91.8	91.8	91.8	85.7	89.8
10	91.8	100	100	87.8	93.9	91.8	85.7	85.7
11	89.8	100	100	87.8	89.8	91.8	87.8	85.7
12	91.8	98.0	95.9	89.8	89.8	91.8	87.8	85.7
13	89.8	98.0	98.0	95.9	87.8	93.9	91.8	87.8
14	93.9	98.0	95.9	100	87.8	93.9	91.8	87.8
15	93.9	98.0	95.9	100	87.8	95.9	87.8	87.8
16	93.9	93.9	98.0	100	89.8	95.9	87.8	87.8
17	93.9	95.9	98.0	100	87.8	89.8	87.8	87.8
18	89.8	93.9	100	100	89.8	89.8	91.8	87.8
19	93.9	93.9	98.0	100	100	95.9	95.9	87.8
20	93.9	100	100	100	100	93.9	95.9	91.8

TABLE 8.3 V-Information on Breast Cancer Data

Gene/η	0.2	0.5	0.8	1.0	1.2	1.5	1.8	2.0
1	85.7	85.7	85.7	85.7	85.7	85.7	85.7	85.7
2	93.9	93.9	93.9	89.8	93.9	93.9	93.9	93.9
3	93.9	93.9	93.9	91.8	93.9	95.9	93.9	93.9
4	93.9	98.0	93.9	95.9	93.9	93.9	91.8	95.9
5	93.9	100	93.9	93.9	93.9	95.9	95.9	95.9
6	98.0	100	87.8	98.0	93.9	95.9	91.8	91.8
7	98.0	100	89.8	91.8	93.9	95.9	85.7	85.7
8	95.9	100	95.9	93.9	93.9	91.8	93.9	85.7
9	98.0	100	95.9	93.9	95.9	91.8	95.9	87.8
10	93.9	100	95.9	100	100	95.9	95.9	87.8
11	95.9	95.9	93.9	98.0	100	95.9	95.9	95.9
12	93.9	95.9	91.8	100	100	95.9	95.9	95.9
13	93.9	95.9	91.8	98.0	100	95.9	93.9	93.9
14	95.9	95.9	89.8	98.0	98.0	93.9	93.9	93.9
15	95.9	93.9	91.8	98.0	98.0	93.9	93.9	91.8
16	95.9	95.9	91.8	95.9	98.0	93.9	98.0	98.0
17	98.0	95.9	95.9	95.9	98.0	95.9	98.0	98.0
18	98.0	95.9	95.9	98.0	98.0	95.9	95.9	95.9
19	98.0	95.9	95.9	98.0	98.0	95.9	95.9	95.9
20	98.0	95.9	98.0	95.9	98.0	95.9	95.9	93.9

TABLE 8.4 χ^2-Information on Breast Cancer Data

Gene/η	0.2	0.5	0.8	1.0	1.2	1.5	1.8	2.0
1	85.7	85.7	85.7	85.7	85.7	85.7	85.7	85.7
2	89.8	89.8	89.8	89.8	89.8	91.8	91.8	91.8
3	91.8	91.8	93.9	91.8	91.8	91.8	91.8	89.8
4	95.9	95.9	95.9	95.9	95.9	91.8	91.8	89.8
5	95.9	95.9	95.9	95.9	95.9	87.8	87.8	87.8
6	93.9	93.9	93.9	93.9	93.9	93.9	87.8	87.8
7	93.9	100	100	93.9	95.9	93.9	91.8	91.8
8	91.8	98.0	100	89.8	93.9	93.9	91.8	91.8
9	93.9	100	100	93.9	91.8	93.9	91.8	89.8
10	93.9	100	98.0	91.8	89.8	93.9	89.8	85.7
11	93.9	100	100	95.9	89.8	89.8	87.8	83.7
12	91.8	98.0	100	100	89.8	87.8	87.8	83.7
13	87.8	98.0	100	98.0	91.8	85.7	87.8	85.7
14	95.9	95.9	100	100	89.8	83.7	85.7	95.9
15	95.9	98.0	100	100	89.8	93.9	95.9	95.9
16	98.0	91.8	100	100	100	98.0	93.9	98.0
17	98.0	100	100	100	100	98.0	93.9	98.0
18	95.9	100	98.0	100	100	100	95.9	98.0
19	95.9	100	98.0	100	98.0	100	95.9	98.0
20	95.9	98.0	98.0	100	100	100	95.9	95.9

TABLE 8.5 Mutual Information on Leukemia Data

Gene/η	0.2	0.5	0.8	1.0	1.2	1.5	1.8	2.0
1	94.4	94.4	94.4	94.4	94.4	94.4	94.4	94.4
2	95.8	95.8	95.8	95.8	95.8	95.8	95.8	95.8
3	95.8	95.8	95.8	95.8	95.8	95.8	97.2	97.2
4	95.8	94.4	95.8	93.1	94.4	94.4	95.8	95.8
5	95.8	95.8	94.4	94.4	95.8	95.8	95.8	95.8
6	95.8	97.2	94.4	95.8	97.2	97.2	95.8	95.8
7	95.8	95.8	94.4	95.8	94.4	94.4	95.8	93.1
8	95.8	95.8	95.8	95.8	95.8	94.4	94.4	94.4
9	95.8	95.8	97.2	98.6	94.4	95.8	95.8	95.8
10	97.2	95.8	95.8	98.6	95.8	95.8	95.8	94.4
11	94.4	95.8	95.8	97.2	94.4	95.8	95.8	93.1
12	94.4	95.8	94.4	95.8	94.4	94.4	94.4	94.4
13	95.8	95.8	93.1	95.8	93.1	93.1	95.8	94.4
14	98.6	95.8	95.8	95.8	95.8	95.8	94.4	94.4
15	97.2	95.8	93.1	93.1	93.1	93.1	94.4	93.1
16	97.2	95.8	93.1	94.4	98.6	97.2	94.4	94.4
17	97.2	95.8	94.4	97.2	97.2	95.8	95.8	95.8
18	94.4	95.8	93.1	94.4	94.4	94.4	95.8	95.8
19	94.4	95.8	93.1	94.4	94.4	94.4	95.8	94.4
20	93.1	95.8	93.1	98.6	94.4	95.8	95.8	94.4

TABLE 8.6 *V*-Information on Leukemia Data

Gene/η	0.2	0.5	0.8	1.0	1.2	1.5	1.8	2.0
1	94.4	90.3	90.3	94.4	94.4	94.4	94.4	94.4
2	95.8	95.8	95.8	98.6	94.4	94.4	94.4	94.4
3	97.2	100	98.6	97.2	95.8	94.4	94.4	94.4
4	95.8	98.6	97.2	97.2	95.8	95.8	95.8	95.8
5	95.8	98.6	97.2	97.2	97.2	95.8	97.2	95.8
6	95.8	100	95.8	97.2	94.4	95.8	97.2	95.8
7	95.8	98.6	97.2	97.2	94.4	97.2	97.2	94.4
8	95.8	98.6	94.4	98.6	95.8	95.8	98.6	95.8
9	98.6	98.6	93.1	98.6	95.8	95.8	98.6	94.4
10	100	98.6	97.2	98.6	97.2	95.8	98.6	97.2
11	97.2	97.2	97.2	98.6	97.2	94.4	98.6	95.8
12	98.6	98.6	97.2	98.6	95.8	94.4	98.6	95.8
13	98.6	98.6	97.2	98.6	94.4	93.1	98.6	97.2
14	98.6	98.6	95.8	97.2	94.4	94.4	95.8	95.8
15	98.6	98.6	97.2	97.2	94.4	94.4	95.8	95.8
16	100	98.6	97.2	95.8	94.4	93.1	97.2	97.2
17	97.2	98.6	95.8	95.8	94.4	95.8	95.8	95.8
18	98.6	97.2	94.4	95.8	94.4	94.4	95.8	95.8
19	100	98.6	95.8	95.8	93.1	94.4	97.2	95.8
20	98.6	98.6	95.8	95.8	91.7	94.4	97.2	94.4

TABLE 8.7 χ^2-Information on Leukemia Data

Gene/η	0.2	0.5	0.8	1.0	1.2	1.5	1.8	2.0
1	94.4	94.4	94.4	94.4	94.4	94.4	94.4	94.4
2	95.8	95.8	95.8	95.8	95.8	94.4	94.4	94.4
3	95.8	97.2	95.8	95.8	95.8	93.1	93.1	93.1
4	94.4	97.2	93.1	93.1	93.1	93.1	93.1	93.1
5	94.4	97.2	93.1	94.4	94.4	94.4	94.4	91.7
6	95.8	97.2	93.1	94.4	94.4	94.4	94.4	94.4
7	95.8	95.8	93.1	97.2	97.2	97.2	94.4	94.4
8	94.4	97.2	95.8	97.2	95.8	97.2	97.2	93.1
9	94.4	97.2	97.2	97.2	97.2	97.2	97.2	97.2
10	94.4	95.8	97.2	95.8	95.8	97.2	95.8	95.8
11	95.8	95.8	97.2	95.8	95.8	95.8	95.8	95.8
12	95.8	95.8	95.8	95.8	95.8	95.8	95.8	94.4
13	98.6	98.6	97.2	97.2	97.2	95.8	94.4	94.4
14	98.6	97.2	97.2	97.2	95.8	95.8	97.2	97.2
15	97.2	98.6	98.6	98.6	95.8	95.8	97.2	97.2
16	95.8	97.2	98.6	98.6	97.2	94.4	97.2	95.8
17	95.8	97.2	98.6	98.6	97.2	95.8	97.2	97.2
18	95.8	95.8	98.6	98.6	97.2	95.8	95.8	97.2
19	95.8	97.2	98.6	98.6	97.2	95.8	94.4	97.2
20	95.8	97.2	97.2	97.2	97.2	97.2	94.4	95.8

TABLE 8.8 Mutual Information on Colon Cancer Data

Gene/η	0.2	0.5	0.8	1.0	1.2	1.5	1.8	2.0
1	75.8	75.8	85.5	85.5	85.5	85.5	85.5	85.5
2	83.9	83.9	83.9	83.9	83.9	83.9	83.9	83.9
3	85.5	85.5	85.5	85.5	85.5	85.5	85.5	85.5
4	87.1	87.1	87.1	87.1	87.1	88.7	87.1	87.1
5	85.5	85.5	87.1	87.1	87.1	88.7	87.1	87.1
6	87.1	85.5	87.1	87.1	87.1	87.1	87.1	87.1
7	88.7	82.3	85.5	90.3	90.3	90.3	90.3	85.5
8	87.1	82.3	88.7	90.3	90.3	90.3	90.3	83.9
9	85.5	88.7	87.1	90.3	90.3	88.7	88.7	83.9
10	85.5	85.5	85.5	90.3	88.7	88.7	88.7	83.9
11	85.5	85.5	85.5	90.3	90.3	87.1	87.1	87.1
12	87.1	85.5	87.1	87.1	87.1	87.1	85.5	85.5
13	85.5	85.5	85.5	85.5	82.3	85.5	85.5	85.5
14	82.3	80.6	80.6	83.9	82.3	80.6	83.9	83.9
15	79.0	85.5	80.6	80.6	80.6	80.6	82.3	83.9
16	79.0	83.9	82.3	82.3	80.6	83.9	79.0	79.0
17	77.4	83.9	82.3	83.9	80.6	83.9	80.6	82.3
18	79.0	80.6	82.3	80.6	82.3	85.5	83.9	80.6
19	80.6	80.6	82.3	82.3	80.6	80.6	80.6	80.6
20	85.5	87.1	82.3	80.6	79.0	80.6	79.0	79.0

TABLE 8.9 V-Information on Colon Cancer Data

Gene/η	0.2	0.5	0.8	1.0	1.2	1.5	1.8	2.0
1	85.5	85.5	85.5	85.5	85.5	85.5	85.5	85.5
2	85.5	83.9	87.1	85.5	85.5	85.5	83.9	83.9
3	87.1	87.1	85.5	85.5	88.7	85.5	88.7	85.5
4	87.1	85.5	85.5	83.9	85.5	85.5	85.5	88.7
5	88.7	90.3	85.5	90.3	87.1	87.1	88.7	88.7
6	90.3	90.3	87.1	85.5	87.1	87.1	88.7	88.7
7	88.7	93.5	87.1	82.3	85.5	85.5	90.3	90.3
8	88.7	88.7	87.1	85.5	83.9	87.1	90.3	90.3
9	85.5	90.3	87.1	87.1	88.7	85.5	90.3	90.3
10	87.1	91.9	82.3	87.1	87.1	85.5	88.7	88.7
11	82.3	90.3	88.7	87.1	83.9	85.5	88.7	88.7
12	85.5	90.3	87.1	85.5	82.3	83.9	85.5	88.7
13	88.7	88.7	88.7	80.6	80.6	83.9	85.5	88.7
14	87.1	87.1	83.9	88.7	80.6	83.9	83.9	88.7
15	85.5	87.1	83.9	88.7	80.6	83.9	83.9	85.5
16	83.9	83.9	83.9	91.9	80.6	80.6	83.9	83.9
17	83.9	80.6	83.9	90.3	80.6	80.6	83.9	80.6
18	83.9	82.3	83.9	90.3	80.6	80.6	85.5	80.6
19	83.9	82.3	83.9	91.9	79.0	80.6	83.9	83.9
20	83.9	83.9	83.9	90.3	83.9	82.3	82.3	83.9

TABLE 8.10 χ^2-Information on Colon Cancer Data

Gene/η	0.2	0.5	0.8	1.0	1.2	1.5	1.8	2.0
1	85.5	85.5	85.5	85.5	85.5	85.5	85.5	85.5
2	83.9	83.9	83.9	83.9	83.9	83.9	83.9	83.9
3	85.5	85.5	88.7	88.7	88.7	85.5	85.5	85.5
4	82.3	87.1	87.1	87.1	85.5	88.7	88.7	88.7
5	88.7	87.1	87.1	85.5	87.1	88.7	88.7	88.7
6	87.1	88.7	90.3	85.5	85.5	88.7	87.1	87.1
7	88.7	88.7	90.3	83.9	83.9	90.3	90.3	88.7
8	85.5	88.7	90.3	87.1	87.1	90.3	90.3	88.7
9	85.5	87.1	90.3	87.1	87.1	90.3	88.7	85.5
10	80.6	87.1	88.7	83.9	83.9	88.7	88.7	88.7
11	80.6	87.1	88.7	88.7	88.7	88.7	83.9	85.5
12	85.5	87.1	87.1	87.1	87.1	82.3	83.9	83.9
13	85.5	85.5	85.5	85.5	82.3	82.3	82.3	83.9
14	83.9	80.6	80.6	80.6	80.6	82.3	83.9	82.3
15	87.1	80.6	80.6	82.3	80.6	80.6	83.9	82.3
16	88.7	83.9	80.6	80.6	80.6	80.6	82.3	83.9
17	87.1	79.0	83.9	83.9	80.6	80.6	82.3	82.3
18	83.9	82.3	83.9	83.9	80.6	80.6	80.6	82.3
19	83.9	80.6	83.9	83.9	80.6	80.6	80.6	80.6
20	85.5	80.6	80.6	91.9	80.6	85.5	80.6	80.6

TABLE 8.11 Mutual Information on RAOA Data

Gene/η	0.2	0.5	0.8	1.0	1.2	1.5	1.8	2.0
1	73.3	93.3	93.3	86.7	93.3	93.3	86.7	86.7
2	86.7	96.7	93.3	93.3	90.0	96.7	93.3	93.3
3	100	100	93.3	90.0	100	93.3	93.3	96.7
4	96.7	96.7	100	90.0	96.7	90.0	93.3	96.7
5	96.7	100	100	96.7	100	93.3	90.0	100
6	100	100	100	100	93.3	90.0	96.7	100
7	96.7	100	100	90.0	90.0	100	96.7	100
8	100	100	100	96.7	96.7	100	100	100
9	96.7	100	96.7	100	100	100	100	96.7
10	100	100	100	100	100	100	96.7	90.0
11	100	100	100	100	100	100	96.7	86.7
12	100	100	100	100	100	100	96.7	100
13	100	100	100	100	100	100	100	100
14	100	100	100	100	100	100	100	100
15	100	100	100	100	100	100	100	100
16	100	100	100	100	100	100	100	100
17	100	100	100	100	100	100	100	96.7
18	100	100	100	100	100	100	100	100
19	100	100	100	100	100	100	100	100
20	100	100	100	100	100	100	96.7	96.7

TABLE 8.12 V-Information on RAOA Data

Gene/η	0.2	0.5	0.8	1.0	1.2	1.5	1.8	2.0
1	73.3	93.3	93.3	93.3	93.3	93.3	93.3	76.7
2	90.0	93.3	90.0	90.0	96.7	100	93.3	90.0
3	90.0	90.0	93.3	93.3	100	100	90.0	93.3
4	96.7	93.3	100	100	100	100	100	96.7
5	96.7	100	100	100	100	100	96.7	96.7
6	93.3	100	100	100	100	96.7	93.3	96.7
7	96.7	100	100	100	100	96.7	96.7	100
8	100	100	100	100	100	100	96.7	100
9	100	100	100	100	100	96.7	100	100
10	100	100	100	100	100	100	96.7	100
11	100	100	100	100	100	100	96.7	100
12	100	100	100	100	100	100	96.7	96.7
13	100	100	100	100	100	100	100	100
14	100	100	100	100	100	100	100	100
15	100	100	100	100	100	100	100	96.7
16	100	100	100	100	100	100	100	96.7
17	100	100	100	100	100	100	100	96.7
18	100	100	100	100	100	100	100	96.7
19	100	100	100	100	100	100	100	96.7
20	100	100	100	100	100	96.7	100	96.7

TABLE 8.13 χ^2-Information on RAOA Data

Gene/η	0.2	0.5	0.8	1.0	1.2	1.5	1.8	2.0
1	73.3	93.3	93.3	93.3	93.3	93.3	73.3	86.7
2	86.7	93.3	90.0	96.7	96.7	93.3	80.0	83.3
3	100	90.0	93.3	96.7	96.7	93.3	96.7	100
4	96.7	93.3	96.7	100	100	100	93.3	86.7
5	100	86.7	90.0	100	100	100	96.7	96.7
6	100	100	100	100	100	96.7	96.7	100
7	100	100	96.7	100	100	100	96.7	100
8	100	100	93.3	100	100	96.7	96.7	100
9	100	100	93.3	100	100	100	100	100
10	100	100	93.3	100	100	100	100	96.7
11	100	100	96.7	100	100	100	93.3	96.7
12	100	100	90.0	100	100	100	96.7	96.7
13	100	100	100	100	100	100	96.7	96.7
14	100	100	100	100	100	100	96.7	93.3
15	100	100	100	100	100	100	96.7	96.7
16	100	100	100	100	100	100	96.7	96.7
17	100	100	100	100	100	100	96.7	96.7
18	100	100	100	100	100	100	96.7	96.7
19	100	100	100	100	100	100	96.7	96.7
20	100	100	100	100	100	100	100	96.7

TABLE 8.14   Mutual Information on RAHC Data

Gene/η	0.2	0.5	0.8	1.0	1.2	1.5	1.8	2.0
1	78.0	78.0	78.0	78.0	78.0	78.0	78.0	78.0
2	84.0	86.0	84.0	86.0	86.0	92.0	86.0	92.0
3	82.0	92.0	90.0	90.0	82.0	92.0	94.0	94.0
4	92.0	90.0	90.0	90.0	92.0	92.0	92.0	96.0
5	90.0	88.0	94.0	94.0	92.0	88.0	92.0	92.0
6	88.0	94.0	94.0	94.0	90.0	92.0	94.0	90.0
7	90.0	100	94.0	94.0	92.0	90.0	92.0	90.0
8	98.0	96.0	96.0	96.0	96.0	90.0	90.0	86.0
9	96.0	94.0	96.0	94.0	96.0	86.0	90.0	88.0
10	92.0	94.0	94.0	92.0	92.0	92.0	90.0	90.0
11	92.0	94.0	94.0	94.0	92.0	90.0	88.0	86.0
12	90.0	90.0	96.0	94.0	94.0	90.0	92.0	94.0
13	92.0	90.0	94.0	94.0	92.0	90.0	88.0	92.0
14	94.0	92.0	100	98.0	94.0	90.0	84.0	92.0
15	92.0	92.0	96.0	96.0	96.0	86.0	86.0	90.0
16	92.0	94.0	96.0	96.0	98.0	84.0	84.0	90.0
17	92.0	92.0	96.0	98.0	98.0	90.0	90.0	88.0
18	92.0	94.0	96.0	98.0	94.0	90.0	88.0	88.0
19	94.0	92.0	98.0	98.0	98.0	90.0	86.0	88.0
20	98.0	94.0	98.0	98.0	98.0	88.0	90.0	90.0

TABLE 8.15 V-Information on RAHC Data

Gene/η	0.2	0.5	0.8	1.0	1.2	1.5	1.8	2.0
1	78.0	84.0	84.0	78.0	78.0	80.0	80.0	76.0
2	86.0	90.0	90.0	92.0	92.0	82.0	82.0	86.0
3	90.0	90.0	96.0	92.0	92.0	86.0	88.0	92.0
4	92.0	96.0	94.0	90.0	90.0	88.0	90.0	92.0
5	98.0	96.0	94.0	84.0	84.0	92.0	90.0	92.0
6	100	96.0	94.0	88.0	96.0	98.0	96.0	94.0
7	100	96.0	94.0	92.0	92.0	98.0	96.0	96.0
8	96.0	98.0	96.0	94.0	94.0	92.0	94.0	94.0
9	94.0	100	100	100	90.0	92.0	92.0	92.0
10	94.0	96.0	100	96.0	96.0	92.0	96.0	94.0
11	98.0	98.0	100	96.0	96.0	96.0	92.0	92.0
12	98.0	94.0	100	100	94.0	96.0	92.0	94.0
13	98.0	94.0	100	100	98.0	96.0	92.0	94.0
14	100	96.0	100	100	98.0	94.0	88.0	94.0
15	100	98.0	100	100	98.0	94.0	90.0	94.0
16	100	98.0	100	100	100	92.0	88.0	96.0
17	100	100	100	100	100	94.0	90.0	96.0
18	100	100	100	100	100	94.0	90.0	96.0
19	100	100	100	100	100	92.0	90.0	96.0
20	100	100	100	100	100	96.0	96.0	100

TABLE 8.16 χ^2-**Information on RAHC Data**

Gene/η	0.2	0.5	0.8	1.0	1.2	1.5	1.8	2.0
1	78.0	78.0	78.0	78.0	78.0	78.0	78.0	78.0
2	84.0	86.0	86.0	86.0	86.0	86.0	86.0	86.0
3	82.0	92.0	92.0	90.0	86.0	92.0	92.0	92.0
4	86.0	90.0	90.0	90.0	92.0	90.0	90.0	92.0
5	88.0	88.0	88.0	90.0	90.0	94.0	94.0	94.0
6	84.0	92.0	94.0	88.0	88.0	88.0	94.0	96.0
7	90.0	96.0	94.0	94.0	86.0	90.0	94.0	96.0
8	90.0	92.0	92.0	92.0	84.0	92.0	92.0	94.0
9	92.0	92.0	92.0	94.0	96.0	92.0	90.0	92.0
10	92.0	94.0	94.0	94.0	92.0	90.0	92.0	92.0
11	92.0	90.0	90.0	92.0	98.0	86.0	92.0	90.0
12	90.0	88.0	98.0	96.0	98.0	88.0	90.0	90.0
13	90.0	88.0	98.0	96.0	96.0	84.0	88.0	92.0
14	90.0	90.0	98.0	98.0	98.0	84.0	92.0	88.0
15	94.0	94.0	98.0	98.0	96.0	82.0	92.0	84.0
16	92.0	94.0	98.0	98.0	100	82.0	92.0	84.0
17	88.0	92.0	98.0	98.0	98.0	84.0	92.0	86.0
18	88.0	94.0	98.0	98.0	98.0	88.0	90.0	88.0
19	90.0	94.0	98.0	100	98.0	88.0	88.0	88.0
20	90.0	96.0	100	98.0	98.0	88.0	86.0	86.0

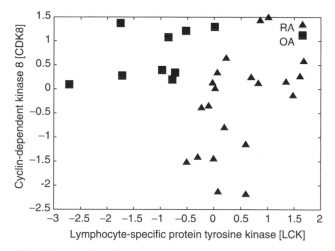

Figure 8.6 Scatter plot of the samples of two classes for the RAOA data.

FEPM-based method achieves consistently better performance for $0.5 \leq \eta \leq 1.5$ with respect to the classification accuracy of the SVM, irrespective of the data sets and information measures used.

From both class separability and classification accuracy analysis reported in Figs. 8.1–8.5 and Tables 8.1–8.16, it can be seen that very large or very small amount of overlapping among three fuzzy sets of the input gene are found to be undesirable. Figure 8.6 represents an example scatter plot of the samples of two classes for the RAOA data set considering two top-ranked genes selected by V-information measure using the FEPM-based method at $\eta = 1.5$. From the figure, it can be seen that the samples of the two classes are linearly separable.

8.5.4   Comparative Performance Analysis

Finally, Tables 8.17 and 8.18 provide the comparative results of different methods with respect to the classification accuracy of the SVM and class separability index. Results are reported for the five microarray data sets and three widely used information measures. From the results reported in Tables 8.17 and 8.18, it can be seen that the FEPM-based method provides better or comparable classification accuracy than that of two existing methods with lower number of selected genes in most of the cases. However, in case of breast cancer data using mutual information and leukemia, colon cancer and RAHC data sets using χ^2-information, the FEPM-based method attains the same accuracy with that of the discrete method with slightly higher number of genes. But, the class separability index of 30 top ranked genes selected using the FEPM-based method is lower than that of the existing two methods, irrespective of the data sets and information measures used. The better performance of the FEPM-based method is achieved because of the fact that the FEPM provides an efficient way to approximate the true marginal and joint distributions of continuous gene expression values than that of the discrete and Parzen-window-based methods.

8.6   CONCLUSION AND DISCUSSION

This chapter deals with the problem of selecting discriminative genes from high dimensional microarray data sets. Several information measures such as entropy, mutual information, and f-information have been shown to be successful for this task. However, for continuous gene expression values, it is very difficult to find the true density functions and to perform the integrations required to compute different information measures. In this regard, the concept of FEPM is used here to approximate the true marginal and joint distributions of continuous gene expression values. The performance of this approach is compared with that of some existing approaches using the class separability index and predictive accuracy of SVM.

For three cancer and two arthritis microarray data sets, significantly better results are found for the FEPM-based method compared to the existing methods,

TABLE 8.17 Comparative Performance Analysis on Cancer Data

Data Set	Measure	Method	Accuracy	Genes	S
		FEPM	100	8	1.889
	I	Discrete	100	6	2.268
		Parzen	95.9	5	2.181
		FEPM	100	5	2.079
Breast	V	Discrete	91.8	10	2.976
		Parzen	98.0	17	3.010
		FEPM	100	7	1.917
	χ^2	Discrete	100	10	2.306
		Parzen	100	11	2.118
		FEPM	98.6	9	1.374
	I	Discrete	98.6	19	1.604
		Parzen	98.6	12	1.613
		FEPM	100	3	1.357
Leukemia	V	Discrete	100	16	1.752
		Parzen	100	7	1.686
		FEPM	98.6	13	1.341
	χ^2	Discrete	98.6	12	1.536
		Parzen	97.2	5	1.407
		FEPM	90.3	7	3.306
	I	Discrete	88.7	10	4.760
		Parzen	90.3	16	4.821
		FEPM	91.9	10	3.334
Colon	V	Discrete	91.9	12	4.850
		Parzen	90.3	8	3.985
		FEPM	91.9	20	3.322
	χ^2	Discrete	91.9	16	4.526
		Parzen	88.7	11	3.527

irrespective of the information measures used. An important finding, however, is that the current approach is shown to be effective for selecting relevant and nonredundant continuous-valued genes from microarray data. All the results reported in this chapter demonstrate the feasibility and effectiveness of the FEPM-based method. It is capable of identifying discriminative genes that may contribute to revealing underlying class structures, providing a useful tool for the exploratory analysis of biological data. The results obtained on gene microarray data sets establish that the FEPM-based method can bring about a remarkable improvement on approximation of the true marginal and joint distributions of continuous feature values. The FEPM-based method is only used for the selection of genes from microarray data sets. This method can be extended to other density approximation tasks depending on the problems at hand.

TABLE 8.18 Comparative Performance Analysis on Arthritis Data

Data Set	Measure	Method	Accuracy	Genes	S
		FEPM	100	3	1.767
	I	Discrete	100	4	2.774
		Parzen	100	3	1.992
		FEPM	100	2	1.683
RAOA	V	Discrete	100	3	3.628
		Parzen	100	3	3.704
		FEPM	100	3	1.616
	χ^2	Discrete	100	8	2.718
		Parzen	100	11	3.008
		FEPM	100	7	2.641
	I	Discrete	100	29	4.169
		Parzen	100	22	4.137
		FEPM	100	6	2.687
RAHC	V	Discrete	98.0	15	6.079
		Parzen	100	13	4.859
		FEPM	100	16	2.576
	χ^2	Discrete	100	8	3.643
		Parzen	100	23	3.892

In Chapters 6–8, we have demonstrated the applications of rough-fuzzy methods to different problems of bioinformatics. The next chapter reports an application of the rough-fuzzy approach to medical imaging, namely, segmentation of brain magnetic resonance (MR) images.

REFERENCES

1. E. Domany. Cluster Analysis of Gene Expression Data. *Journal of Statistical Physics*, 110:1117–1139, 2003.

2. T. R. Golub, D. K. Slonim, P. Tamayo, C. Huard, M. Gaasenbeek, J. P. Mesirov, H. Coller, M. L. Loh, J. R. Downing, M. A. Caligiuri, C. D. Bloomfield, and E. S. Lander. Molecular Classification of Cancer: Class Discovery and Class Prediction by Gene Expression Monitoring. *Science*, 286:531–537, 1999.

3. G. J. McLachlan, K.-A. Do, and C. Ambroise. *Analyzing Microarray Gene Expression Data*. John Wiley & Sons, Inc., Hoboken, NJ, 2004.

4. P. A. Devijver and J. Kittler. *Pattern Recognition: A Statistical Approach*. Prentice Hall, Englewood Cliffs, NJ, 1982.

5. C. Ding and H. Peng. Minimum Redundancy Feature Selection from Microarray Gene Expression Data. *Journal of Bioinformatics and Computational Biology*, 3(2):185–205, 2005.

6. J. Li, H. Su, H. Chen, and B. W. Futscher. Optimal Search-Based Gene Subset Selection for Gene Array Cancer Classification. *IEEE Transactions on Information Technology in Biomedicine*, 11(4):398–405, 2007.

7. D. Jiang, C. Tang, and A. Zhang. Cluster Analysis for Gene Expression Data: A Survey. *IEEE Transactions on Knowledge and Data Engineering*, 16(11):1370–1386, 2004.

8. L. J. Heyer, S. Kruglyak, and S. Yooseph. Exploring Expression Data: Identification and Analysis of Coexpressed Genes. *Genome Research*, 9:1106–1115, 1999.

9. H. Peng, F. Long, and C. Ding. Feature Selection Based on Mutual Information: Criteria of Max-Dependency, Max-Relevance, and Min-Redundancy. *IEEE Transactions on Pattern Analysis and Machine Intelligence*, 27(8):1226–1238, 2005.

10. R. Battiti. Using Mutual Information for Selecting Features in Supervised Neural Net Learning. *IEEE Transactions on Neural Networks*, 5(4):537–550, 1994.

11. D. Huang and T. W. S. Chow. Effective Feature Selection Scheme Using Mutual Information. *Neurocomputing*, 63:325–343, 2004.

12. P. Maji. f-Information Measures for Efficient Selection of Discriminative Genes from Microarray Data. *IEEE Transactions on Biomedical Engineering*, 56(4):1063–1069, 2009.

13. X. Liu, A. Krishnan, and A. Mondry. An Entropy Based Gene Selection Method for Cancer Classification Using Microarray Data. *BMC Bioinformatics*, 6(76):1–14, 2005.

14. J. R. Quinlan. *C4.5: Programs for Machine Learning*. Morgan Kaufmann, San Francisco, CA, USA, 1993.

15. N. Kwak and C.-H. Choi. Input Feature Selection by Mutual Information Based on Parzen Window. *IEEE Transactions on Pattern Analysis and Machine Intelligence*, 24(12):1667–1671, 2002.

16. A. M. Fraser and H. L. Swinney. Independent Coordinates for Strange Attractors from Mutual Information. *Physical Review A*, 33(2):1134–1140, 1986.

17. M. Young, B. Rajagopalan, and U. Lall. Estimation of Mutual Information Using Kernel Density Estimators. *Physical Review E*, 52(3B):2318–2321, 1995.

18. J.-H. Chiang and S.-H. Ho. A Combination of Rough-Based Feature Selection and RBF Neural Network for Classification Using Gene Expression Data. *IEEE Transactions on NanoBioscience*, 7(1):91–99, 2008.

19. J. Fang and J. W. G. Busse. Mining of MicroRNA Expression Data: A Rough Set Approach. In *Proceedings of the 1st International Conference on Rough Sets and Knowledge Technology*, pages 758–765. Springer, Berlin, 2006.

20. A. Gruzdz, A. Ihnatowicz, and D. Slezak. Interactive Gene Clustering: A Case Study of Breast Cancer Microarray Data. *Information Systems Frontiers*, 8:21–27, 2006.

21. P. Maji and S. Paul. Rough Set Based Maximum Relevance-Maximum Significance Criterion and Gene Selection from Microarray Data. *International Journal of Approximate Reasoning*, 52(3):408–426, 2011.

22. D. Slezak. Rough Sets and Few-Objects-Many-Attributes Problem: The Case Study of Analysis of Gene Expression Data Sets. In *Proceedings of the Frontiers in the Convergence of Bioscience and Information Technologies*, pages 233–240, 2007.

23. D. Slezak and J. Wroblewski. Roughfication of Numeric Decision Tables: The Case Study of Gene Expression Data. In *Proceedings of the 2nd International Conference on Rough Sets and Knowledge Technology*, pages 316–323. Springer, Berlin, 2007.

24. J. J. Valdes and A. J. Barton. Relevant Attribute Discovery in High Dimensional Data: Application to Breast Cancer Gene Expressions. In *Proceedings of the 1st International Conference on Rough Sets and Knowledge Technology*, pages 482–489. Springer, Berlin, 2006.

25. Q. Hu, D. Yu, Z. Xie, and J. Liu. Fuzzy Probabilistic Approximation Spaces and Their Information Measures. *IEEE Transactions on Fuzzy Systems*, 14(2):191–201, 2007.

26. Y. Tang, Y.-Q. Zhang, Z. Huang, X. Hu, and Y. Zhao. Recursive Fuzzy Granulation for Gene Subsets Extraction and Cancer Classification. *IEEE Transactions on Information Technology in Biomedicine*, 12(6):723–730, 2008.

27. P. Maji and S. K. Pal. Feature Selection Using f-Information Measures in Fuzzy Approximation Spaces. *IEEE Transactions on Knowledge and Data Engineering*, 22(6):854–867, 2010.

28. P. Maji and S. K. Pal. Fuzzy-Rough Sets for Information Measures and Selection of Relevant Genes from Microarray Data. *IEEE Transactions on Systems Man and Cybernetics Part B-Cybernetics*, 40(3):741–752, 2010.

29. C. Shannon and W. Weaver. *The Mathematical Theory of Communication*. University Illinois Press, Champaign, IL, 1964.

30. J. P. W. Pluim, J. B. A. Maintz, and M. A. Viergever. f-Information Measures in Medical Image Registration. *IEEE Transactions on Medical Imaging*, 23(12):1508–1516, 2004.

31. I. Vajda. *Theory of Statistical Inference and Information*. Kluwer Academic, Dordrecht, The Netherlands, 1989.

32. M. J. Beynon. Stability of Continuous Value Discretisation: An Application within Rough Set Theory. *International Journal of Approximate Reasoning*, 35:29–53, 2004.

33. D. Huang, T. W. S. Chow, E. W. M. Ma, and J. Li. Efficient Selection of Discriminative Genes from Microarray Gene Expression Data for Cancer Diagnosis. *IEEE Transactions on Circuits and Systems-I*, 52(9):1909–1918, 2005.

34. E. Parzen. On Estimation of a Probability Density Function and Mode. *Annals of Mathematical Statistics*, 33:1065–1076, 1962.

35. B. W. Silverman. *Density Estimation for Statistics and Data Analysis*. Statistics and Applied Probability. Chapman & Hall, London, 1986.

36. V. Vapnik. *The Nature of Statistical Learning Theory*. Springer-Verlag, New York, 1995.

37. M. West, C. Blanchette, H. Dressman, E. Huang, S. Ishida, R. Spang, H. Zuzan, J. A. Olson, J. R. Marks, and J. R. Nevins. Predicting the Clinical Status of Human Breast Cancer by Using Gene Expression Profiles. *Proceedings of the National Academy Of Sciences of the United States of America*, 98(20):11462–11467, 2001.

38. U. Alon, N. Barkai, D. A. Notterman, K. Gish, S. Ybarra, D. Mack, and A. J. Levine. Broad Patterns of Gene Expression Revealed by Clustering Analysis of Tumor and Normal Colon Tissues Probed by Oligonucleotide Arrays. *Proceedings of the National Academy Of Sciences of the United States of America*, 96(12):6745–6750, 1999.

39. T. C. T. M. van der Pouw Kraan, F. A. van Gaalen, P. V. Kasperkovitz, N. L. Verbeet, T. J. M. Smeets, M. C. Kraan, M. Fero, P.-P. Tak, T. W. J. Huizinga, E. Pieterman, F. C. Breedveld, A. A. Alizadeh, and C. L. Verweij. Rheumatoid Arthritis is a Heterogeneous Disease: Evidence for Differences in the Activation of the STAT-1 Pathway Between Rheumatoid Tissues. *Arthritis and Rheumatism*, 48(8):2132–2145, 2003.

40. T. C. T. M. van der Pouw Kraan, C. A. Wijbrandts, L. G. M. van Baarsen, A. E. Voskuyl, F. Rustenburg, J. M. Baggen, S. M. Ibrahim, M. Fero, B. A. C. Dijkmans, P. P. Tak, and C. L. Verweij. Rheumatoid Arthritis Subtypes Identified by Genomic Profiling of Peripheral Blood Cells: Assignment of a Type I Interferon Signature in a Subpopulation of Patients. *Annals of the Rheumatic Diseases*, 66:1008–1014, 2007.

9

SEGMENTATION OF BRAIN MAGNETIC RESONANCE IMAGES

9.1 INTRODUCTION

Segmentation is a process of partitioning an image space into some nonoverlapping meaningful homogeneous regions. The success of an image analysis system depends on the quality of segmentation [1]. If the domain of the image is given by Ω, then the segmentation problem is to determine the sets $S_k \subset \Omega$, whose union is the entire domain Ω. The sets that make up a segmentation must satisfy

$$\Omega = \bigcup_{k=1}^{K} S_k, \text{ where } S_k \cap S_j = \emptyset \text{ for } k \neq j, \tag{9.1}$$

and each S_k is connected. Hence, a segmentation method is supposed to find those sets that correspond to distinct anatomical structures or regions of interest in the image. In the analysis of medical images for computer-aided diagnosis and therapy, segmentation is often required as a preliminary stage. However, medical image segmentation is a complex and challenging task owing to the intrinsic nature of the images. The brain has a particularly complicated structure and its precise segmentation is very important for detecting tumors, edema, and necrotic tissues, in order to prescribe appropriate therapy [2].

In medical imaging technology, a number of complementary diagnostic tools such as X-ray computer tomography, magnetic resonance imaging (MRI), and

Rough-Fuzzy Pattern Recognition: Applications in Bioinformatics and Medical Imaging,
First Edition. Pradipta Maji and Sankar K. Pal.
© 2012 John Wiley & Sons, Inc. Published 2012 by John Wiley & Sons, Inc.

positron emission tomography are available. The MRI is an important diagnostic imaging technique for the early detection of abnormal changes in tissues and organs. Its unique advantage over other modalities is that it can provide multi-spectral images of tissues with a variety of contrasts on the basis of the three MR parameters ρ, T1, and T2. Therefore, the majority of research in medical image segmentation concerns MR images [2].

Conventionally, brain MR images are interpreted visually and qualitatively by radiologists. Advanced research requires quantitative information such as the size of the brain ventricles after a traumatic brain injury or the relative volume of ventricles to brain. Fully automatic methods sometimes fail, producing incorrect results and requiring the intervention of a human operator. This is often true owing to the restrictions imposed by image acquisition, pathology, and biological variation. So, it is important to have a faithful method to measure various structures in the brain. One such method is the segmentation of images to isolate objects and regions of interest.

Many image-processing techniques have been proposed for MR image segmentation [3, 4], most notably thresholding [5–7], region-growing [8], edge detection [9], pixel classification [10, 11], and clustering [12–14]. Some algorithms using the neural network approach have also been investigated in MR image segmentation problems [15, 16]. One of the main problems in medical image segmentation is uncertainty. Some of the sources of this uncertainty include imprecision in computations and vagueness in class definitions. In this background, the possibility concept introduced by the fuzzy set theory and rough set theory have gained popularity in modeling and propagating uncertainty. Both fuzzy set and rough set provide a mathematical framework to capture uncertainties associated with the human cognition process. The segmentation of MR images using fuzzy c-means (FCM) has been reported [12, 16–20]. Image segmentation using rough sets has also been done [21–28]. Recently, a review is reported in Hassanien et al. [29] on the application of rough sets and near sets in medical imaging.

In this chapter, the application of different rough-fuzzy clustering algorithms is presented for segmentation of brain MR images. Details of these algorithms have been reported in Chapter 3 and Maji and Pal [22, 30]. One of the important issues of the partitive-clustering-algorithm-based brain MR image segmentation method is the selection of initial prototypes of different classes or categories. The concept of discriminant analysis, based on the maximization of class separability, is used to circumvent the initialization and local minima problems of the partitive clustering algorithms. In effect, it enables efficient segmentation of brain MR images [23]. The effectiveness of different rough-fuzzy clustering algorithms, along with a comparison with other c-means algorithms, is demonstrated on a set of real and synthetic brain MR images using some standard validity indices.

The rest of this chapter is organized as follows: While Section 9.2 deals with the pixel classification problem, Section 9.3 gives an overview of the feature extraction techniques employed in segmentation of brain MR images, along with the initialization method of c-means algorithm based on the maximization of class separability. Implementation details, experimental results, and a comparison among different c-means are presented in Section 9.4. Concluding remarks are given in Section 9.5.

9.2 PIXEL CLASSIFICATION OF BRAIN MR IMAGES

In this section, the results of different c-means algorithms are presented on pixel classification of brain MR images, that is, the results of clustering based on only gray value of pixels. The performance of three hybrid algorithms, namely, rough-fuzzy c-means (RFCM), rough-possibilistic c-means (RPCM), and rough-fuzzy-possibilistic c-means (RFPCM), is compared extensively with that of different c-means algorithms. The algorithms compared are hard c-means (HCM) [31], FCM [12], possibilistic c-means (PCM) [32], FPCM [33], FPCM of Masulli and Rovetta (FPCM^{MR}) [34], kernel-based HCM (KHCM) [35], kernel-based FCM (KFCM) [36, 37], kernel-based PCM (KPCM) [37], kernel-based FPCM (KFPCM) [38], rough c-means (RCM) [39], and RFCM of Mitra et al. (RFCM^{MBP}) [40].

All the algorithms are implemented in C language and run in LINUX environment with the machine configuration Pentium IV, 3.2 GHz, 1 MB cache, and 1 GB RAM. In all the experiments, the input parameters of different c-means algorithms used are as follows:

Fuzzifiers : $\acute{m}_1 = 2.0$ and $\acute{m}_2 = 2.0$
Constants : $a = 0.5$ and $b = 0.5$
Weight parameter : $w = 0.95$ and $\tilde{w} = 0.05$

The parameters are held constant across all runs. The values of δ for the RCM, RFCM, RPCM, and RFPCM algorithms are calculated using the procedure reported in Chapter 3. The final prototypes of the FCM are used to initialize the PCM, FPCM, KFCM, KPCM, and KFPCM, while the Gaussian function is used as the kernel.

The experimentation is done in two parts. In the first part, some real brain MR images are used. Above 100 brain MR images with different sizes and 16-bit gray levels are tested with different c-means algorithms. All the brain MR images are collected from Advanced Medicare and Research Institute, Salt Lake, Kolkata, India. In the second part, the segmentation results are presented on some benchmark images obtained from *BrainWeb: Simulated Brain Database* (http://www.bic.mni.mcgill.ca/brainweb/). The comparative performance of different c-means is reported with respect to α, ϱ, $\alpha^\star$, γ, DB, and Dunn indices reported in Chapter 3, as well as the β index [19] described in Chapter 4.

9.2.1   Performance on Real Brain MR Images

Figure 9.1 presents some examples of real brain MR images. Each image is of the size 256×180 with 16-bit gray levels. So, the number of objects in each data set is 46,080. Consider Fig. 9.2 as an example, which represents an MR image (I-20497774) along with the segmented images obtained using different c-means algorithms. Table 9.1 depicts the values of DB index, Dunn index, and β index of both FCM and RFPCM for different values of c on the data set of I-20497774. The results reported here with respect to DB and Dunn index confirm that both FCM and RFPCM achieve their best results for $c = 4$ corresponding to four classes or categories such as background, gray matter, white matter, and cerebrospinal fluid. Also, the value of β index, as expected, increases with an increase in the value of c. For a particular value of c, the performance of the RFPCM is better than that of the FCM.

Figure 9.3 shows the scatter plots of the highest and second highest memberships of all the objects in the data set of I-20497774 at first and final iterations, respectively, for the RFPCM considering $c = 4$. The diagonal line represents the zone where two highest memberships of objects are equal. From Fig. 9.3, it is observed that although the average difference between two highest memberships of the objects is very low at first iteration ($\delta = 0.145$), they become ultimately very high at the final iteration ($\delta = 0.652$). In Fig. 9.4, the variations of different indices such as ϱ, $\alpha^\star$, and γ over different iterations are reported for the I-20497774 data set. All the results reported in Fig. 9.4 clearly establish the fact that as the iteration increases, the value of ϱ decreases and the values of $\alpha^\star$ and γ increase. Ultimately, all the values are saturated after 20 iterations. That is, the rough-sets-based indices reported in Chapter 3 provide good quantitative measures for rough-fuzzy clustering.

Finally, Table 9.2 provides the comparative results of different c-means algorithms on I-20497774 with respect to the values of DB, Dunn, β, and CPU time (in milliseconds). The corresponding segmented images along with the original are presented in Fig. 9.2. The results reported in Fig. 9.2 and

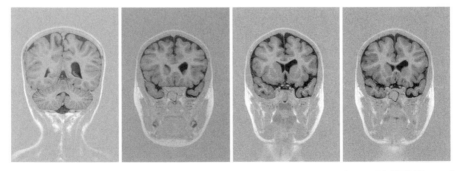

Figure 9.1 Example of brain MRI: I-20497774, I-20497761, I-20497763, and I-20497777.

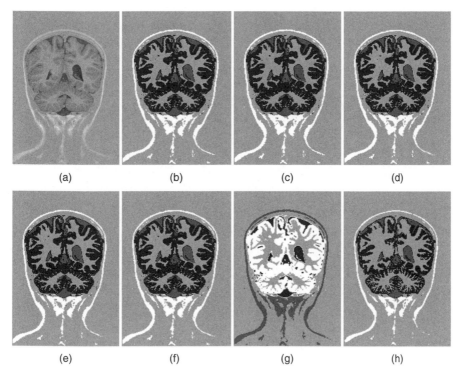

Figure 9.2 I-20497774: original and segmented versions of different c-means. (a) Original, (b) FCM, (c) KFCM, (d) RCM, (e) RFCM^{MBP}, (f) RFCM, (g) RPCM, and (h) RFPCM.

TABLE 9.1   Performance of FCM and RFPCM on I-20497774

Value of c	DB Index		Dunn Index		β Index	
	FCM	RFPCM	FCM	RFPCM	FCM	RFPCM
2	0.51	0.20	2.30	6.14	2.15	2.17
3	0.25	0.18	1.11	1.66	3.55	3.60
4	0.16	0.15	1.50	1.54	9.08	10.01
5	0.39	0.17	0.10	0.64	10.45	10.97
6	0.20	0.18	0.66	1.09	16.93	17.00
7	0.23	0.27	0.98	0.10	21.63	19.46
8	0.34	0.25	0.09	0.32	25.82	25.85
9	0.32	0.28	0.12	0.11	31.75	29.50
10	0.30	0.26	0.08	0.11	38.04	38.17

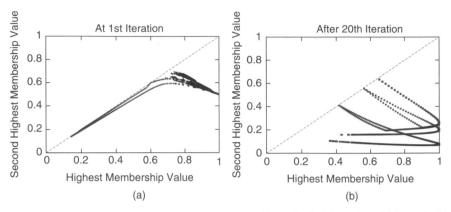

Figure 9.3   Scatter plots of two highest memberships of all objects in I-20497774. (a) $\delta = 0.145$ and (b) $\delta = 0.652$.

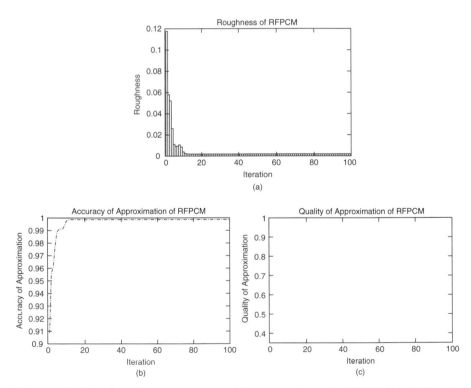

Figure 9.4   Variation of ϱ, $\alpha^{\star}$, and γ over different iterations for I-20497774. (a) Roughness: ϱ, (b) accuracy of approximation: $\alpha^{\star}$, and (c) quality of approximation: γ.

TABLE 9.2 Performance of Different c-Means Algorithms

Algorithms	DB Index	Dunn Index	β Index	Time
HCM	0.18	1.17	8.11	817
FCM	0.16	1.50	9.08	1742
FPCM^{MR}	0.16	1.51	9.11	1019
KHCM	0.17	1.21	8.59	1599
KFCM	0.17	1.49	9.06	1904
RCM	0.17	1.51	9.10	912
RFCM^{MBP}	0.15	1.51	9.02	1035
RFCM	0.15	1.64	9.68	1028
RPCM	0.15	1.53	9.79	1138
RFPCM	0.15	1.54	10.01	1109

Table 9.2 confirm that the rough-fuzzy clustering algorithms produce more promising segmented images than do the conventional methods. Some of the existing algorithms such as the PCM, FPCM, KPCM, and KFPCM have failed to produce multiple segments as they generate coincident clusters even when they have been initialized with the final prototypes of the FCM.

Table 9.3 compares the performance of different c-means algorithms on other brain MR images such as I-20497761, I-20497763, and I-20497777 with respect to DB, Dunn, β, and CPU time considering $c = 4$. The original images along with the segmented versions of different c-means are shown in Figs. 9.5, 9.6, and 9.7. All the results reported in Table 9.3 and Figs. 9.5, 9.6, and 9.7 confirm that although each c-means algorithm, except the PCM, FPCM, KPCM, and KFPCM, generates good segmented images, the values of DB, Dunn, and β index of the RFPCM are better compared to other c-means algorithms.

9.2.2 Performance on Simulated Brain MR Images

Extensive experimentation is done to evaluate the performance of the RFPCM algorithm on simulated brain MR images obtained from *BrainWeb: Simulated Brain Database*.

Figures 9.8, 9.9, and 9.10 present the original and segmented images obtained using the RFPCM algorithm for different slice thicknesses and noise levels. The noise is calculated relative to the brightest tissue. The results are reported for three different slice thicknesses: 1, 3, and 5 mm, and the noise varies from 0 to 9%. Finally, Tables 9.4, 9.5, and 9.6 compare the values of DB, Dunn, and β indices of different c-means algorithms for different slice thicknesses and noise levels. All the results reported in Figs. 9.8, 9.9, and 9.10 and Tables 9.4, 9.5, and 9.6 confirm that the RFPCM algorithm generates good segmented images irrespective of the slice thickness and noise level. Also, the performance of the RFPCM algorithm in terms of DB, Dunn, and β indices is significantly better compared to other c-means algorithms.

TABLE 9.3   Performance of Different c-Means Algorithms

Data Set	Algorithms	DB Index	Dunn Index	β Index	Time
	HCM	0.16	2.13	12.07	719
	FCM	0.14	2.26	12.92	1813
	FPCM^{MR}	0.13	2.22	12.56	1026
	KHCM	0.16	2.17	12.11	1312
I-20497761	KFCM	0.14	2.27	12.78	2017
	RCM	0.15	2.31	11.68	806
	RFCM^{MBP}	0.14	2.34	9.99	1052
	RFCM	0.13	2.39	13.06	1033
	RPCM	0.11	2.61	12.98	1179
	RFPCM	0.10	2.73	13.93	1201
	HCM	0.18	1.88	12.02	706
	FCM	0.16	2.02	12.63	1643
	FPCM^{MR}	0.15	2.03	12.41	1132
	KHCM	0.17	2.01	12.10	1294
I-20497763	KFCM	0.15	2.08	12.71	1983
	RCM	0.15	2.14	12.59	751
	RFCM^{MBP}	0.15	2.08	10.59	1184
	RFCM	0.11	2.12	13.30	1017
	RPCM	0.12	2.71	13.14	1105
	RFPCM	0.10	2.86	13.69	1057
	HCM	0.17	2.01	8.68	765
	FCM	0.16	2.16	9.12	1903
	FPCM^{MR}	0.16	2.21	8.81	992
	KHCM	0.16	2.08	8.92	1294
I-20497777	KFCM	0.16	2.19	9.08	1907
	RCM	0.15	2.34	9.28	779
	RFCM^{MBP}	0.15	2.33	9.69	1127
	RFCM	0.14	2.39	9.81	1058
	RPCM	0.15	2.42	9.77	1079
	RFPCM	0.14	2.49	9.82	1063

9.3   SEGMENTATION OF BRAIN MR IMAGES

In this section, the feature extraction methodology for segmentation of brain MR images is first described. Next, the methodology to select initial centroids for different c-means algorithms is provided on the basis of the concept of maximization of class separability [23].

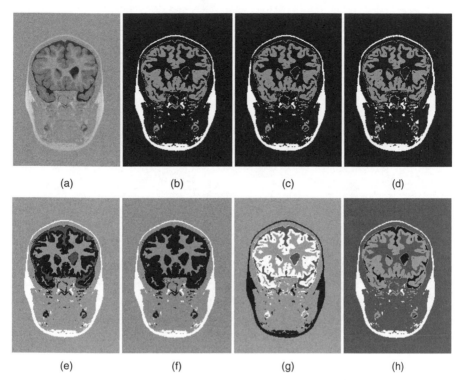

Figure 9.5 I-20497761: original and segmented versions of different c-means. (a) Original, (b) KHCM, (c) FCM, (d) RCM, (e) RFCM$^{\mathbf{MBP}}$, (f) RFCM, (g) RPCM, and (h) RFPCM.

9.3.1 Feature Extraction

Statistical texture analysis derives a set of statistics from the distribution of pixel values or blocks of pixel values. There are different types of statistical texture, first-order, second-order, and higher order statistics, based on the number of pixel combinations used to compute the textures. The first-order statistics such as mean, standard deviation, range, entropy, and the qth moment about the mean, are calculated using the histogram formed by the gray scale value of each pixel. These statistics consider the properties of the gray scale values, but not their spatial distribution. The second-order statistics are based on pairs of pixels. These take into account the spatial distribution of the gray scale distribution. In the present work, only first- and second-order statistical textures are considered.

A set of 13 input features is used for clustering the brain MR images. These include gray value of the pixel, two recently introduced first-order statistics, namely, homogeneity and edge value of the pixel [23], and 10 Haralick's textural

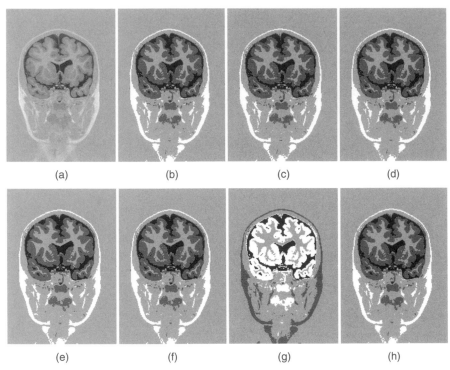

Figure 9.6 I-20497763: original and segmented versions of different c-means. (a) Original, (b) HCM, (c) FCM, (d) FPCM^{MR}, (e) RCM, (f) RFCM, (g) RPCM, and (h) RFPCM.

features [41], which are second-order statistics, namely, angular second moment, contrast, correlation, inverse difference moment, sum average, sum variance, sum entropy, second-order entropy, difference variance, and difference entropy. They are useful in characterizing images, and can be used as features of a pixel. Hence, these features have promising application in clustering-based brain MR image segmentation.

9.3.1.1 Homogeneity   If H is the homogeneity of a pixel $I_{m,n}$ within a 3×3 neighborhood, then

$$
H = 1 - \frac{1}{6(I_{\max} - I_{\min})} \left\{ |I_{m-1,n-1} + I_{m+1,n+1} - I_{m-1,n+1} - I_{m+1,n-1}| \right.
$$
$$
\left. + |I_{m-1,n-1} + 2I_{m,n-1} + I_{m+1,n-1} - I_{m-1,n+1} - 2I_{m,n+1} - I_{m+1,n+1}| \right\}
$$

$$(9.2)$$

where $I_{\max}$ and $I_{\min}$ represent the maximum and minimum gray values of the image, respectively. The region that is entirely within an organ will have a high

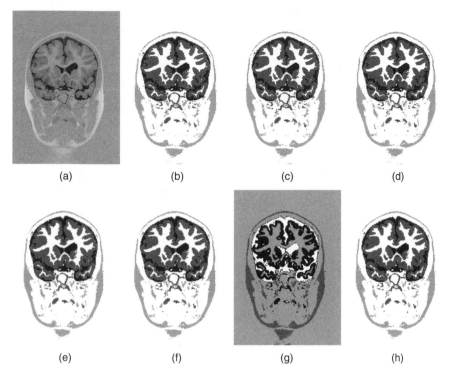

Figure 9.7 I-20497777: original and segmented versions of different c-means. (a) Original, (b) HCM, (c) FCM, (d) RCM, (e) RFCM^{MBP}, (f) RFCM, (g) RPCM, and (h) RFPCM.

H value. On the other hand, the regions that contain more than one organ will have lower H values [23].

9.3.1.2 Edge Value In MR imaging, the histogram of the given image is in general unimodal. One side of the peak may display a shoulder or slope change, or one side may be less steep than the other, reflecting the presence of two peaks that are close together or that differ greatly in height. The histogram may also contain a third, usually smaller, population corresponding to points on the object–background border. These points have gray levels intermediate between those of the object and the background; their presence raises the level of the valley floor between the two peaks, or if the peaks are already close together, makes it harder to detect the fact that they are not a single peak. As the histogram peaks are close together and very unequal in size, it may be difficult to detect the valley between them. In determining how each point of the image should contribute to the segmentation method, the current method takes into account the rate of change of gray level at the point, as well as the point's gray level or edge value, that is, the maximum of differences of average gray levels in pairs of horizontally and vertically adjacent 2×2 neighborhoods [6, 42]. If Δ is the

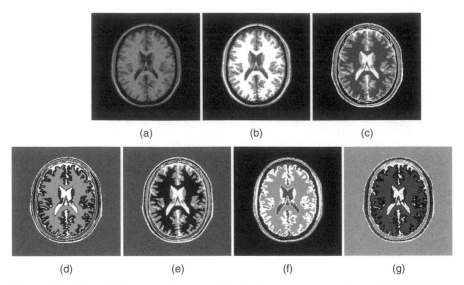

Figure 9.8   Slice thickness = 1 mm: original and segmented versions of RFPCM algorithm for different noise levels. (a) Original, (b) noise = 0%, (c) noise = 1%, (d) noise = 3%, (e) noise = 5%, (f) noise = 7%, and (g) noise = 9%.

edge value at a given point $I_{m,n}$, then

$$\Delta = \tfrac{1}{4} \max \left\{ |I_{m-1,n} + I_{m-1,n+1} + I_{m,n} + I_{m,n+1} - I_{m+1,n} - I_{m+1,n+1} \right.$$
$$-I_{m+2,n} - I_{m+2,n+1}|, |I_{m,n-1} + I_{m,n} + I_{m+1,n-1} + I_{m+1,n}$$
$$\left. -I_{m,n+1} - I_{m,n+2} - I_{m+1,n+1} - I_{m+1,n+2}| \right\}. \tag{9.3}$$

According to the image model, points interior to the object and background should generally have low edge values, since they are highly correlated with their neighbors, while those on the object–background border should have high edge values [6].

9.3.1.3   Haralick's Textural Feature
Texture is one of the important features used in identifying objects or regions of interest in an image. It is often described as a set of statistical measures of the spatial distribution of gray levels in an image. This scheme has been found to provide a powerful input feature representation for various recognition problems. Haralick et al. [41] proposed different textural properties for image classification. Haralick's textural measures are based on the moments of a joint probability density function that is estimated as the joint co-occurrence matrix or gray level co-occurrence matrix [41, 43]. A gray level co-occurrence matrix $P_{d,\theta}(i, j)$ reflects the distribution of the probability of occurrence of a pair of gray levels (i, j) separated by a given distance d at angle θ. On the basis of normalized gray level co-occurrence matrix, Haralick

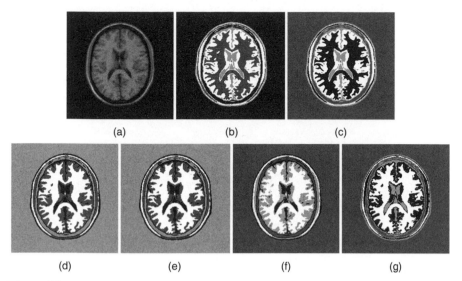

Figure 9.9   Slice thickness = 3 mm: original and segmented versions of RFPCM algorithm for different noise levels. (a) Original, (b) noise = 0%, (c) noise = 1%, (d) noise = 3%, (e) noise = 5%, (f) noise = 7%, (g) noise = 9%.

proposed several quantities as a measure of texture such as energy, contrast, correlation, sum of squares, inverse difference moments, sum average, sum variance, sum entropy, entropy, difference variance, difference entropy, information measure of correlation 1, and correlation 2. In order to define these texture measures, let us introduce some entities used in the derivation of Haralick's texture measures.

$$p(i, j) = \frac{P_{d,\theta}(i, j)}{\sum_{i=I_{\min}}^{I_{\max}} \sum_{j=I_{\min}}^{I_{\max}} P_{d,\theta}(i, j)}; \quad p_x(i) = \sum_{j=I_{\min}}^{I_{\max}} p(i, j) \tag{9.4}$$

$$p_{x+y}(k) = \sum_{i=I_{\min}}^{I_{\max}} \sum_{j=I_{\min}}^{I_{\max}} p(i, j); \quad i + j = k; \quad k = 2I_{\min}, 2I_{\min} + 1, \ldots, 2I_{\max}. \tag{9.5}$$

$$p_{x-y}(k) = \sum_{i=I_{\min}}^{I_{\max}} \sum_{j=I_{\min}}^{I_{\max}} p(i, j); \quad |i - j| = k; \quad k = 0, 1, \ldots, I_{\max} - I_{\min}. \tag{9.6}$$

where $I_{\max}$ and $I_{\min}$ represent the maximum and minimum gray values of the image, respectively. On the basis of the above entities, the Haralick's texture

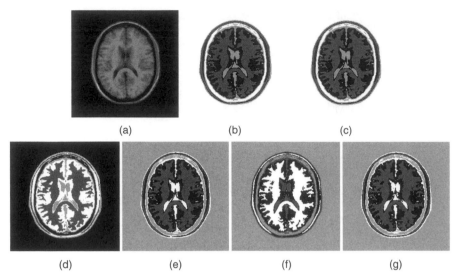

Figure 9.10 Slice thickness = 5 mm: original and segmented versions of RFPCM algorithm for different noise levels. (a) Original, (b) noise = 0%, (c) noise = 1%, (d) noise = 3%, (e) noise = 5%, (f) noise = 7%, (g) noise = 9%.

measures are defined as follows:

$$\text{Angular second moment or energy: } F_1 = \sum_{i=I_{\min}}^{I_{\max}} \sum_{j=I_{\min}}^{I_{\max}} p^2(i, j). \tag{9.7}$$

$$\text{Contrast: } F_2 = \sum_k k^2 \sum_{i=I_{\min}}^{I_{\max}} \sum_{j=I_{\min}}^{I_{\max}} p(i, j); |i - j| = k. \tag{9.8}$$

$$\text{Correlation: } F_3 = \frac{\sum_i \sum_j ijp(i, j) - \mu_x \mu_y}{\sigma_x \sigma_y} \tag{9.9}$$

where μ_x and μ_y are the means, and σ_x and σ_y are the standard deviations of p_x and p_y, respectively.

$$\text{Sum of squares: } F_4 = \sum_{i=I_{\min}}^{I_{\max}} \sum_{j=I_{\min}}^{I_{\max}} (i - \mu_f)^2 p(i, j), \tag{9.10}$$

TABLE 9.4 Value of DB Index for Simulated Brain MRI

Slice Thickness	Algorithms/ Methods	Noise (%)					
		0	1	3	5	7	9
	HCM	0.34	0.35	0.35	0.35	0.36	0.38
	FCM	0.29	0.29	0.29	0.31	0.30	0.30
	FPCM^{MR}	0.27	0.26	0.26	0.26	0.28	0.27
	KHCM	0.30	0.30	0.31	0.31	0.30	0.31
1	KFCM	0.29	0.28	0.28	0.28	0.29	0.29
	RCM	0.17	0.18	0.18	0.17	0.18	0.19
	RFCM^{MBP}	0.26	0.26	0.25	0.27	0.27	0.27
	RFCM	0.12	0.12	0.12	0.13	0.13	0.14
	RPCM	0.13	0.12	0.13	0.13	0.12	0.12
	RFPCM	0.12	0.12	0.12	0.13	0.13	0.13
	HCM	0.41	0.41	0.40	0.42	0.42	0.42
	FCM	0.31	0.31	0.30	0.32	0.32	0.32
	FPCM^{MR}	0.29	0.30	0.31	0.31	0.31	0.32
	KHCM	0.36	0.36	0.36	0.37	0.36	0.35
3	KFCM	0.29	0.29	0.29	0.29	0.30	0.31
	RCM	0.23	0.19	0.20	0.19	0.19	0.18
	RFCM^{MBP}	0.31	0.32	0.28	0.28	0.28	0.29
	RFCM	0.14	0.13	0.14	0.14	0.14	0.14
	RPCM	0.13	0.13	0.13	0.14	0.15	0.12
	RFPCM	0.13	0.13	0.13	0.13	0.13	0.13
	HCM	0.45	0.46	0.45	0.45	0.46	0.46
	FCM	0.32	0.32	0.32	0.32	0.30	0.33
	FPCM^{MR}	0.30	0.31	0.31	0.31	0.31	0.31
	KHCM	0.34	0.33	0.33	0.33	0.33	0.33
5	KFCM	0.31	0.30	0.31	0.30	0.30	0.30
	RCM	0.19	0.21	0.24	0.21	0.18	0.18
	RFCM^{MBP}	0.28	0.29	0.29	0.29	0.30	0.29
	RFCM	0.15	0.15	0.15	0.15	0.15	0.15
	RPCM	0.15	0.15	0.14	0.14	0.14	0.14
	RFPCM	0.14	0.14	0.14	0.14	0.14	0.14

where μ_f is the mean gray level of the image.

$$\text{Inverse difference moment}: F_5 = \sum_{i=I_{\min}}^{I_{\max}} \sum_{j=I_{\min}}^{I_{\max}} \frac{p(i, j)}{1 + (i - j)^2}. \tag{9.11}$$

$$\text{Sum average}: F_6 = \sum_k k p_{x+y}(k). \tag{9.12}$$

TABLE 9.5   Value of Dunn Index for Simulated Brain MRI

Slice Thickness	Algorithms/ Methods	Noise (%)					
		0	1	3	5	7	9
	HCM	2.87	3.11	3.14	3.05	3.19	2.97
	FCM	3.63	3.71	3.66	3.78	3.92	3.99
	FPCM^{MR}	3.70	3.81	3.55	3.78	3.49	3.96
	KHCM	3.18	3.21	3.29	3.34	3.30	3.37
1	KFCM	3.64	3.68	3.68	3.89	4.11	4.01
	RCM	5.74	5.51	5.58	6.30	5.65	7.43
	RFCM^{MBP}	6.13	6.74	6.58	6.91	7.03	6.95
	RFCM	8.34	8.33	8.79	8.88	8.90	8.33
	RPCM	8.53	8.28	8.91	9.07	9.11	8.70
	RFPCM	8.64	8.16	9.12	9.58	9.27	9.27
	HCM	2.51	2.51	2.56	2.67	2.48	2.70
	FCM	3.41	3.37	3.43	3.52	3.53	3.59
	FPCM^{MR}	3.49	3.50	3.51	3.44	3.41	3.75
	KHCM	3.01	3.04	3.19	3.19	3.23	3.24
3	KFCM	3.41	3.43	3.43	3.57	3.59	3.59
	RCM	3.89	5.22	4.92	5.19	5.73	5.47
	RFCM^{MBP}	6.33	6.31	6.57	6.48	6.90	6.77
	RFCM	7.24	7.27	7.39	7.45	7.53	7.47
	RPCM	7.39	7.52	7.17	7.42	7.88	7.60
	RFPCM	8.42	8.22	8.29	8.44	8.89	9.25
	HCM	2.14	2.19	2.23	2.30	2.27	2.36
	FCM	3.02	3.09	3.09	3.15	3.22	3.28
	FPCM^{MR}	3.07	3.11	3.28	3.28	3.21	3.34
	KHCM	2.85	2.86	2.90	2.89	3.01	3.06
5	KFCM	3.03	3.09	3.09	3.18	3.24	3.32
	RCM	5.34	4.52	3.74	5.01	5.95	5.18
	RFCM^{MBP}	5.31	5.37	5.02	4.99	4.71	4.95
	RFCM	5.76	5.74	6.06	5.95	6.03	6.05
	RPCM	6.12	6.04	6.11	6.28	6.23	6.29
	RFPCM	7.48	7.77	8.24	7.89	7.88	7.52

$$\text{Sum variance} : F_7 = \sum_k (k - F_6)^2 p_{x+y}(k). \tag{9.13}$$

$$\text{Sum entropy} : F_8 = -\sum_k p_{x+y}(k)\log_2[p_{x+y}(k)]. \tag{9.14}$$

$$\text{Entropy} : F_9 = -\sum_{i=I_{\min}}^{I_{\max}} \sum_{j=I_{\min}}^{I_{\max}} p(i, j)\log_2[p(i, j)]. \tag{9.15}$$

TABLE 9.6   Value of β Index for Simulated Brain MRI

Slice Thickness	Algorithms/ Methods	Noise (%)					
		0	1	3	5	7	9
	HCM	36.2	35.8	35.3	34.7	34.4	33.8
	FCM	41.6	41.1	40.4	39.5	37.8	35.9
	FPCM^{MR}	41.8	41.0	40.6	39.8	38.1	36.9
	KHCM	40.3	39.1	38.3	37.0	35.4	34.1
1	KFCM	41.7	41.1	40.6	39.4	38.0	36.1
	RCM	45.0	44.0	42.7	42.7	39.5	38.2
	RFCM^{MBP}	46.1	45.7	45.2	43.8	40.1	39.5
	RFCM	56.9	56.1	56.0	53.2	50.3	45.2
	RPCM	56.5	55.4	55.1	52.0	49.4	44.5
	RFPCM	56.7	55.7	55.7	52.8	49.7	44.9
	HCM	32.7	30.6	29.1	27.8	24.1	22.3
	FCM	35.9	34.3	32.7	32.0	30.5	28.6
	FPCM^{MR}	35.7	34.6	33.1	32.2	31.4	29.5
	KHCM	34.3	32.9	31.0	29.7	28.8	26.1
3	KFCM	36.1	34.4	32.9	31.8	30.7	29.5
	RCM	39.9	41.1	40.6	40.9	40.2	39.3
	RFCM^{MBP}	40.3	40.5	38.1	35.9	32.6	31.1
	RFCM	51.4	50.8	50.4	49.9	47.8	45.2
	RPCM	50.1	49.5	50.2	48.7	46.1	44.2
	RFPCM	50.4	49.9	49.7	49.0	46.8	44.5
	HCM	26.1	25.4	24.0	22.9	22.4	21.8
	FCM	32.7	31.8	30.2	29.9	26.5	27.6
	FPCM^{MR}	33.0	32.3	30.8	29.1	27.6	27.3
	KHCM	31.3	31.5	29.8	29.4	27.3	24.1
5	KFCM	32.7	31.9	30.4	30.1	27.2	27.8
	RCM	40.2	39.5	38.3	38.5	40.3	38.0
	RFCM^{MBP}	43.6	41.7	40.2	39.5	39.0	38.8
	RFCM	48.1	48.5	48.3	47.3	46.4	43.0
	RPCM	45.4	44.7	43.8	42.0	41.3	41.0
	RFPCM	46.6	46.8	47.9	45.8	45.0	42.0

Difference variance : F_{10} = variance of p_{x-y} (9.16)

Difference entropy: $F_{11} = -\sum_{k} p_{x-y}(k)\log_2[p_{x-y}(k)]$.   (9.17)

Information measure of correlation 1 : $F_{12} = \dfrac{H_{xy} - H_{xy1}}{\max\{H_x, H_y\}}$ (9.18)

Information measure of correlation 2 : $F_{13} = \{1 - \exp[-2(H_{xy2} - H_{xy})]\}^2$ (9.19) where $H_{xy} = F_9$; H_x and H_y are the entropies of p_x and p_y, respectively;

$$H_{xy1} = -\sum_{i=I_{min}}^{I_{max}} \sum_{j=I_{min}}^{I_{max}} p(i,j)\log_2[p_x(i)p_y(j)]; \tag{9.20}$$

$$H_{xy2} = -\sum_{i=I_{min}}^{I_{max}} \sum_{j=I_{min}}^{I_{max}} p_x(i)p_y(j)\log_2[p_x(i)p_y(j)]. \tag{9.21}$$

In Haralick et al. [41] these properties were calculated for large blocks in aerial photographs. Every pixel within each of these large blocks was then assigned the same texture values. This leads to a significant loss of resolution that is unacceptable in medical imaging. However, in the present work, the texture values are assigned to a pixel by using a 3×3 sliding window centered about that pixel. The gray level co-occurrence matrix is constructed by mapping the gray level co-occurrence probabilities on the basis of spatial relations of pixels in different angular directions ($\theta = 0°, 45°, 90°, 135°$) with unit pixel distance, while scanning the window (centered about a pixel) from left to right and top to bottom [41, 43]. Ten texture measures, namely, angular second moment, contrast, correlation, inverse difference moment, sum average, sum variance, sum entropy, second-order entropy, difference variance, and difference entropy, are computed for each window. For four angular directions, a set of four values is obtained for each of 10 measures. The mean of each of the 10 measures, averaged over four values, along with gray value, homogeneity, and edge value of the pixel, comprises the set of 13 features, which is used as feature vector of the corresponding pixel.

9.3.2 Selection of Initial Prototypes

A limitation of the c-means algorithm is that it can only achieve a local optimum solution that depends on the initial choice of the centroids. Consequently, computing resources may be wasted in that some initial centroids get stuck in regions of the input space with a scarcity of data points and may therefore never have the chance to move to new locations where they are needed. To overcome this limitation of the c-means algorithm, a method is described to select initial centroids, which is based on discriminant analysis maximizing some measures of class separability [44]. It enables the algorithm to converge to an optimum or near optimum solution [23].

Before describing the method for selecting initial centroids, a quantitative measure of class separability [44] is provided, which is given by

$$J(T) = \frac{P_1(T)P_2(T)[m_1(T) - m_2(T)]^2}{P_1(T)\sigma_1^2(T) + P_2(T)\sigma_2^2(T)}, \tag{9.22}$$

where

$$P_1(T) = \sum_{z=0}^{T} h(z); \quad P_2(T) = \sum_{z=T+1}^{L-1} h(z) = 1 - P_1(T); \tag{9.23}$$

$$m_1(T) = \frac{1}{P_1(T)} \sum_{z=0}^{T} z h(z); \quad m_2(T) = \frac{1}{P_2(T)} \sum_{z=T+1}^{L-1} z h(z); \tag{9.24}$$

$$\sigma_1^2(T) = \frac{1}{P_1(T)} \sum_{z=0}^{T} [z - m_1(T)]^2 h(z);$$

$$\sigma_2^2(T) = \frac{1}{P_2(T)} \sum_{z=T+1}^{L-1} [z - m_2(T)]^2 h(z). \tag{9.25}$$

Here, L is the total number of discrete values ranging between $[0, L - 1]$, T is the threshold value, which maximizes $J(T)$, and h(z) represents the percentage of data having feature value z over the total number of discrete values of the corresponding feature. To maximize $J(T)$, the means of the two classes should be as well separated as possible and the variances in both classes should be as small as possible.

On the basis of the concept of maximization of class separability, the method for selecting initial centroids is described next. The main steps of this method proceed as follows:

1. The data set $X = \{x_1, \ldots, x_i, \ldots, x_n\}$ with $x_i \in \Re^m$ are first discretized to facilitate class separation method. Suppose the possible value range of a feature F_j in the data set is $(F_{\min, j}, F_{\max, j})$, and the real value that the data element x_i takes at F_j is F_{ij}, then the discretized value of F_{ij} is

$$\text{Discretized}(F_{ij}) = (L - 1) \times \left\{ \frac{F_{ij} - F_{\min, j}}{F_{\max, j}} - F_{\min, j} \right\}, \tag{9.26}$$

where L is the total number of discrete values ranging between $[0, L - 1]$.

2. For each feature F_j, calculate $h(z)$ for $0 \leq z < L$.

3. Calculate the threshold value T_j for the feature F_j, which maximizes class separability along that feature.

4. On the basis of the threshold T_j, discretize the corresponding feature F_j of the data element x_i as follows:

$$\overline{F}_{ij} = \begin{cases} 1, & \text{if discretized } (F_{ij}) \geq T_j \\ 0, & \text{otherwise} \end{cases} \tag{9.27}$$

5. Repeat steps 2 to 4 for all the features and generate the set of discretized objects $\overline{X} = \{\overline{x}_1, \ldots, \overline{x}_i, \ldots, \overline{x}_n\}$.

6. Calculate total number of similar discretized objects $N(x_i)$ and mean of similar objects $\overline{v}(x_i)$ of x_i as

$$N(x_i) = \sum_{j=1}^{n} \delta_j, \qquad (9.28)$$

and

$$\overline{v}(x_i) = \frac{1}{N(x_i)} \sum_{j=1}^{n} \delta_j \times x_j \qquad (9.29)$$

where

$$\delta_j = \begin{cases} 1 & \text{if } \overline{x}_j = \overline{x}_i \\ 0 & \text{otherwise} \end{cases}. \qquad (9.30)$$

7. Sort n objects according to their values of $N(x_i)$ such that $N(x_1) > N(x_2) > \cdots > N(x_n)$.

8. If $\overline{x}_i = \overline{x}_j$, then $N(x_i) = N(x_j)$ and $\overline{v}(x_j)$ should not be considered as a centroid or mean, resulting in a reduced set of objects to be considered for initial centroids.

9. Let there be $\acute{n}$ objects in the reduced set having $N(x_i)$ values such that $N(x_1) > N(x_2) > \cdots > N(x_{\acute{n}})$. A heuristic threshold function can be defined as follows [45]:

$$\text{Tr} = \frac{R}{\tilde{\epsilon}}, \quad \text{where } R = \sum_{i=1}^{\acute{n}} \frac{1}{N(x_i) - N(x_{i+1})} \qquad (9.31)$$

where $\tilde{\epsilon}$ is a constant ($= 0.5$, say), so that all the means $\overline{v}(x_i)$ of the objects in reduced set having $N(x_i)$ value higher than it are regarded as the candidates for initial centroids or means.

The value of Tr is high if most of the $N(x_i)$'s are large and close to each other. The above condition occurs when a small number of large clusters are present. On the other hand, if the $N(x_i)$'s have wide variation among them, then the number of clusters with smaller size increases. Accordingly, Tr attains a lower value automatically. Note that the main motive of introducing this threshold function lies in reducing the number of centroids. Actually, it attempts to eliminate noisy centroids, that is, data representatives having lower values of $N(x_i)$, from the whole data set. The whole approach is, therefore, data dependent.

9.4 EXPERIMENTAL RESULTS

In this section, the performance of different c-means algorithms in segmentation of brain MR images is presented. The algorithms compared are HCM [31], FCM[12], PCM [32], FPCM [33], RCM [39], and RFCM. Details of the experimental setup, data collection, and objective of the experiments are the same as those of Section 9.2.

9.4.1 Illustrative Example

Consider Fig. 9.11 as an example that represents the segmented images of an MR image (I-20497774) obtained using different c-means algorithms. Each image is of size 256×180 with 16-bit gray levels. So, the number of objects in the data set of I-20497774 is 46,080. The parameters generated in the discriminant-analysis-based initialization method are shown in Table 9.7 only for I-20497774 data set along with the values of input parameters. The threshold values for 13 features of the given data set are also reported in this table. Table 9.8 depicts the values of DB index, Dunn index, and β index of both FCM and RFCM for different values of c on the data set of I-20497774. The results reported here with respect to DB and Dunn indices confirm that both FCM and RFCM achieve their best results for $c = 4$. Also, the value of β index, as expected, increases with increase in the value of c. For a particular value of c, the performance of the RFCM is better than that of the FCM.

Finally, Table 9.9 provides the comparative results of different c-means algorithms on I-20497774 with respect to the values of DB index, Dunn index, and β index. The corresponding segmented images are presented in Fig. 9.11. The results reported in Fig. 9.11 and Table 9.9 confirm that the RFCM algorithm produces more promising segmented image than do the conventional c-means algorithms. Some of the existing algorithms such as the PCM and FPCM fail to produce multiple segments as they generate coincident clusters even when they are initialized with final prototypes of the FCM.

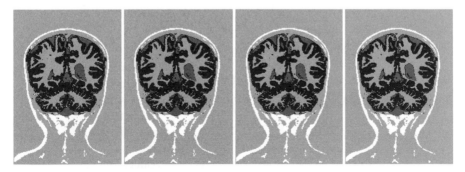

Figure 9.11   I-20497774: segmented versions of HCM, FCM, RCM, and RFCM.

TABLE 9.7   Values of Different Parameters

Size of image $= 256 \times 180$
Minimum gray value $= 1606$, maximum gray value $= 2246$
Samples per pixel $= 1$, bits allocated $= 16$, bits stored $= 12$

Number of objects $= 46080$
Number of features $= 13$, value of L $= 101$

Threshold values:
Gray value $= 1959$, homogeneity $= 0.17$, edge value $= 0.37$
Angular second moment $= 0.06$, contrast $= 0.12$
Correlation $= 0.57$, Inverse difference moment $= 0.18$
Sum average $= 0.17$, sum variance $= 0.14$, sum entropy $= 0.87$
Entropy $= 0.88$, difference variance $= 0.07$, difference entropy $= 0.79$

TABLE 9.8   Performance of FCM and RFCM on I-20497774 Data Set

Value of c	DB Index		Dunn Index		β Index	
	FCM	RFCM	FCM	RFCM	FCM	RFCM
2	0.38	0.19	2.17	3.43	3.62	4.23
3	0.22	0.16	1.20	1.78	7.04	7.64
4	0.15	0.13	1.54	1.80	11.16	13.01
5	0.29	0.19	0.95	1.04	11.88	14.83
6	0.24	0.23	0.98	1.11	19.15	19.59
7	0.23	0.21	1.07	0.86	24.07	27.80
8	0.31	0.21	0.46	0.95	29.00	33.02
9	0.30	0.24	0.73	0.74	35.06	40.07
10	0.30	0.22	0.81	0.29	41.12	44.27

TABLE 9.9   Performance of Different c-Means on I-20497774 Data Set

Algorithms	DB Index	Dunn Index	β Index
HCM	0.17	1.28	10.57
FCM	0.15	1.54	11.16
RCM	0.16	1.56	11.19
RFCM	0.13	1.80	13.01

9.4.2   Importance of Homogeneity and Edge Value

Table 9.10 presents the comparative results of different c-means for Haralick's features and features reported in Maji and Pal [23] on I-20497774 data set. While P-2 and H-13 stand for the set of two new features [23] and 13 Haralick's features, H-10 represents that of the 10 Haralick's features that are used in the current study. Two new features, namely, homogeneity and edge value, are found as important as Haralick's 10 features for clustering-based segmentation of brain

TABLE 9.10   Different Image Features on I-20497774 Data Set

Algorithms	Features	DB Index	Dunn Index	β Index	Time
HCM	H-13	0.19	1.28	10.57	4308
	H-10	0.19	1.28	10.57	3845
	P-2	0.18	1.28	10.57	1867
	H-10 ∪ P-2	0.17	1.28	10.57	3882
FCM	H-13	0.15	1.51	10.84	36,711
	H-10	0.15	1.51	10.84	34,251
	P-2	0.15	1.51	11.03	14,622
	H-10 ∪ P-2	0.15	1.54	11.16	43,109
RCM	H-13	0.19	1.52	11.12	5204
	H-10	0.19	1.52	11.12	5012
	P-2	0.17	1.51	11.02	1497
	H-10 ∪ P-2	0.16	1.56	11.19	7618
RFCM	H-13	0.13	1.76	12.57	15,705
	H-10	0.13	1.76	12.57	15,414
	P-2	0.13	1.77	12.88	6866
	H-10 ∪ P-2	0.13	1.80	13.01	17,084

MR images. The sct of 13 features, comprised of gray value, two new features, and 10 Haralick's features, improves the performance of all c-means with respect to DB, Dunn, and β indices. It is also observed that the three Haralick's features, namely, sum of squares, information measure of correlation 1, and correlation 2, do not contribute any extra information for segmentation of brain MR images.

9.4.3   Importance of Discriminant Analysis-Based Initialization

Table 9.11 provides comparative results of different c-means algorithms with random initialization of centroids and the discriminant-analysis-based initialization method described in Section 9.3.2 for the data sets I-20497761, I-20497763, and I-20497777 (Fig. 9.1). The discriminant-analysis-based initialization method is found to improve the performance in terms of DB index, Dunn index, and β index as well as reduce the time requirement of all c-means algorithms. It is also observed that the HCM with this initialization method performs similar to the RFCM with random initialization, although it is expected that the RFCM is superior to the HCM in partitioning the objects. While in random initialization, the c-means algorithms get stuck in local optimums, the discriminant-analysis-based initialization method enables the algorithms to converge to an optimum or near optimum solution. In effect, the execution time required for different c-means algorithms is lesser in this scheme compared to random initialization.

TABLE 9.11   Performance of Discriminant-Analysis-Based Initialization

Data	Algorithms	Initialization	DB	Dunn	β Index	Time
	HCM	Random	0.23	1.58	9.86	8297
		Discriminant	0.15	2.64	12.44	4080
	FCM	Random	0.19	1.63	12.73	40,943
I-20497761		Discriminant	0.12	2.69	13.35	38,625
	RCM	Random	0.19	1.66	10.90	9074
		Discriminant	0.14	2.79	12.13	6670
	RFCM	Random	0.15	2.07	11.89	19,679
		Discriminant	0.11	2.98	13.57	16,532
	HCM	Random	0.26	1.37	10.16	3287
		Discriminant	0.16	2.03	13.18	3262
	FCM	Random	0.21	1.54	10.57	46,157
I-20497763		Discriminant	0.15	2.24	13.79	45,966
	RCM	Random	0.21	1.60	10.84	10,166
		Discriminant	0.14	2.39	13.80	6770
	RFCM	Random	0.17	1.89	11.49	19,448
		Discriminant	0.10	2.38	14.27	15,457
	HCM	Random	0.33	1.52	6.79	4322
		Discriminant	0.16	2.38	8.94	3825
	FCM	Random	0.28	1.67	7.33	42,284
I-20497777		Discriminant	0.15	2.54	10.02	40,827
	RCM	Random	0.27	1.71	7.47	8353
		Discriminant	0.13	2.79	9.89	7512
	RFCM	Random	0.19	1.98	8.13	18,968
		Discriminant	0.11	2.83	11.04	16,930

9.4.4   Comparative Performance Analysis

Table 9.12 compares the performance of different c-means algorithms on some brain MR images with respect to DB, Dunn, and β indices. The segmented versions of different c-means are shown in Figs. 9.12, 9.13, and 9.14. All the results reported in Table 9.12 and Figs. 9.12, 9.13, and 9.14 confirm that although each c-means algorithm, except the PCM and FPCM, generates good segmented images, the values of DB, Dunn, and β indices of the RFCM are better compared to other c-means algorithms. Both the PCM and FPCM fail to produce multiple segments of the brain MR images as they generate coincident clusters even when they are initialized with the final prototypes of other c-means algorithms.

Table 9.12 also provides execution time (in milliseconds) of different c-means. The execution time required for the RFCM is significantly lesser compared to that for FCM. For the HCM and RCM, although the execution time is less, the performance is considerably poorer than that of the RFCM.

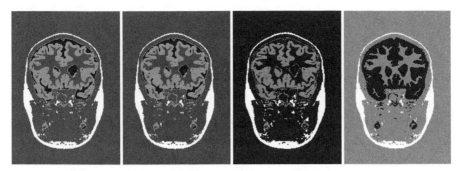

Figure 9.12   I-20497761: segmented versions of HCM, FCM, RCM, and RFCM.

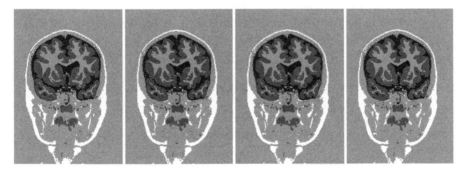

Figure 9.13   I-20497763: segmented versions of HCM, FCM, RCM, and RFCM.

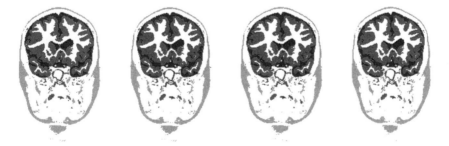

Figure 9.14   I-20497777: segmented versions of HCM, FCM, RCM, and RFCM.

The following conclusions can be drawn from the results reported in this chapter:

1. It is observed that the RFCM is superior to other c-means algorithms. However, the RFCM requires higher time compared to the HCM and RCM

**TABLE 9.12   Performance of Different c-Means
Algorithms**

Data	Algorithms	DB	Dunn	β Index	Time
	HCM	0.15	2.64	12.44	4080
I-20497761	FCM	0.12	2.69	13.35	38,625
	RCM	0.14	2.79	12.13	6670
	RFCM	0.11	2.98	13.57	16,532
	HCM	0.16	2.03	13.18	3262
I-20497763	FCM	0.15	2.24	13.79	45,966
	RCM	0.14	2.39	13.80	6770
	RFCM	0.10	2.38	14.27	15,457
	HCM	0.16	2.38	8.94	3825
I-20497777	FCM	0.15	2.54	10.02	40,827
	RCM	0.13	2.79	9.89	7512
	RFCM	0.11	2.83	11.04	16,930

and lesser time compared to the FCM. But, the performance of the RFCM with respect to DB, Dunn, and β is significantly better than that of all other c-means. The performance of the FCM and RCM is intermediate between the RFCM and HCM.

2. The discriminant-analysis-based initialization is found to improve the values of DB, Dunn, and β indices as well as reduce the time requirement substantially for all c-means algorithms.

3. Two new features reported in Maji and Pal [23], namely, homogeneity and edge value, are as important as Haralick's 10 features for clustering-based segmentation of brain MR images.

4. Use of rough sets and fuzzy memberships adds a small computational load to the HCM algorithm; however, the corresponding integrated method, that is, the RFCM algorithm, shows a definite increase in Dunn index and decrease in DB index.

The best performance of the current segmentation method in terms of DB, Dunn, and β indices is achieved because of the following reasons:

1. The discriminant-analysis-based initialization of centroids enables the algorithm to converge to an optimum or near optimum solutions.

2. Membership of the RFCM handles efficiently overlapping partitions.

3. The concept of crisp lower bound and fuzzy boundary of the rough-fuzzy clustering algorithm deals with uncertainty, vagueness, and incompleteness in class definition.

In effect, promising segmented brain MR images are obtained using the rough-fuzzy clustering algorithm.

9.5 CONCLUSION AND DISCUSSION

The problem of segmenting brain MR images is considered in rough-fuzzy computing framework. A robust segmentation technique is presented in this chapter, integrating the merits of rough sets, fuzzy sets, and c-means algorithm, for brain MR images. Some new measures are reported, based on the local properties of MR images, for accurate segmentation. The method, based on the concept of maximization of class separability, is found to be successful in effectively circumventing the initialization and local minima problems of iterative refinement clustering algorithms such as c-means. The extensive experimental results on a set of real and benchmark brain MR images show that the rough-fuzzy clustering algorithm produces a segmented image more promising than do the conventional algorithms.

REFERENCES

1. A. Rosenfeld and A. C. Kak. *Digital Picture Processing*. Academic Press, Inc., Orlando, Florida, 1982.
2. P. Suetens. *Fundamentals of Medical Imaging*. Cambridge University Press, Cambridge, UK, 2002.
3. J. C. Bezdek, L. O. Hall, and L. P. Clarke. Review of MR Image Segmentation Techniques Using Pattern Recognition. *Medical Physics*, 20(4):1033–1048, 1993.
4. R. Haralick and L. Shapiro. Survey: Image Segmentation Techniques. *Computer Vision Graphics and Image Processing*, 29:100–132, 1985.
5. C. Lee, S. Hun, T. A. Ketter, and M. Unser. Unsupervised Connectivity Based Thresholding Segmentation of Midsaggital Brain MR Images. *Computers in Biology and Medicine*, 28:309–338, 1998.
6. P. Maji, M. K. Kundu, and B. Chanda. Second Order Fuzzy Measure and Weighted Co-Occurrence Matrix for Segmentation of Brain MR Images. *Fundamenta Informaticae*, 88(1–2):161–176, 2008.
7. P. K. Sahoo, S. Soltani, A. K. C. Wong, and Y. C. Chen. A Survey of Thresholding Techniques. *Computer Vision Graphics and Image Processing*, 41:233–260, 1988.
8. I. N. Manousakes, P. E. Undrill, and G. G. Cameron. Split and Merge Segmentation of Magnetic Resonance Medical Images: Performance Evaluation and Extension to Three Dimensions. *Computers and Biomedical Research*, 31(6):393–412, 1998.
9. H. R. Singleton and G. M. Pohost. Automatic Cardiac MR Image Segmentation Using Edge Detection by Tissue Classification in Pixel Neighborhoods. *Magnetic Resonance in Medicine*, 37(3):418–424, 1997.
10. N. R. Pal and S. K. Pal. A Review on Image Segmentation Techniques. *Pattern Recognition*, 26(9):1277–1294, 1993.
11. J. C. Rajapakse, J. N. Giedd, and J. L. Rapoport. Statistical Approach to Segmentation of Single Channel Cerebral MR Images. *IEEE Transactions on Medical Imaging*, 16:176–186, 1997.
12. J. C. Bezdek. *Pattern Recognition with Fuzzy Objective Function Algorithm*. Plenum, New York, 1981.

13. K. V. Leemput, F. Maes, D. Vandermeulen, and P. Suetens. Automated Model-Based Tissue Classification of MR Images of the Brain. *IEEE Transactions on Medical Imaging*, 18(10):897–908, 1999.

14. W. M. {Wells} III, W. E. L. Grimson, R. Kikinis, and F. A. Jolesz. Adaptive Segmentation of MRI Data. *IEEE Transactions on Medical Imaging*, 15(4):429–442, 1996.

15. S. Cagnoni, G. Coppini, M. Rucci, D. Caramella, and G. Valli. Neural Network Segmentation of Magnetic Resonance Spin Echo Images of the Brain. *Journal of Biomedical Engineering*, 15(5):355–362, 1993.

16. L. O. Hall, A. M. Bensaid, L. P. Clarke, R. P. Velthuizen, M. S. Silbiger, and J. C. Bezdek. A Comparison of Neural Network and Fuzzy Clustering Techniques in Segmenting Magnetic Resonance Images of the Brain. *IEEE Transactions on Neural Networks*, 3(5):672–682, 1992.

17. M. E. Brandt, T. P. Bohan, L. A. Kramer, and J. M. Fletcher. Estimation of CSF, White and Gray Matter Volumes in Hydrocephalic Children Using Fuzzy Clustering of MR Images. *Computerized Medical Imaging and Graphics*, 18:25–34, 1994.

18. C. L. Li, D. B. Goldgof, and L. O. Hall. Knowledge-Based Classification and Tissue Labeling of MR Images of Human Brain. *IEEE Transactions on Medical Imaging*, 12(4):740–750, 1993.

19. S. K. Pal, A. Ghosh, and B. U. Shankar. Segmentation of Remotely Sensed Images with Fuzzy Thresholding and Quantitative Evaluation. *International Journal of Remote Sensing*, 21(11):2269–2300, 2000.

20. K. Xiao, S. H. Ho, and A. E. Hassanien. Automatic Unsupervised Segmentation Methods for MRI Based on Modified Fuzzy C-Means. *Fundamenta Informaticae*, 87(3–4):465–481, 2008.

21. A. E. Hassanien. Fuzzy Rough Sets Hybrid Scheme for Breast Cancer Detection. *Image and Vision Computing*, 25(2):172–183, 2007.

22. P. Maji and S. K. Pal. Rough Set Based Generalized Fuzzy C-Means Algorithm and Quantitative Indices. *IEEE Transactions on Systems Man and Cybernetics Part B-Cybernetics*, 37(6):1529–1540, 2007.

23. P. Maji and S. K. Pal. Maximum Class Separability for Rough-Fuzzy C-Means Based Brain MR Image Segmentation. *LNCS Transactions on Rough Sets*, 9:114–134, 2008.

24. M. M. Mushrif and A. K. Ray. Color Image Segmentation: Rough-Set Theoretic Approach. *Pattern Recognition Letters*, 29(4):483–493, 2008.

25. S. K. Pal and P. Mitra. Multispectral Image Segmentation Using the Rough Set-Initialized-EM Algorithm. *IEEE Transactions on Geoscience and Remote Sensing*, 40(11):2495–2501, 2002.

26. S. Widz, K. Revett, and D. Slezak. A Hybrid Approach to MR Imaging Segmentation Using Unsupervised Clustering and Approximate Reducts. In *Proceedings of the 10th International Conference on Rough Sets, Fuzzy Sets, Data Mining, and Granular Computing*, Regina, Canada, pages 372–382, 2005.

27. S. Widz, K. Revett, and D. Slezak. A Rough Set-Based Magnetic Resonance Imaging Partial Volume Detection System. In *Proceedings of the 1st International Conference on Pattern Recognition and Machine Intelligence*, Kolkata, India, pages 756–761, 2005.

28. S. Widz and D. Slezak. Approximation Degrees in Decision Reduct-Based MRI Segmentation. In *Proceedings of the Frontiers in the Convergence of Bioscience and Information Technologies*, Jeju Island, Korea, pages 431–436, 2007.

29. A. E. Hassanien, A. Abraham, J. F. Peters, G. Schaefer, and C. Henry. Rough Sets and Near Sets in Medical Imaging: A Review. *IEEE Transactions on Information Technology in Biomedicine*, 13(6):955–968, 2009.

30. P. Maji and S. K. Pal. RFCM: A Hybrid Clustering Algorithm Using Rough and Fuzzy Sets. *Fundamenta Informaticae*, 80(4):475–496, 2007.

31. A. K. Jain and R. C. Dubes. *Algorithms for Clustering Data*. Prentice Hall, Englewood Cliffs, NJ, 1988.

32. R. Krishnapuram and J. M. Keller. A Possibilistic Approach to Clustering. *IEEE Transactions on Fuzzy Systems*, 1(2):98–110, 1993.

33. N. R. Pal, K. Pal, J. M. Keller, and J. C. Bezdek. A Possibilistic Fuzzy C-Means Clustering Algorithm. *IEEE Transactions on Fuzzy Systems*, 13(4):517–530, 2005.

34. F. Masulli and S. Rovetta. Soft Transition from Probabilistic to Possibilistic Fuzzy Clustering. *IEEE Transactions on Fuzzy Systems*, 14(4):516–527, 2006.

35. M. Girolami. Mercer Kernel-Based Clustering in Feature Space. *IEEE Transactions on Neural Networks*, 13(3):780–784, 2002.

36. S. Miyamoto and D. Suizu. Fuzzy C-Means Clustering Using Kernel Functions in Support Vector Machines. *Journal of Advanced Computational Intelligence and Intelligent Informatics*, 7(1):25–30, 2003.

37. D.-Q. Zhang and S.-C. Chen. Kernel Based Fuzzy and Possibilistic C-Means Clustering. In *Proceedings of the International Conference on Artificial Neural Network, Turkey*, pages 122–125, 2003.

38. X.-H. Wu and J.-J. Zhou. Possibilistic Fuzzy C-Means Clustering Model Using Kernel Methods. In *Proceedings of the International Conference on Computational Intelligence for Modelling, Control and Automation*, volume 2, pages 465–470. IEEE Computer Society, Los Alamitos, CA, 2005.

39. P. Lingras and C. West. Interval Set Clustering of Web Users with Rough K-Means. *Journal of Intelligent Information Systems*, 23(1):5–16, 2004.

40. S. Mitra, H. Banka, and W. Pedrycz. Rough-Fuzzy Collaborative Clustering. *IEEE Transactions on Systems Man and Cybernetics Part B-Cybernetics*, 36:795–805, 2006.

41. R. M. Haralick, K. Shanmugam, and I. Dinstein. Textural Features for Image Classification. *IEEE Transactions on Systems Man and Cybernetics*, 3(6):610–621, 1973.

42. J. S. Weszka and A. Rosenfeld. Histogram Modification for Threshold Selection. *IEEE Transactions on Systems Man and Cybernetics*, 9(1):38–52, 1979.

43. R. M. Rangayyan. *Biomedical Image Analysis*. CRC Press, Boca Raton, Florida, 2004.

44. N. Otsu. A Threshold Selection Method from Gray Level Histogram. *IEEE Transactions on Systems Man and Cybernetics*, 9(1):62–66, 1979.

45. M. Banerjee, S. Mitra, and S. K. Pal. Rough-Fuzzy MLP: Knowledge Encoding and Classification. *IEEE Transactions on Neural Networks*, 9(6):1203–1216, 1998.

INDEX

Rough-Fuzzy Pattern Recognition: Applications in Bioinformatics and Medical Imaging,
First Edition. Pradipta Maji and Sankar K. Pal.
© 2012 John Wiley & Sons, Inc. Published 2012 by John Wiley & Sons, Inc.